Advanced Mathematics

A UNIFIED COURSE: BOOK ONE INCLUDING INTRODUCTION TO VECTORS AND MECHANICS

L. K. Turner

8.7.77

CONTENTS OF BOOK TWO

9 Further calculus

9.1 The logarithmic function, ln x
9.2 The exponential function
9.3 Interlude: infinite series
9.4 Maclaurin series
9.5 Integration by parts: reduction formulae
9.6 Interlude: integration by parts
9.7 General integration
9.8 Differential equations
9.9 Hyperbolic functions
9.10 Lengths of curves and areas of surfaces
9.11 Areas and lengths in polar coordinates

10 Complex numbers

10.1 Introduction
10.2 The Argand diagram
10.3 Complex conjugates
10.4 de Moivre's theorem
10.5 Euler's equation: cis $\theta = e^{i\theta}$
10.6 Transformations in the Argand diagram
10.7 Loci in the Argand diagram
10.8 Functions of a complex variable

11 Further vectors and mechanics

11.1 Points of subdivision
11.2 Centres of mass
11.3 Scalar products
11.4 Coordinate geometry of three dimensions
11.5 Vector products
11.6 Plane kinematics
11.7 Impulse and momentum
11.8 Work, energy and power
11.9 Oscillations
11.10 General statics: moments, couples, and equilibrium
11.11 General dynamics; rotation about a fixed axis
11.12 Dimensions

12 Probability distributions and further statistics

12.1 Probability distributions and generators
12.2 Continuous probability distributions
12.3 The Normal distribution
12.4 The Poisson distribution
12.5 Samples
12.6 Significance and confidence
12.7 Correlation and regression

13 Matrices, determinants, and linear algebra

13.1 Introduction: matrices and matrix addition
13.2 Linear transformations and their matrices
13.3 Products of transformations and matrices
13.4 Inverse transformations and matrices
13.5 Determinants
13.6 Systems of linear equations
13.7 Systematic reduction
13.8 Further applications

14 Introduction to algebraic structures

14.1 The idea of structure
14.2 Binary operations
14.3 Groups
14.4 Isomorphisms

Epilogue

Answers to exercises

LONGMAN

ADVANCED MATHEMATICS

A UNIFIED COURSE: BOOK ONE INCLUDING INTRODUCTION TO VECTORS AND MECHANICS

L. K. Turner

HEADMASTER, WATFORD GRAMMAR SCHOOL

Longman Group Limited
London
Associated companies, branches and representatives throughout the world

First published 1976

ISBN 0 582 35237 1

The Pitman Press, England

Acknowledgements

Figures 1–4, page 181 are based on photographs by Professor Dayton C. Miller by permission of the Case Institute of Technology, Cleveland, Ohio.

We are grateful to the following for permission to reproduce copyright material which appears in Advanced Mathematics Book 1, and Advanced Mathematics Book 2: The Associated Examining Board for the General Certificate of Education for questions from 'A' Level Mathematics Papers:— 1970; Statistics Paper No. 2, 1970: Statistics Paper 10 No. 2, 1972: Applied Mathematics Paper 3 No. 8 and 1972: Modern Syllabus Paper 6 No. 10: Joint Matriculation Board for questions set by the Board in previous examinations in Mathematics; Oxford Delegacy of Local Examinations for questions from 1972 'A' Level Mathematics II Nos. 21 and 22 and Oxford Scholarship Examinations 1969 Mathematics III No. D.14 and 1968 Natural Science I (Mathematics) No. 17; University of Cambridge Local Examinations Syndicate for questions from Entrance & Schools Examination Papers:— 1969 Maths III 4A and Further Maths 2A, 1970 Maths III 3A and Further Maths IIB and for questions from four 'A' Level Mathematics Papers 1971 and questions from four 'A' Level Mathematics Papers 1972: University of London for four questions from London 'GCE' Mathematics Examinations Papers 1968 and 1969 and the Welsh Joint Education Committee for questions from July 1972 Paper:— Mathematics A2 and Special Paper No. 11.

Contents

Sections marked * may be delayed until a second reading.

Preface ix

Notation xi

O **Prologue** 1

0.1 Logical symbols 1
0.2 Sets and Boolean algebra 2
0.3 Coordinates 6
0.4 Straight lines 10
*0.5 Coordinate geometry 14
0.6 Linear transformations and matrices 16
0.7 Multiplication of matrices 19
*0.8 The inverse of a matrix 21

I **Functions and their graphs** 25

1.1 Functions 25
1.2 Graphs of functions 29
1.3 Continuity 33
1.4 Limits 34
1.5 Interlude $f: x \mapsto x^2$ 39
*1.6 Quadratic functions 42

*1.7 Polynomial functions 48
*1.8 Rational functions 55
*1.9 Functions of several variables 58

2 **Power functions and logarithmic functions** 62

2.1 Power functions 62
2.2 The power function, 10^x 64
2.3 Increasing and decreasing power functions 65
2.4 The logarithmic function, $\log_{10}x$ 68
*2.5 Properties of logarithmic functions 70
*2.6 Graphs of logarithmic functions 73
*2.7 Logarithmic scales 73

3 **Rates of change: differentiation** 79

3.1 Average velocity and instantaneous velocity 79
3.2 Differentiation 84
3.3 Standard methods of differentiation 87
3.4 Higher derivatives 89
3.5 Alternative notation 92
3.6 Kinematics 96
3.7 Reverse processes 99
*3.8 Stationary values: maxima and minima 102
*3.9 Further differentiation 111
*3.10 Rates of change 116
*3.11 Small changes 121
3.12 Further kinematics 124

4 **Areas: integration** 130

4.1 The area beneath a curve: the integral 131
4.2 The two staircases 132
4.3 The generation of area 134
4.4 The fundamental theorem of calculus 136
4.5 The calculation of definite integrals 139
*4.6 Negative areas 142
*4.7 Numerical integration 145
*4.8 Other summations 149
*4.9 Mean values 154
*4.10 Integration by substitution 156

5 Trigonometric functions 162

5.1 Sine, cosine, and tangent 162
5.2 Simple equations 166
5.3 Cotangent, secant, and cosecant and the use of Pythagoras' theorem 169
*5.4 Sine and cosine rules 172
5.5 Functions of compound angles 174
5.6 Multiple angles 178
*5.7 The combination of waves 180
5.8 Factor formulae 185
5.9 Radians. Small angles 188
*5.10 Oscillations: amplitude, period, and frequency 193
5.11 Differentiation of trigonometric functions 195
5.12 Integration of trigonometric functions 200
*5.13 Inverse trigonometric functions 202
*5.14 Polar coordinates 206

6 Sequences and series 210

6.1 Sequences and series 210
6.2 Arithmetic progressions 212
6.3 Geometric progressions 215
*6.4 Finite series: the method of differences 218
*6.5 Mathematical induction 221
*6.6 Iteration: recurrence relations 224
*6.7 Interlude: Fibonacci numbers and prime numbers 226
6.8 Permutations and combinations 229
6.9 Pascal's triangle: the binomial theorem 233

7 Probability and statistics 238

7.1 Trials, events, and probabilities 238
7.2 Compound events 241
*7.3 Conditional probability 247
*7.4 Probability distributions: three standard types 251
*7.5 Expectation 256
7.6 Location and spread 258
*7.7 Frequency distributions 266

8 **Introduction to vectors and mechanics** 275

8.1 Vectors and vector addition 275
8.2 Position vectors and components 281
8.3 Differentiation of vectors 287
8.4 Motion with constant acceleration: projectiles 290
8.5 Circular motion 295
*8.6 Relative motion 299
8.7 Introduction to mechanics: Newton's first law 302
8.8 Force, mass and weight: Newton's second law 305
8.9 Reactions: Newton's third law 315
8.10 Friction 321
*8.11 Universal gravitation 328

***** **Appendix: parabolas, ellipses, and hyperbolas** 334

A.1 The parabola 337
A.2 The ellipse 343
A.3 The hyperbola 347
A.4 The sections of a cone 353

Answers to exercises 355

Preface

This is the first of two volumes written primarily for those in sixth forms and colleges. Directness, clarity, and unification have been principal aims, so that it has been possible to cover, with minor exceptions, the work of all the new Advanced Level syllabuses within the space of two books.

Knowing where to start is an author's first difficulty, which has here been solved by means of a prologue. This serves to introduce for some readers, and to revise for others, a number of topics which will be needed in the main body of the book: Boolean algebra, in readiness for use with probabilities; elementary coordinate geometry; and, for the great benefit they bring to the chapter on trigonometric functions, a preliminary discussion of 2×2 matrices.

In Boolean algebra I have followed the practice of a growing number of writers in replacing $A \cup B$ and $A \cap B$ by $A + B$ and AB. It is a great gain, for instance, to replace $(A \cap B) \cup (C \cap D)$ by $AB + CD$, whilst any reader who needs to use the former notation will easily make the required translation.

There is one place where I have been bold enough to introduce a new term which I have found valuable in the classroom. After the functions x^2, x^3 and polynomials, one naturally moves to a discussion of 10^x and a^x, and here one is suddenly at a loss for a name. Commonly they are called exponential functions; and so they are, but this is daunting for newcomers and a trifle embarrassing for those already familiar with $\exp x$. So here I use the term *power functions* and, once it is clear that 10^x and 2^x are utterly different from such simple functions as x^2 and x^3, the name is very descriptive.

Familiarity with the idea of a limit is, of course, necessary at an early stage, and this has been introduced in an informal and visual manner. A fuller discussion of ε and δ or of neighbourhoods is quite inappropriate at this level and can only reduce momentum and delay the introduction of calculus.

The individual chapters are considerably longer than is usual in a book of this kind and can perhaps best be considered as individual booklets, each a collection of separate sections. It is hoped that this arrangement, as well as increasing ease of reference, will make for flexibility in use. There is, for instance, no intention that chapters 1 and 2 should be completed before chapter 3 is begun, and there is much to be said for a quick reading of sections 1.1 to 1.4 being followed by the earlier sections of chapter 3. Similarly the first ten sections of chapter 5 and the whole of chapter 6 could, if it were desired, be taken at a much earlier stage. Those using the book will be able to decide for themselves on the most suitable order, though they may wish to delay the sections which have been marked with an asterisk.

Considerable care has been taken over the selection and arrangement of exercises, and it is hoped that abler students will pay particular attention to the miscellaneous problems at the end of each chapter. I am grateful for permission to use examination questions from the following Boards: Oxford and Cambridge (O.C., S.M.P., M.E.I.), the Joint Matriculation Board (J.M.B.), the Cambridge Local Examination Syndicate (C), the University of London (L), the Associated Examining Board (A.E.B.) and the Welsh Joint Education Committee (W); and also to the Clarendon Press and the Cambridge University Press for permission to use questions set in Entrance Scholarship Examinations at Oxford and Cambridge (OS and CS).

Figures 1–4, page 181 are based on photographs by Professor Dayton C. Miller by permission of the Case Institute of Technology, Cleveland, Ohio.

It is a particular pleasure to record my thanks to Mr F. J. Budden, who has read the entire manuscript and made many valuable suggestions; to my colleague, Mr D. Knighton, for his helpful comments on chapter 7; to various former pupils who have kindly assisted with the checking of answers, and to Mr D. J. Simmons, for his great help in reading proofs. Responsibility for faults which remain is wholly mine.

Finally, my thanks are due to the Warden and Fellows of New College, Oxford, without whose generous provision of a Schoolmaster Studentship this book would never have been started; and pre-eminently to my wife and sons for their forbearance, without which it could never have been finished.

L. K. T.

Notation

Logical symbols

$p, q, r \cdots$	statements, or propositions
p' (or $\sim p$)	negation of p
$\Rightarrow$	implies
$\Leftarrow$	is implied by
$\Leftrightarrow$	implies and is implied by
$\not\Rightarrow$ (etc.)	does not imply (etc.)

Equalities and inequalities

$=$	is equal to
$\neq$	is not equal to
$\approx$	is approximately equal to
$\equiv$	is always equal (or identical) to
$<$	is less than
$>$	is greater than

Sets

$x, y, \ldots$	elements of a set
$A, B, \ldots$	sets of elements
$n(A)$	number of elements in A
$\in$	belongs to
$\notin$	does not belong to
$\{x : x \in A\}$	set of elements x such that $x \in A$

$\varnothing$	empty set
$\mathscr{E}$	universal set
A'	complement of A within $\mathscr{E}$ $= \{x : x \in \mathscr{E} \text{ and } x \notin A\}$
$A + B$ (or $A \cup B$)	union of A and B $= \{x : x \in A \text{ and/or } x \in B\}$
AB (or $A \cap B$)	intersection of A and B $= \{x : x \in A \text{ and } x \in B\}$
Z	{all integers}
Q	{all rational numbers}
R	{all real numbers}
Z^+ (etc.)	{all positive integers} (etc.)

Matrices and determinants

$\boldsymbol{A} = \begin{pmatrix} a & b \\ c & d \end{pmatrix}$	matrix $\boldsymbol{A}$
Δ	determinant of $\boldsymbol{A}$ $(= ad - bc)$
$\boldsymbol{A}^{-1}$	inverse matrix of $\boldsymbol{A}$ (when $\Delta \neq 0$)

Functions

$f : x \mapsto y$	function f maps x on to y
$f(x)$	image of x under mapping f (or output of f, when input is x)
f^{-1}	inverse function of f
gf	function f followed by function g
$\lvert x \rvert$	magnitude of x
$[x]$	integral part of x (greatest integer not greater than x)
$\log_a x$	logarithm to base a of x

Limits, derivatives and integrals

$\rightarrow$	tends to
$\rightarrow \infty$	becomes unlimitedly large (tends to infinity)
$\lim_{x \to a} f(x)$	limit of $f(x)$ as $x \rightarrow a$
$\lim_{x \to a+} f(x)$	limit of $f(x)$ as $x \rightarrow a$ from above
$\lim_{x \to a-} f(x)$	limit of $f(x)$ as $x \rightarrow a$ from below
$\delta x, \delta y \ldots$	slight increments in $x, y \ldots$
$f'(x)$ or $\dfrac{\mathrm{d}y}{\mathrm{d}x}$	derivative

$f''(x)$ or $\dfrac{d^2y}{dx^2}$	second derivative
$\int_a^b f(x)\,dx$	definite integral, or area beneath $f(x)$ from $x = a$ to $x = b$
$\int f(x)\,dx$	indefinite integral, or primitive

Trigonometric functions

rad	radian $\left(1 \text{ rad} = \dfrac{180°}{\pi}\right)$
$\sin^{-1}$	inverse sine $(-\frac{1}{2}\pi \leqslant \sin^{-1} x \leqslant \frac{1}{2}\pi)$
$\cos^{-1}$	inverse cosine $(0 \leqslant \cos^{-1} x \leqslant \pi)$
$\tan^{-1}$	inverse tangent $(-\frac{1}{2}\pi < \tan^{-1} x < \frac{1}{2}\pi)$

Series

$\sum_{r=1}^{n} u_r$ or $\sum_{1}^{n} u_r$	$u_1 + u_2 + \cdots + u_n$
$a + (a + d) + (a + 2d) + \cdots$	arithmetic progression (a.p.)
$a + ap + ap^2 + \cdots$	geometric progression (g.p.)

Permutations and combinations

$n!$	n factorial $(= n \times (n - 1) \times \cdots \times 3 \times 2 \times 1)$
${}^nP_r$	number of permutations of n different objects, taken r at a time
${}^nC_r$	number of combinations of n different objects, taken r at a time

Probability

$\mathscr{E}$	universal set of equally likely occurrences in a given trial
$\boldsymbol{E}$	a particular event
E	set of occurrences corresponding to $\boldsymbol{E}$
$p(\boldsymbol{E})$	probability of $\boldsymbol{E} = \dfrac{n(E)}{n(\mathscr{E})}$
$p(\boldsymbol{A} + \boldsymbol{B})$	probability of event $\boldsymbol{A}$ and/or event $\boldsymbol{B}$
$p(\boldsymbol{AB})$	probability of event $\boldsymbol{A}$ and event $\boldsymbol{B}$
$p(\boldsymbol{A}/\boldsymbol{B})$	probability of $\boldsymbol{A}$ conditional upon $\boldsymbol{B}$
$\boldsymbol{A}$, $\boldsymbol{B}$ exclusive	$AB = \varnothing$

$\boldsymbol{A}, \boldsymbol{B}$ exhaustive	$A + B = \mathscr{E}$
$\boldsymbol{A}, \boldsymbol{B}$ independent	$p(\boldsymbol{AB}) = p(\boldsymbol{A})p(\boldsymbol{B})$
p_r	probability of event $\boldsymbol{E}_r$
x_r	random variable associated with $\boldsymbol{E}_r$
$E[x]$	expectation of $x = \sum p_r x_r$

Statistics

$x_1, x_2, \ldots$	values
m	mean $= \frac{1}{n}\sum x_r$
s^2	variance $= \frac{1}{n}\sum (x_r - m)^2$ $= \frac{1}{n}\sum x_r^2 - m^2$
s	standard deviation
$f_1, f_2, \ldots$	frequencies of $x_1, x_2 \ldots$ $\Rightarrow m = \frac{1}{n}\sum f_r x_r$ and $s^2 = \frac{1}{n}\sum f_r(x_r - m)^2$ $= \frac{1}{n}\sum f_r x_r^2 - m^2$

Vectors

$\boldsymbol{AB}, \boldsymbol{CD} \ldots$	displacement vectors
$\boldsymbol{a}, \boldsymbol{b} \ldots$	free vectors, or position vectors, with moduli $a, b, c \ldots$
$\boldsymbol{a} + \boldsymbol{b}, \boldsymbol{a} - \boldsymbol{b}$	vector sum and vector difference
$\boldsymbol{i}, \boldsymbol{j}, \boldsymbol{k}$	unit vectors, with right-hand set of mutually perpendicular axes
$\boldsymbol{r} = x\boldsymbol{i} + y\boldsymbol{j} + z\boldsymbol{k}$	vector $\boldsymbol{r}$ with components $x\boldsymbol{i}, y\boldsymbol{j}, z\boldsymbol{k}$
$\boldsymbol{v} = \frac{\mathrm{d}\boldsymbol{r}}{\mathrm{d}t} = \dot{\boldsymbol{r}}$ $\nearrow \boldsymbol{v}$ or $\nearrow \boldsymbol{v}$	velocity
$\boldsymbol{a} = \frac{\mathrm{d}\boldsymbol{v}}{\mathrm{d}t} = \ddot{\boldsymbol{r}}$ $\nearrow\!\!\nearrow \boldsymbol{a}$	acceleration
$\boldsymbol{v}_{\mathrm{P}}(\mathrm{Q}) = \frac{\mathrm{d}}{\mathrm{d}t}(\boldsymbol{PQ})$	velocity of Q relative to P

$\boldsymbol{a}_{\mathrm{P}}(\mathrm{Q}) = \dfrac{\mathrm{d}}{\mathrm{d}t}\boldsymbol{v}_{\mathrm{P}}(\mathrm{Q})$	acceleration of Q relative to P
$\dot{\theta} = \omega$	angular velocity
$\ddot{\theta} = \dot{\omega}$	angular acceleration

Mechanics

$\boldsymbol{g}$	acceleration due to gravity ($g \approx 9.8$ m s^{-2} on Earth)

force $\boldsymbol{F}$

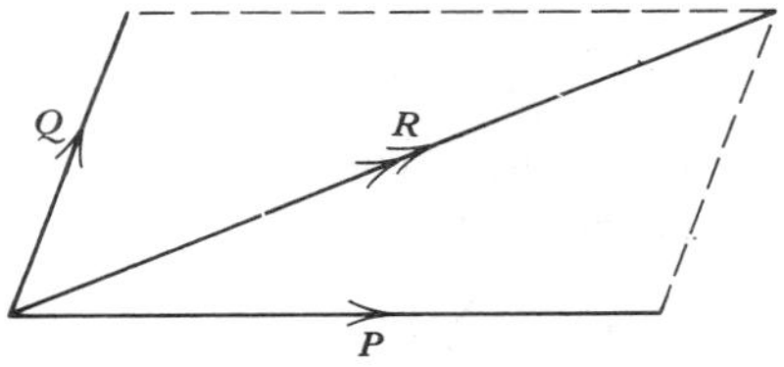

resultant $\boldsymbol{R}$, with components $\boldsymbol{P}$, $\boldsymbol{Q}$

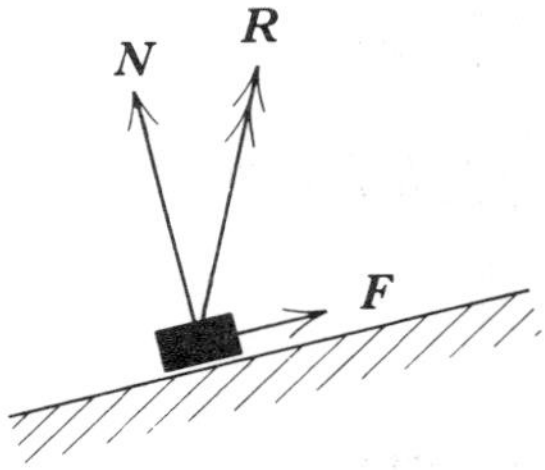

resultant reaction $\boldsymbol{R}$, with normal component $\boldsymbol{N}$ and frictional component $\boldsymbol{F}$

μ	coefficient of friction
λ	angle of friction
$G(\approx 6.67 \times 10^{-11})$	constant of gravitation

Unless otherwise stated, SI units are used throughout, and in questions involving gravitation the value of g has generally been taken as 10 m s^{-2}, the 2 per cent error being for present purposes unimportant.

0

Prologue

0.1 Logical symbols

Every piece of mathematics is about the consequences of a particular set of rules or axioms. As with the basic laws of a country or the rules of a game, some sets of axioms prove to be more useful or more interesting than others; but their only essential quality is to be clear and consistent, without contradictions.

From these axioms we proceed to other mathematical statements. This usually begins with a process of speculation, frequently needing considerable imagination and insight. If our speculations can be proved correct, we call them *deductions* from our axioms, and one of our first needs is for a set of symbols to denote this process of *implication*.

Suppose that p and q are two such statements. We shall use the symbols $\Rightarrow$, $\Leftarrow$, and $\Leftrightarrow$ to denote implications as follows:

$p \Rightarrow q$ means 'p implies q'
$p \Leftarrow q$ means 'p is implied by q' or 'q implies p'
$p \Leftrightarrow q$ means 'p implies q' and 'q implies p'.

For example, if p is the statement 'ABCD is a square'
and q is the statement 'ABCD is a rectangle',
then $p \Rightarrow q$.

Similarly $x^2 = 1 \quad \Leftarrow \quad x = 1$
and $\triangle$ABC is equilateral $\Leftrightarrow$ $\triangle$ABC is equiangular.

In this last case, the two statements are said to be *equivalent*.

A line striking through any of these symbols inserts the word 'not'.

So $p \nRightarrow q$ means 'p does not imply q'. This is not to say that q cannot be true, but simply that it is not a consequence of p.

For example

$$x^2 = 1 \nRightarrow x = 1$$

(for though x might be 1, it might equally well be -1. Such a case which shows the implication to be invalid is usually called a *counter-example.*)

If p is a given statement, we shall use p' to indicate its opposite, or *negation.*†

So if p denotes 'John is English',
then p' denotes 'John is not English'.

If, further, q denotes 'John is a Yorkshireman',
then q' denotes 'John is not a Yorkshireman' and we see that
$p \nRightarrow q, \quad p \Leftarrow q, \quad p' \Rightarrow q'$, etc.

Exercise 0.1

1 Use the symbols $\Rightarrow$, $\Leftarrow$, and $\Leftrightarrow$ to connect the statements p and q, where:

(*i*) p: Jock lives in Scotland
q: Jock lives in Glasgow

(*ii*) p: ABCD is a square
q: ABCD is a rhombus

(*iii*) p: ABCD is a parallelogram
q: ABCD is a rectangle

(*iv*) p: In $\triangle$ABC, $\hat{A} = 90°$
q: In $\triangle$ABC, $AB^2 + AC^2 = BC^2$

(*v*) p: Hexagon ABCDEF is equilateral
q: Hexagon ABCDEF is equiangular

2 Use the symbols $\nRightarrow$, $\nLeftarrow$, $\nLeftrightarrow$ to connect the above statements, giving counter-examples to show why the implication is not valid.

3 Repeat no. 1 for the statements p' and q'.

4 What do you know about p' and q' if (*i*) $p \Rightarrow q$. (*ii*) $p \Leftarrow q$, (*iii*) $p \Leftrightarrow q$?

5 Of three men only two always tell the truth. One day the first said to the second about the third, 'He's always truthful'. Shortly afterwards the second said the same to the third about the first. Which wouldn't you trust?

0.2 Sets and Boolean algebra

We now look briefly at the mathematics of collections, or sets. This may already be familiar to the reader and is of interest to us for two reasons: firstly because it leads to an algebra, named after its inventor George Boole (1815–64), which provides both striking similarities and also sharp contrasts with the algebra of ordinary numbers; and secondly because it gives

† $\sim p$ is sometimes used instead of p'.

us a language and a notation which will be particularly useful when we investigate (as in chapter 7) questions of probability.

Sets may be of any kind: the pupils of a school, the species of animals in a zoo, or the numbers which are multiples of 5. The only requirement is that we should always be able to decide whether or not an object belongs to the set: those which do are called its *elements*. 'Belongs to' is abbreviated as $\in$, whilst the set itself is usually written in braces, $\{\ \}$, or denoted by a capital letter $A, B, C, \ldots$.

So if A is the set of even integers from 2 to 10, we write

$$A = \{\text{Even integers from 2 to 10}\} = \{2, 4, 6, 8, 10\},$$

and $2 \in A$, $4 \in A$, but $3 \notin A$.

When our attention is entirely confined within a particular set of objects, we usually call it the *universal set*, which we shall represent by $\mathscr{E}$ (ensemble). Other sets of objects under consideration can then be regarded as *sub-sets* of $\mathscr{E}$.

If, for instance, we are interested only in the positive integers from 1 to 10, we can take

$$\mathscr{E} = \{1, 2, 3, 4, 5, 6, 7, 8, 9, 10\};$$

so $A = \{2, 4, 6, 8, 10\}$ is a sub-set of $\mathscr{E}$, and we write $A \subset \mathscr{E}$, or $\mathscr{E} \supset A$.

Using the symbol $\in$, we can also write

$$A = \{x : x \in \mathscr{E}, x \text{ is even}\}$$

Here the colon means 'such that'; so the whole sentence should be read:

'A is the set of every x belonging to $\mathscr{E}$ such that x is even.'

The *complement of A within $\mathscr{E}$* is the set of all elements of $\mathscr{E}$ which do not belong to A, and is written A'.

So if $\mathscr{E} = \{1, 2, 3, 4, 5, 6, 7, 8, 9, 10\}$
and $A = \{2, 4, 6, 8, 10\}$,
we see that $A' = \{1, 3, 5, 7, 9\}$.

Sometimes a particular set has no members,

e.g. $\{x : x \in \mathscr{E}, x \text{ is a multiple of } 11\}$

Such a set is called *empty*, or a *null-set*, and written $\varnothing$.

When two sets A and B contain exactly the same elements they are said to be *equal*, and we write $A = B$.

It is also useful to have special names for three important sets of numbers:

$$N = \{\text{natural numbers}\} = \{1, 2, 3, 4, 5 \ldots\}$$

$$Z = \{\text{all integers}\} = \{0, \pm 1, \pm 2, \pm 3 \ldots\}$$

$Q = \{\text{all rational numbers}\dagger\}$

$R = \{\text{all real numbers}\dagger\}$

The sets of all positive integers, positive rational numbers and positive real numbers are denoted by Z^+, Q^+, R^+ respectively.

So $Z^+ = \{x : x \in Z, x > 0\} = N$, etc.

Exercise 0.2a

1 Simplify (*i*) $\{x : x \in Z, -1 \leqslant x \leqslant 3\}$
(*ii*) $\{x : x \in Z^+, -1 \leqslant x \leqslant 3\}$
(*iii*) $\{x : x \in Z, x > 4\}$
(*iv*) $\{x : x \in Z^+, x \leqslant 2\}$
(*v*) $\{x : x \in Z^+, x \leqslant 0\}$.

2 Simplify (*i*) $\varnothing'$; (*ii*) $\mathscr{E}'$; (*iii*) $(A')'$.

Union and Intersection

If A and B are two sets, we now define two new sets:

The set of all elements belonging to A *or* B *or both* is called the *union* of A and B, and written $A \cup B$ or $A + B$.

The set of all elements belonging to A *and* B is called the *intersection* of A and B, and written $A \cap B$ or AB.

So if $A = \{a, b, c, d, e\}$

and $B = \{a, e, i, o, u\}$

$$A + B = A \cup B = \{a, b, c, d, e, i, o, u\}$$

and $$AB = A \cap B = \{a, e\}.$$

The use of two notations may seem to the reader unnecessarily confusing, particularly as he should become familiar with both. Each has its advantages. The symbols $\cup$ and $\cap$ ('cup' and 'cap') emphasise that union and intersection are completely new operations and have nothing to do with addition and multiplication of ordinary arithmetic. But once this is firmly recognised there is considerable advantage to be gained by use of the notation $A + B$ and AB. As A and B are sets, not numbers, there will be very little chance of confusion with addition and multiplication, and their use is more convenient: the expression $(A \cap B) \cup (C \cap D)$, for instance, looks considerably less forbidding when written $AB + CD$. For this reason we shall adopt it as our standard notation.

It will have been noticed that the operations of union and intersection are independent of the order of A and B, so that $A + B = B + A$ and $AB = BA$. Such operations are said to be *commutative*.

† See page 41.

They can also be illustrated very conveniently in *Venn diagrams:*

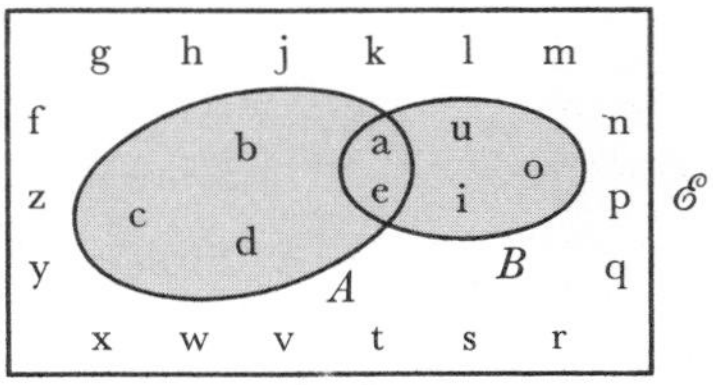

$A+B$

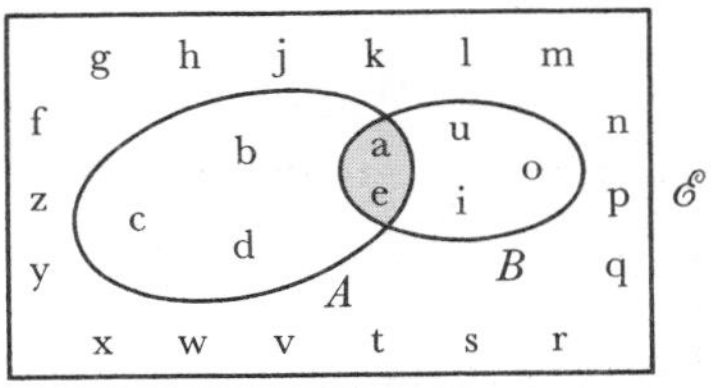

AB

$\mathscr{E}$ = {The 26 letters of the alphabet}

A = {The first 5 letters of the alphabet}

B = {The vowels}.

Such diagrams can also be used to depict the complement of A:

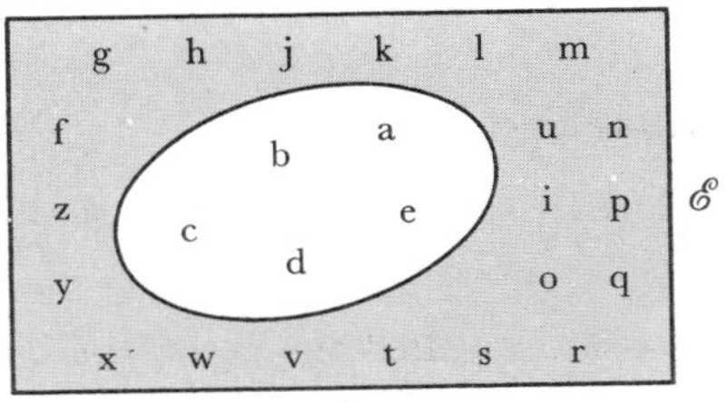

A'

Exercise 0.2b

1 Simplify:
(*i*) $A + \varnothing$; (*ii*) $A\varnothing$; (*iii*) $A + \mathscr{E}$; (*iv*) $A\mathscr{E}$; (*v*) $A + A$; (*iv*) AA; (*vii*) $A + AB$; (*viii*) $A(A + B)$; (*ix*) $A + A'$; (*x*) AA'.

2 Use Venn diagrams to show that union and intersection are *associative*: that is,
(*i*) $A + (B + C) = (A + B) + C$; (*ii*) $A(BC) = (AB)C$
(so that they can be written without ambiguity as $A + B + C$ and ABC).

3 Use Venn diagrams to show that *intersection is distributive over union*, and that *union is distributive over intersection*: that is,
(*i*) $A(B + C) = AB + AC$, (*ii*) $A + BC = (A + B)(A + C)$.

[The symmetry, or *duality*, of these two equations is best displayed in the other notation:
(*i*) $A \cap (B \cup C) = (A \cap B) \cup (A \cap C)$,
(*ii*) $A \cup (B \cap C) = (A \cup B) \cap (A \cup C)$.

It will also be noticed that whilst the first is similar to

$a(b + c) = ab + ac$

in ordinary algebra and arithmetic, the second has no such equivalent.]

4 Use Venn diagrams to verify *de Morgan's laws*:
(*i*) $(A + B)' = A'B'$; (*ii*) $(AB)' = A' + B'$.

5 Simplify:
(*i*) $A(A' + B)$; (*ii*) $AB(A' + B')$; (*iii*) $(A + B)' + B'$;
(*iv*) $(A + B)(A + B')$.

6 Simplify:
(*i*) $(B + C)(C + A)(A + B)$; (*ii*) $(A + B')(B + C')(C + A')$
(*iii*) $(A + B)(A + B' + C)$; (*iv*) $(A + B)(A + C)(B + C)C'$.

0.3 Coordinates

Suppose that a point O is marked on a plane, together with a pair of perpendicular lines, Ox and Oy, each with a uniform scale. The reader will be used to defining other points in the plane by means of *ordered pairs* of numbers (x, y), known as their *Cartesian* coordinates (after the French mathematician and philosopher, René Descartes (1596–1650)). O is called the *origin* and Ox, Oy are the *coordinate axes*.

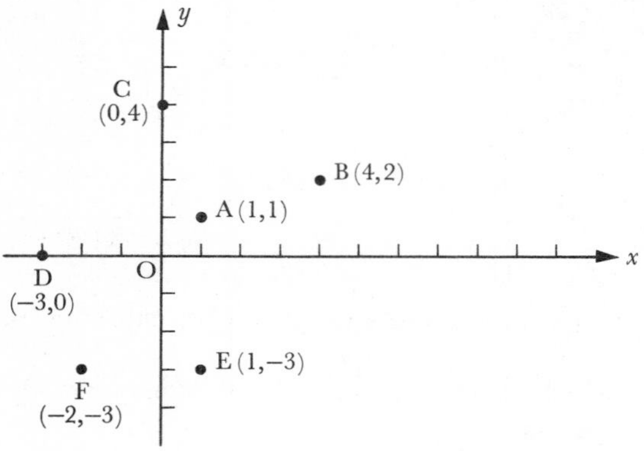

Mid-point, distance and gradient

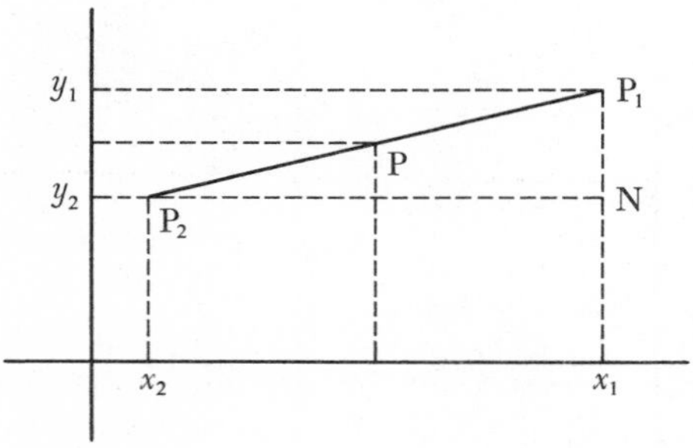

Suppose that P_1 is (x_1, y_1) and P_2 is (x_2, y_2). Hence the coordinates of the mid-point are given by

$$x = \tfrac{1}{2}(x_1 + x_2)$$

$$y = \tfrac{1}{2}(y_1 + y_2).$$

It is also seen that

$$P_1P_2^2 = P_2N^2 + NP_1^2$$

$$= (x_1 - x_2)^2 + (y_1 - y_2)^2$$

So $\quad P_1P_2 = \sqrt{[(x_1 - x_2)^2 + (y_1 - y_2)^2]}.$

Further the gradient of P_1P_2 is defined as

$$\frac{y_1 - y_2}{x_1 - x_2}$$

So, in our figure, the mid-points, lengths and gradients of AB, DE, EF, and AE are as follows

	mid-point	*length*	*gradient*
AB:	$\left(\frac{1+4}{2}, \frac{1+2}{2}\right) = \left(\frac{5}{2}, \frac{3}{2}\right)$	$\sqrt{[(4-1)^2 + (2-1)^2]} = \sqrt{10}$	$\frac{2-1}{4-1} = \frac{1}{3}$
DE:	$\left(\frac{-3+1}{2}, \frac{0-3}{2}\right) = \left(-1, -\frac{3}{2}\right)$	$\sqrt{[(1+3)^2 + (-3)^2]} = 5$	$\frac{-3-0}{1-(-3)} = \frac{3}{4}$
EF:	$\left(\frac{-2+1}{2}, \frac{-3-3}{2}\right) = \left(-\frac{1}{2}, -3\right)$	$\sqrt{[(1+2)^2 + 0^2]} = 3$	$\frac{-3+3}{1+2} = 0$
AE:	$\left(\frac{1+1}{2}, \frac{1-3}{2}\right) = (1, -1)$	$\sqrt{[(1-1)^2 + (1+3)^2]} = 4$	$\frac{1-(-3)}{1-1} = \frac{4}{0}$ which is undefined.

Perpendicular lines

Suppose that P is a point, not on a coordinate axis, with coordinates (h, k).

If OP is rotated anti-clockwise through a right angle, then P moves to Q, $(-k, h)$, and the gradients of OP and OQ are

$$m_1 = \frac{k}{h} \quad \text{and} \quad m_2 = -\frac{h}{k}$$

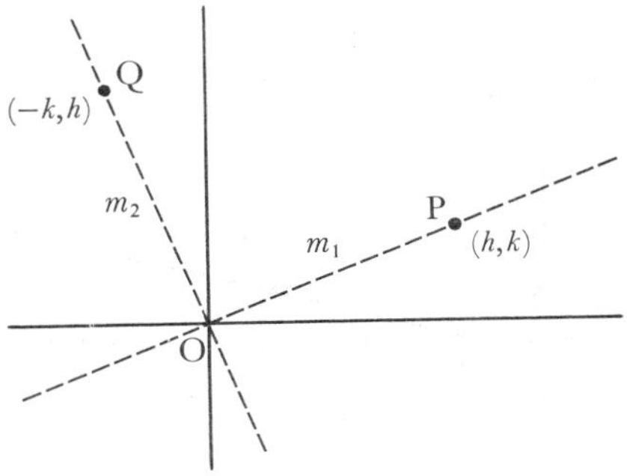

So OP, OQ perpendicular

$$\Rightarrow m_1m_2 = \frac{k}{h} \times -\frac{h}{k}$$

$$\Rightarrow \boxed{m_1m_2 = -1}$$

Conversely, if $m_1m_2 = -1$ it is readily seen that the two lines must be perpendicular.

Example

In our original figure (p. 6), gradient of AB $= \dfrac{2 - 1}{4 - 1} = \dfrac{1}{3}$

gradient of OE $= \dfrac{-3 - 0}{1 - 0} = -3$

So AB and OE are perpendicular.

Exercise 0.3a

For 1–3, draw axes Ox, Oy and label them from $x = -5$ to $+5$ and $y = -5$ to $+5$.

1 Mark the points
O (0, 0); A (4, 5); B (3, 0); C (0, −2); D (−2, 1).

2 Find the mid-points, lengths and gradients of OA, OB, OC, OD.
Name any two lines which are perpendicular.

3 Find the mid-points, lengths, and gradients of the six lines joining A, B, C, and D.
Name any pairs which are parallel or perpendicular.

4 M is the mid-point of side BC of △ABC. What do you know about $AB^2 + AC^2$ and $2(AM^2 + BM^2)$?
[Let M, B, C, A be $(0, 0)$, $(a, 0)$, $(-a, 0)$, (x, y) respectively.]

Equations and inequalities: curves and regions

Sometimes we do not know the exact position of a point P but merely that its coordinates x, y are restricted by some equation or inequality. There is then a corresponding restriction on the position of P and the set of all possible positions is a sub-set of all the points of the plane.

Example 1

If $y = x^2$, possible pairs (x, y) are $(0, 0)$, $(1, 1)$, $(2, 4)$, $(\frac{1}{2}, \frac{1}{4})$, $(-0.3, +0.09)$, etc, [but *not* $(1, 2)$, $(2, 3)$, $(-1, -1)$ etc.], and it is readily seen that the set of all points is indicated by:

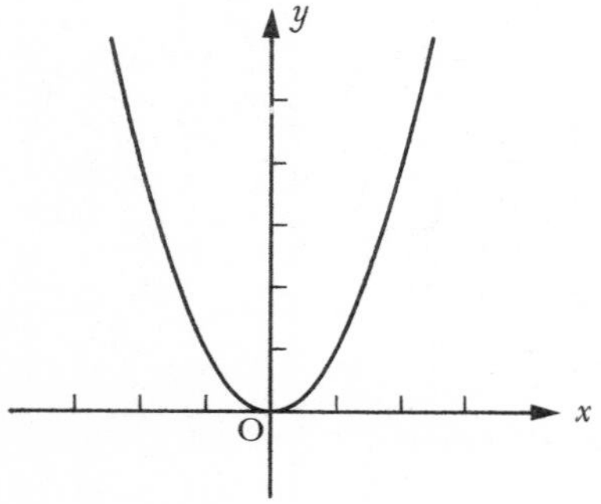

Example 2

Similarly, if $x + y > 2$, possible positions of (x, y) are $(2\frac{1}{2}, 0)$, $(1\frac{1}{4}, 1)$, $(-1, 4)$, $(5, -2.9)$, etc., [but *not* $(0, 1.9)$, $(-1, -2)$, $(1, 1)$, etc.]. So the set of all such points can be indicated by:

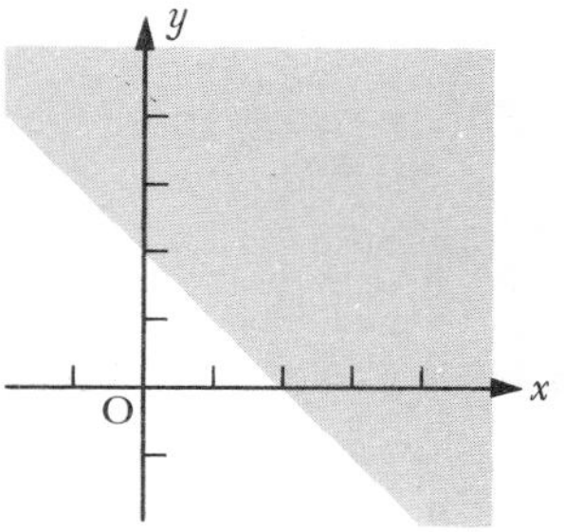

If we were told that $x + y \geqslant 2$, the region would also include the set of points for which $x + y = 2$ (i.e., the line $x + y = 2$):

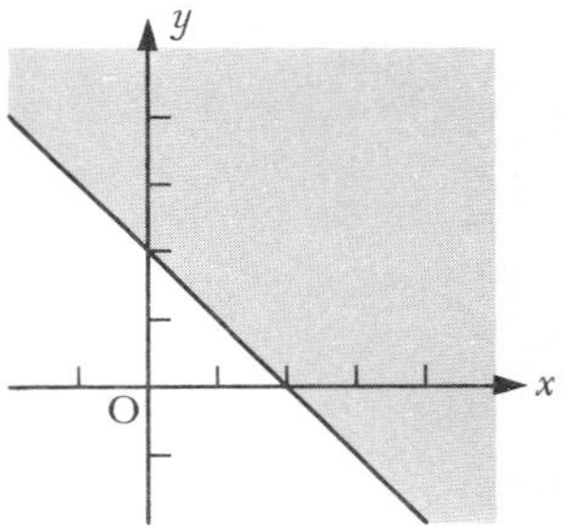

Example 3

Suppose (x, y) is restricted by $x + 3y = 9$ *and* $2x - y = 4$.

$$\left.\begin{array}{ll}\text{Then} & 2x + 6y = 18 \\ \text{and} & 2x - y = 4\end{array}\right\} \Rightarrow 7y = 14 \Rightarrow y = 2 \Rightarrow x = 3.$$

So $(3, 2)$ is the only possible position of (x, y) and this is the point of intersection of $x + 3y = 9$ and $2x - y = 4$.

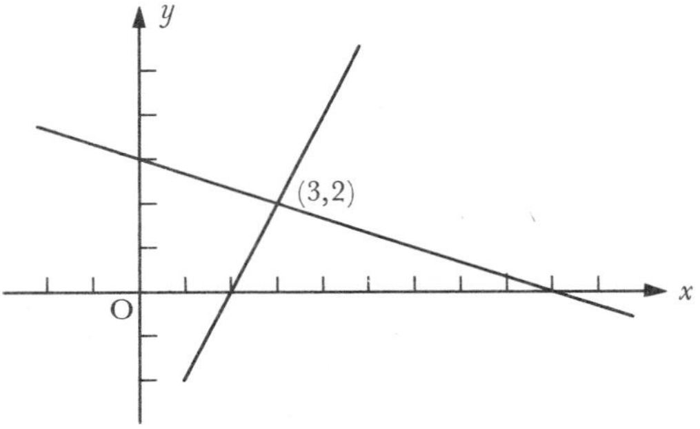

Exercise 0.3b

In each of numbers 1–10 draw axes Ox, Oy and label them from $x = -5$ to $+5$ and from $y = -5$ to $+5$. Then for each of the given equations or inequalities

(*a*) name four points which satisfy the restriction;
(*b*) name two points which do not satisfy the restriction;
(*c*) indicate in your figure the set of points defined by the restriction, i.e., the corresponding curve or region.

1 (*i*) $x + y < 4$; (*ii*) $x + y = 4$; (*iii*) $x + y > 4$.

2 (*i*) $y \leqslant 2x + 3$; (*ii*) $y \geqslant 2x + 3$.

3 (*i*) $x^2 + y^2 \leqslant 25$; (*ii*) $x^2 + y^2 > 25$.

4 (*i*) $xy = 12$; (*ii*) $xy \neq 12$.

5 (*i*) $xy > 0$; (*ii*) $xy = 0$.

6 (*i*) $x^2 + y^2 < 0$; (*ii*) $x^2 + y^2 = 0$; (*iii*) $x^2 + y^2 \geqslant 0$.

7 (*i*) $xy + 20 = 0$; (*ii*) $xy + 20 > 0$.

8 (*i*) $y = 2x$; (*ii*) $y = x$; (*iii*) $y = \frac{1}{2}x$; (*iv*) $y = 0x$; (*v*) $y = -x$.

9 (*i*) $x^2 - y^2 > 0$; (*ii*) $x^2 - y^2 = 0$; (*iii*) $x^2 - y^2 < 0$.

10 (*i*) $y = x^3$; (*ii*) $4x^2 + 9y^2 = 36$.

11 Illustrate the following, and in each case find and show in your illustration the points of intersection:

(*i*) $x + y = 4$ and $y = 2x + 1$
(*ii*) $x + y = 4$ and $7x + y + 8 = 0$
(*iii*) $y = 2x + 1$ and $7x + y + 8 = 0$
(*iv*) $y = x + 2$ and $y = x^2$
(*v*) $y = x^3$ and $y = 4x$.

12 Illustrate the region of the plane where

(*i*) $0 \leqslant y < x$
(*ii*) $x^2 + y^2 \leqslant 1$ and $xy > 0$
(*iii*) $x + 1 < y < 2x - 3$.

0.4 Straight lines

We saw in the last section that when the coordinates (x, y) of a point P were restricted by an equation, the point P was usually confined to a particular curve. So we refer to the original restriction as the *equation of the curve*. There is therefore a two-way problem: to investigate the curve which arises from a given equation, and to find the equation of a given curve. We shall begin with the case of a 'curve' which is a straight line.

Example 1

Find the equation of the straight line with gradient $\frac{1}{3}$ which passes through the point (1, 2).

P (x, y) is on this line

$$\Rightarrow \quad \frac{y-2}{x-1} = \frac{1}{3}$$

$$\Rightarrow 3y - 6 = x - 1$$

$$\Rightarrow \quad x - 3y + 5 = 0$$

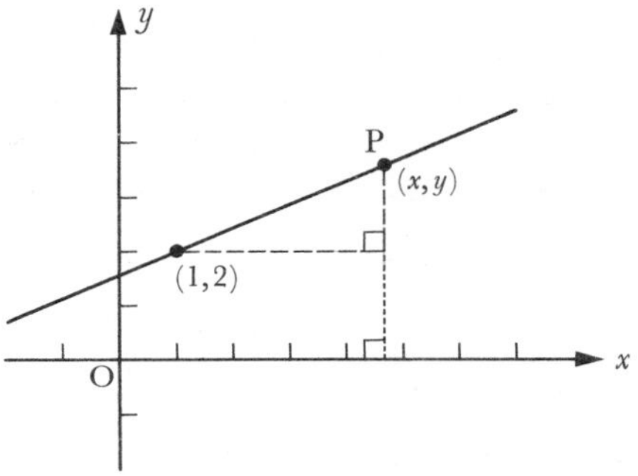

which is therefore the equation of the line.

Example 2

Find the equation of the line which passes through (1, 3) and (3, 6).

P (x, y) is on this line

$$\Rightarrow \quad \frac{y-6}{x-3} = \frac{6-3}{3-1} = \frac{3}{2}$$

$$\Rightarrow \quad 2(y-6) = 3(x-3)$$

$$\Rightarrow \quad 3x - 2y + 3 = 0$$

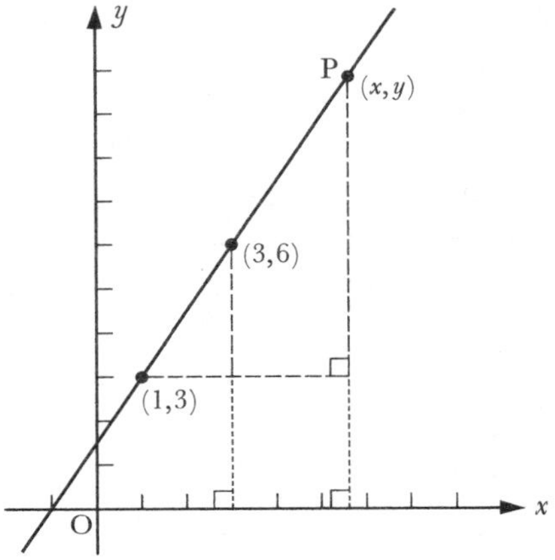

Example 3

A is (3, 4); B is (7, 5); C is (5, 1).

Find the equation of the line through A which is perpendicular to BC.

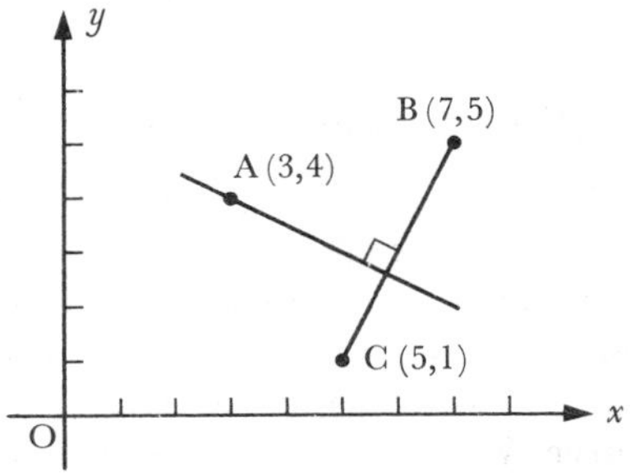

Gradient of BC $= \dfrac{5 - 1}{7 - 5} = \dfrac{4}{2} = 2$

$\Rightarrow$ Required gradient $= -\frac{1}{2}$.

So required line is $\dfrac{y - 4}{x - 3} = -\frac{1}{2} \Rightarrow x + 2y - 11 = 0.$

Exercise 0.4a

In 1–6, O is (0, 0); A is (3, 1); B is (5, 5); C is (9, 1).

Find the equations of:

1 (*i*) OA; (*ii*) OB; (*iii*) OC.

2 (*i*) BC; (*ii*) CA; (*iii*) AB.

3 the three *medians* (which join the vertices of $\triangle$ABC to the mid-points of the opposite sides). Show that they all meet in a point G (the *centroid* of $\triangle$ABC), and find the coordinates of G.

4 the three *altitudes* of $\triangle$ABC (which pass through the vertices and are perpendicular to the opposite sides). Show that they all meet in a point H (the *orthocentre* of $\triangle$ABC) and find the coordinates of H.

5 the three perpendicular bisectors of BC, CA, AB. Show that they all meet in a point X (the *circumcentre* of $\triangle$ABC), and find the coordinates of X. Verify by coordinate geometry that XA = XB = XC.

6 Show that the centroid, the orthocentre and the circumcentre all lie on a straight line, and find its equation.

7 Show that the equation of the straight line with gradient m which passes through the point $(0, c)$ is $y = mx + c$.

The general equation of a straight line

In all these examples the equation of a straight line has been of the form

$$ax + by + c = 0,$$

where a, b, c are constants and a, b are not both zero (since otherwise $c = 0$ and the equation would entirely collapse).

To examine this equation, we shall distinguish two cases:

Case 1 $b = 0$. Then $ax + c = 0$

$\Leftrightarrow x = -\dfrac{c}{a}$ (since $a \neq 0$)

and this is the equation of a straight line parallel to Oy.

Case 2 $b \neq 0$. Then $ax + by + c = 0$

$$\Leftrightarrow y = -\frac{a}{b}x - \frac{c}{b} \quad \text{(since } b \neq 0\text{)},$$

and (by No. 7 of the last exercise) this is the equation of a straight line of gradient $-\frac{a}{b}$ which goes through $\left(0, -\frac{c}{b}\right)$.

So again the given equation represents a straight line.

Hence in every case $ax + by + c = 0$ represents a straight line.

The plotting of straight lines is frequently helped by concentrating on particular features:

Example 1

$3x + 4y - 12 = 0$

$x = 0 \Rightarrow 4y - 12 = 0 \Rightarrow y = 3$

$y = 0 \Rightarrow 3x - 12 = 0 \Rightarrow x = 4.$

So the line passes through (0, 3) and (4, 0) and is therefore

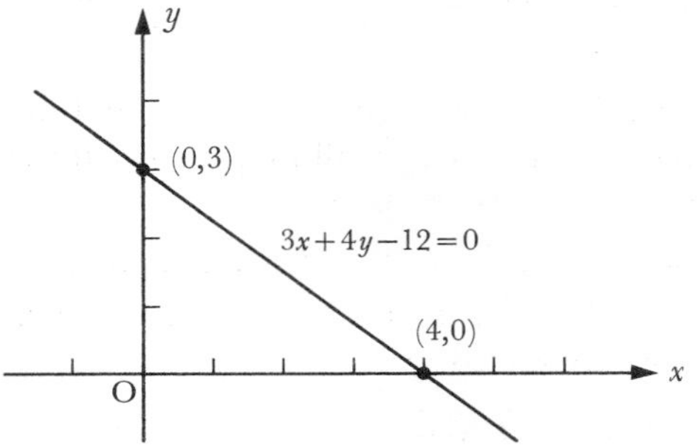

Example 2

$x + 2y = 0$

This line passes through (0, 0) and has gradient $-\frac{1}{2}$ (since $y = -\frac{1}{2}x$). So the line is

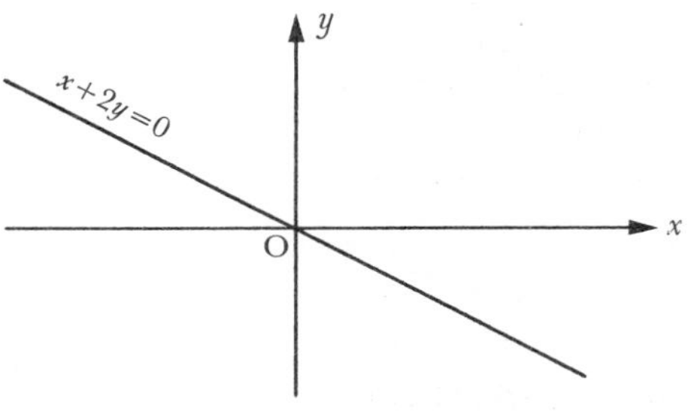

Exercise 0.4b

Sketch the lines:

1 (*i*) $y = x$; (*ii*) $y = x + 1$; (*iii*) $y = x + 2$; (*iv*) $y = x - 1$.
2 (*i*) $y = -2x$; (*ii*) $y = -2x + 1$; (*iii*) $y = -2x + 2$; (*iv*) $y = -2x - 2$.

Sketch the triangles defined by the following sets of lines and find the co-ordinates of their vertices:

3 (*i*) $x - 4 = 0$; (*ii*) $x + y - 4 = 0$; (*iii*) $x - y - 2 = 0$.

4 (*i*) $x - 2y + 3 = 0$; (*ii*) $x + y = 0$; (*iii*) $y + 2 = 0$.

5 (*i*) $2x + y - 4 = 0$; (*ii*) $3x - y - 6 = 0$; (*iii*) $x + y = 0$.

*0.5 Coordinate geometry

It is now clear that we are able to translate geometrical statements about points into algebraic statements about their coordinates, and vice-versa. We are therefore in a position to prove geometrical results by algebraic methods, or *coordinate geometry*, as it is usually called. This will not be pursued very far, but we shall illustrate how it can be developed for circles:

Example 1
Find the equation of the circle with centre C (2, 3) and radius 4

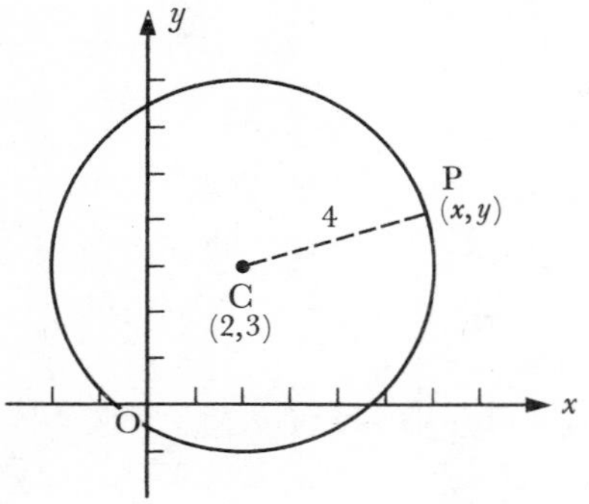

$$P(x, y) \text{ is on this circle} \Rightarrow CP^2 = 16$$

$$\Rightarrow \quad (x - 2)^2 + (y - 3)^2 = 16$$

$$\Rightarrow \quad x^2 + y^2 - 4x - 6y - 3 = 0$$

which is the equation of the circle.

Example 2
Find the curve represented by $x^2 + y^2 + 4x - 2y - 4 = 0$.

This can be written

$(x + 2)^2 + (y - 1)^2 = 9.$

So if P is (x, y) and C is $(-2, 1)$,

$$CP^2 = 9$$

$$\Rightarrow \quad CP = 3 \quad (\text{since } CP > 0)$$

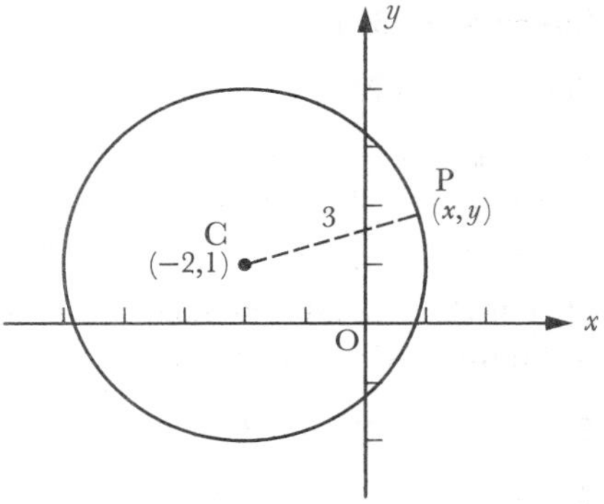

Hence the equation represents a circle whose centre is $(-2, 1)$ and radius is 3.

Example 3

A moving point P remains equidistant from the point A $(0, 1)$ and the line $y = -1$. Find the curve on which it moves.

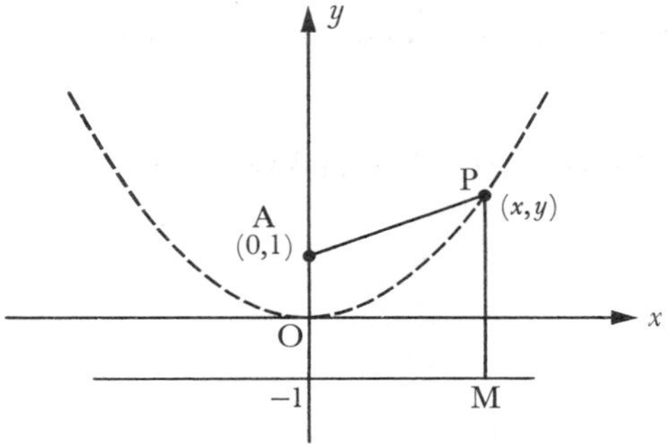

$AP^2 = x^2 + (y - 1)^2$ and $PM = y + 1$.

So $x^2 + (y - 1)^2 = (y + 1)^2$

$\Rightarrow \quad x^2 + y^2 - 2y + 1 = y^2 + 2y + 1$

$\Rightarrow \quad y = \frac{1}{4}x^2$

This is the equation of the path, or *locus*, of P, and is called a *parabola*.

Exercise 0.5

1 Find the equation of the circle

(*i*) with centre $(0, 0)$ and radius 4

(*ii*) with centre $(1, 2)$ and radius 3

(*iii*) with centre $(3, -4)$ and radius 5

(*iv*) with centre (a, b) and radius r.

2 What is the curve represented by

(*i*) $x^2 + y^2 = 10$
(*ii*) $x^2 + y^2 - 4x - 2y + 4 = 0$
(*iii*) $x^2 + y^2 + 6x - 8y = 0$
(*iv*) $x^2 + y^2 + 2x + 6y - 6 = 0$
(*v*) $x^2 - y = 0$
(*vi*) $xy = 12$.

3 Show that, if g, f, and c are constants,

$$x^2 + y^2 + 2gx + 2fy + c = 0$$

represents a circle, provided that $g^2 + f^2 > c$.
What is its centre and radius?

4 If A is $(1, 0)$ and B is $(-1, 0)$, find the curve on which P lies if

(*i*) $AP = 2BP$; (*ii*) $AP = BP$; (*iii*) $AP^2 + BP^2 = 4$; (*iv*) $AP^2 - BP^2 = 4$; (*v*) P is equidistant from A and the y-axis.

0.6 Linear transformations and matrices

Consider the equations

$$x' = 2x + y$$

$$y' = 3x - 2y$$

We can use these equations to transform any point P with coordinates (x, y) into another point P′ with coordinates (x', y'). For instance,

$(2, 1)$ is transformed into $(5, 4)$

$(0, 2)$ is transformed into $(2, -4)$

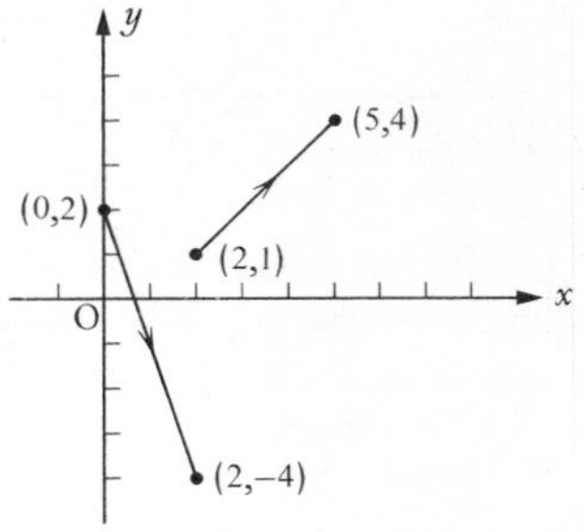

In this way we obtain a transformation of the entire plane, and this transformation is called 'linear' on account of the original equations. More generally,

$$x' = ax + by$$
$$y' = cx + dy$$

where a, b, c, d are constants, is a *linear transformation*.

We shall now investigate the effects of a number of simple linear transformations.

Example 1

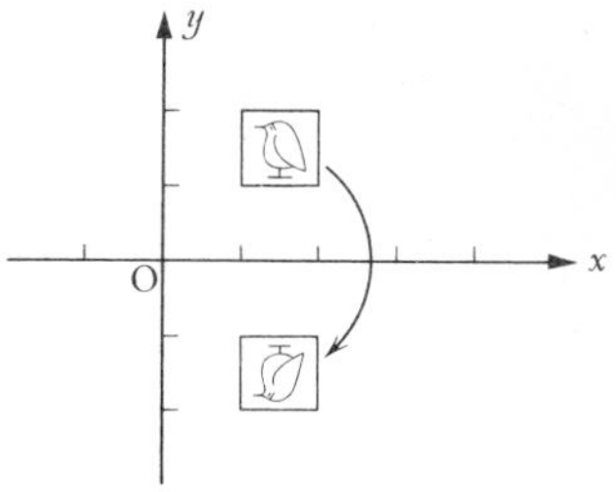

$x' = x$
$y' = -y$

transforms (1, 1) into (1, −1)

(1, 2) into (1, −2), etc.

So we see that this linear transformation represents a *reflection* in the x-axis.

Example 2

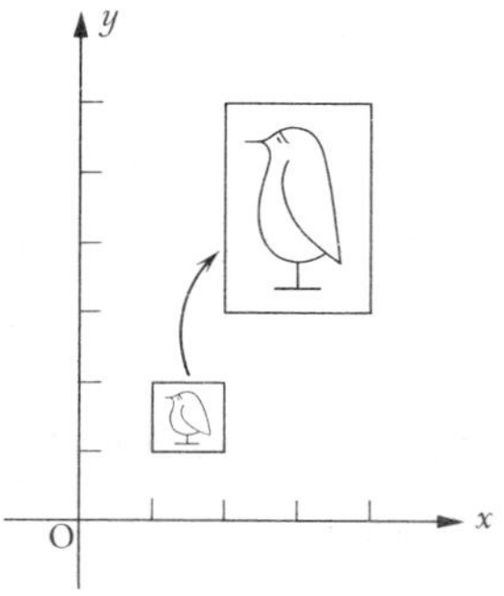

$x' = 2x$
$y' = 3y$

transforms (1, 1) into (2, 3)

(2, 2) into (4, 6), etc.,

so represents

a stretch $\times 2$ parallel Ox

and a stretch $\times 3$ parallel Oy

Example 3

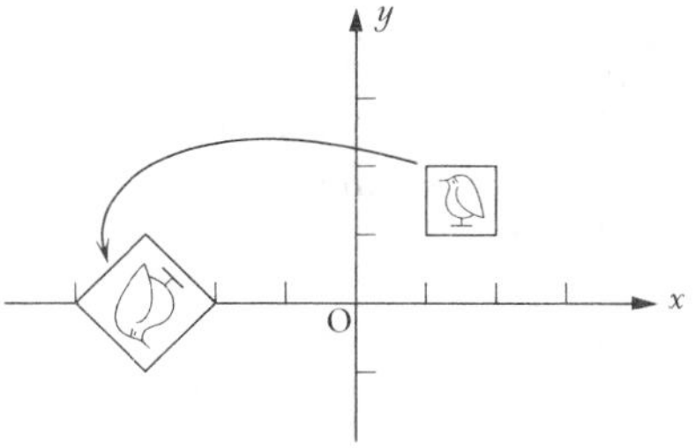

$x' = -x - y$
$y' = x - y$

The square ABCD, where

A is (1, 1), B (2, 1), C (2, 2), D (1, 2)

is transformed into A′B′C′D′ where

A′ is (−2, 0), B′ (−3, 1), C′ (−4, 0), D′ (−3, −1).

So the transformation represents an anticlockwise rotation through 135° combined with a 2-way stretch $\times\sqrt{2}$.

Example 4

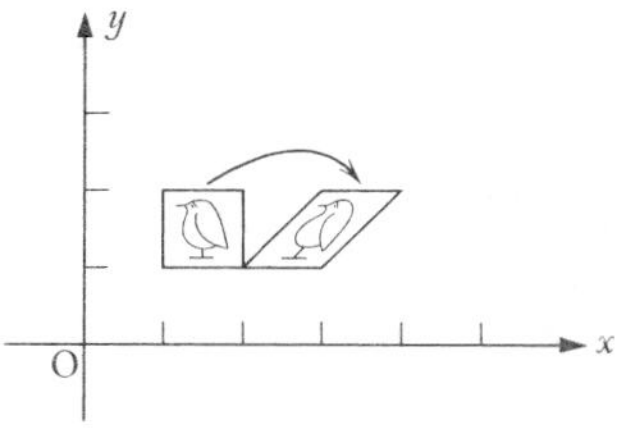

$x' = x + y$
$y' = y$

transforms A, B, C, D into

A′ (2, 1), B′ (3, 1), C′ (4, 2), D′ (3, 2).

Such a transformation is called a *shear*.

It is clear from these examples that the linear transformation

$x' = ax + by$
$y' = cx + dy$

is defined by the values of the four elements in the square arrangement:

$$\begin{pmatrix} a & b \\ c & d \end{pmatrix}.$$

This is called the *matrix* of the transformation, which can be written as

$$\begin{pmatrix} x' \\ y' \end{pmatrix} = \begin{pmatrix} a & b \\ c & d \end{pmatrix}\begin{pmatrix} x \\ y \end{pmatrix},$$

where $\begin{pmatrix} a & b \\ c & d \end{pmatrix}\begin{pmatrix} x \\ y \end{pmatrix} = \begin{pmatrix} ax + by \\ cx + dy \end{pmatrix}.$

Example 5

What are the transformations defined by

(*i*) $\begin{pmatrix} 3 & 0 \\ 0 & 1 \end{pmatrix}$ (*ii*) $\begin{pmatrix} 0 & 1 \\ -1 & 0 \end{pmatrix}$ (*iii*) $\begin{pmatrix} 2 & -2 \\ -1 & 1 \end{pmatrix}$?

(*i*) $\begin{pmatrix} 3 & 0 \\ 0 & 1 \end{pmatrix}$ is the matrix of the transformation

$$\begin{pmatrix} x' \\ y' \end{pmatrix} = \begin{pmatrix} 3 & 0 \\ 0 & 1 \end{pmatrix}\begin{pmatrix} x \\ y \end{pmatrix} = \begin{pmatrix} 3x + 0y \\ 0x + 1y \end{pmatrix} = \begin{pmatrix} 3x \\ y \end{pmatrix}$$

and so represents a stretch ×3 parallel to Ox:

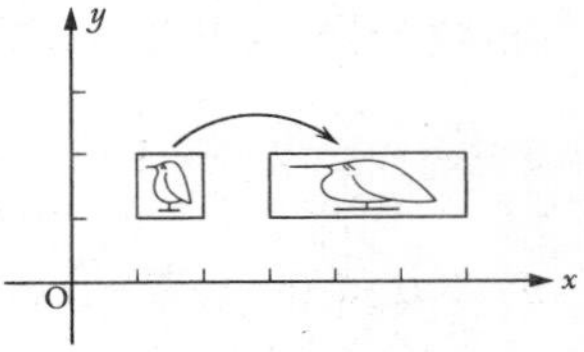

(*ii*) $\begin{pmatrix} 0 & 1 \\ -1 & 0 \end{pmatrix}$ is the matrix of

$$\begin{pmatrix} x' \\ y' \end{pmatrix} = \begin{pmatrix} 0 & 1 \\ -1 & 0 \end{pmatrix}\begin{pmatrix} x \\ y \end{pmatrix} = \begin{pmatrix} 0x + 1y \\ -1x + 0y \end{pmatrix}$$
$$= \begin{pmatrix} y \\ -x \end{pmatrix}$$

and so represents a clockwise rotation through 90°:

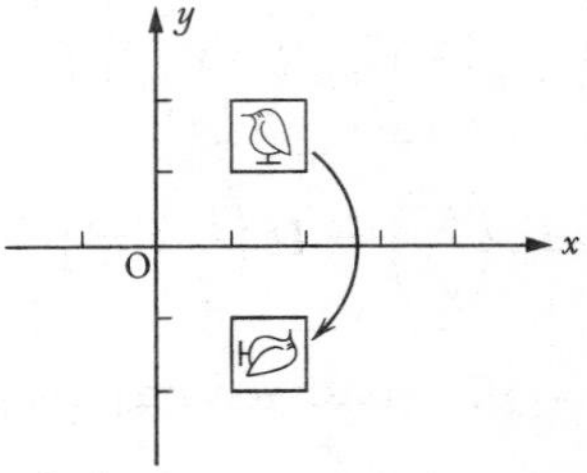

(*iii*) $\begin{pmatrix} 2 & -2 \\ -1 & 1 \end{pmatrix}$ is the matrix of

$$\begin{pmatrix} x' \\ y' \end{pmatrix} = \begin{pmatrix} 2 & -2 \\ -1 & 1 \end{pmatrix}\begin{pmatrix} x \\ y \end{pmatrix} = \begin{pmatrix} 2x - 2y \\ -x + y \end{pmatrix}$$

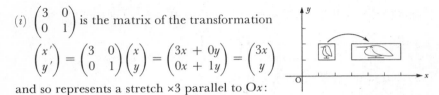

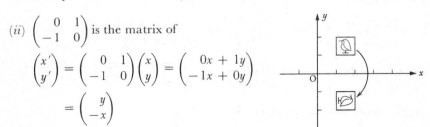

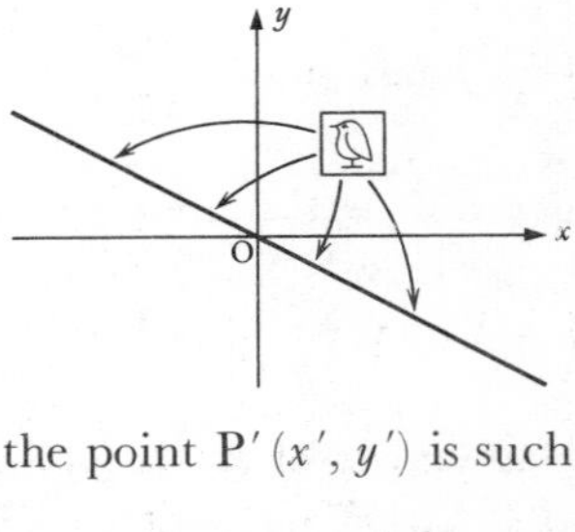

from which we see that for every point P(x, y), the point P$'(x', y')$ is such that $x' = -2y'$.

Hence every point P is transformed into a point of the line $x' = -2y'$, and our figure becomes squashed and unrecognisable.

Such a transformation and its corresponding matrix are called *singular*.

Exercise 0.6

Transform the points

O (0, 0), U (1, 0), V (0, 1), W (1, 1)

by means of the following matrices, and in each case illustrate the effect of the transformation:

1 $\begin{pmatrix} 1 & 0 \\ 0 & 1 \end{pmatrix}$, usually called $\boldsymbol{I}$ (the *identity* or *unit* matrix).

2 $\begin{pmatrix} -1 & 0 \\ 0 & 1 \end{pmatrix}$ 3 $\begin{pmatrix} -1 & 0 \\ 0 & -1 \end{pmatrix}$ 4 $\begin{pmatrix} 0 & 1 \\ 1 & 0 \end{pmatrix}$ 5 $\begin{pmatrix} 2 & 0 \\ 0 & 2 \end{pmatrix}$

6 $\begin{pmatrix} 0 & 0 \\ 0 & 0 \end{pmatrix}$ 7 $\begin{pmatrix} 0 & -1 \\ 1 & 0 \end{pmatrix}$ 8 $\begin{pmatrix} 0 & 2 \\ -2 & 0 \end{pmatrix}$ 9 $\begin{pmatrix} 1 & 1 \\ 1 & 1 \end{pmatrix}$ 10 $\begin{pmatrix} 1 & 0 \\ 1 & 1 \end{pmatrix}$.

0.7 Multiplication of matrices

Suppose that a transformation with matrix $\boldsymbol{T} = \begin{pmatrix} a & b \\ c & d \end{pmatrix}$ is followed by a transformation with matrix $\boldsymbol{U} = \begin{pmatrix} \alpha & \beta \\ \gamma & \delta \end{pmatrix}$.

If $\boldsymbol{T}$ transforms P (x, y) into P′ (x', y')
and $\boldsymbol{U}$ transforms P′ (x', y') into P″ (x'', y'')

then $\begin{pmatrix} x' \\ y' \end{pmatrix} = \boldsymbol{T}\begin{pmatrix} x \\ y \end{pmatrix} = \begin{pmatrix} a & b \\ c & d \end{pmatrix}\begin{pmatrix} x \\ y \end{pmatrix} \Leftrightarrow \begin{aligned} x' &= ax + by \\ y' &= cx + dy \end{aligned}$

and $\begin{pmatrix} x'' \\ y'' \end{pmatrix} = \boldsymbol{U}\begin{pmatrix} x' \\ y' \end{pmatrix} = \begin{pmatrix} \alpha & \beta \\ \gamma & \delta \end{pmatrix}\begin{pmatrix} x' \\ y' \end{pmatrix} \Leftrightarrow \begin{aligned} x'' &= \alpha x' + \beta y' \\ y'' &= \gamma x' + \delta y' \end{aligned}$

So the combined transformation is

$$x'' = \alpha(ax + by) + \beta(cx + dy) = (\alpha a + \beta c)\,x + (\alpha b + \beta d)\,y$$

$$y'' = \gamma(ax + by) + \delta(cx + dy) = (\gamma a + \delta c)\,x + (\gamma b + \delta d)\,y,$$

which has matrix $\begin{pmatrix} \alpha a + \beta c & \alpha b + \beta d \\ \gamma a + \delta c & \gamma b + \delta d \end{pmatrix}$.

But this combination can also be written

$$\begin{pmatrix} x'' \\ y'' \end{pmatrix} = \boldsymbol{U}\begin{pmatrix} x' \\ y' \end{pmatrix} = \boldsymbol{UT}\begin{pmatrix} x \\ y \end{pmatrix}.$$

The matrix of the combined transformation is therefore called the *product* of $\boldsymbol{U}$ and $\boldsymbol{T}$ and written $\boldsymbol{UT}$.

So $\boldsymbol{UT} = \begin{pmatrix} \alpha & \beta \\ \gamma & \delta \end{pmatrix}\begin{pmatrix} a & b \\ c & d \end{pmatrix} = \begin{pmatrix} \alpha a + \beta c & \alpha b + \beta d \\ \gamma a + \delta c & \gamma b + \delta d \end{pmatrix}$.

The reader should note very clearly that transformation $\boldsymbol{T}$ followed by $\boldsymbol{U}$ has matrix $\boldsymbol{UT}$, and $\boldsymbol{U}$ followed by $\boldsymbol{T}$ has matrix $\boldsymbol{TU}$, and can easily show that these two matrices are normally different.

Example 1

If $\boldsymbol{A} = \begin{pmatrix} 1 & 2 \\ 3 & 4 \end{pmatrix}$, $\boldsymbol{B} = \begin{pmatrix} 5 & 6 \\ 7 & 8 \end{pmatrix}$, find $\boldsymbol{AB}$ and $\boldsymbol{BA}$

$$\boldsymbol{AB} = \begin{pmatrix} 1 & 2 \\ 3 & 4 \end{pmatrix}\begin{pmatrix} 5 & 6 \\ 7 & 8 \end{pmatrix} = \begin{pmatrix} 1\times5 + 2\times7 & 1\times6 + 2\times8 \\ 3\times5 + 4\times7 & 3\times6 + 4\times8 \end{pmatrix} = \begin{pmatrix} 19 & 22 \\ 43 & 50 \end{pmatrix}$$

and $$\boldsymbol{BA} = \begin{pmatrix} 5 & 6 \\ 7 & 8 \end{pmatrix}\begin{pmatrix} 1 & 2 \\ 3 & 4 \end{pmatrix} = \begin{pmatrix} 5\times1 + 6\times3 & 5\times2 + 6\times4 \\ 7\times1 + 8\times3 & 7\times2 + 8\times4 \end{pmatrix} = \begin{pmatrix} 23 & 34 \\ 31 & 46 \end{pmatrix}$$

So $\boldsymbol{AB} \neq \boldsymbol{BA}$

and we see that in general the multiplication of matrices is *not* commutative.

Example 2

If $\boldsymbol{I} = \begin{pmatrix} 1 & 0 \\ 0 & 1 \end{pmatrix}$, $\boldsymbol{A} = \begin{pmatrix} a & b \\ c & d \end{pmatrix}$, calculate $\boldsymbol{IA}$ and $\boldsymbol{AI}$.

$$\boldsymbol{IA} = \begin{pmatrix} 1 & 0 \\ 0 & 1 \end{pmatrix}\begin{pmatrix} a & b \\ c & d \end{pmatrix} = \begin{pmatrix} 1a + 0c & 1b + 0d \\ 0a + 1c & 0b + 1d \end{pmatrix} = \begin{pmatrix} a & b \\ c & d \end{pmatrix}$$

$$\boldsymbol{AI} = \begin{pmatrix} a & b \\ c & d \end{pmatrix}\begin{pmatrix} 1 & 0 \\ 0 & 1 \end{pmatrix} = \begin{pmatrix} a1 + b0 & a0 + b1 \\ c1 + d0 & c0 + d1 \end{pmatrix} = \begin{pmatrix} a & b \\ c & d \end{pmatrix}$$

So $\boldsymbol{IA} = \boldsymbol{AI} = \boldsymbol{A}$,

and we call $\boldsymbol{I}$ the *unit* (or *identity*) matrix.

Example 3

If $\boldsymbol{A} = \begin{pmatrix} 1 & 0 \\ 0 & -1 \end{pmatrix}$, $\boldsymbol{B} = \begin{pmatrix} -1 & 0 \\ 0 & 1 \end{pmatrix}$, $\boldsymbol{C} = \begin{pmatrix} -1 & 0 \\ 0 & -1 \end{pmatrix}$

calculate $\boldsymbol{A}^2$ ($=\boldsymbol{AA}$), $\boldsymbol{B}^2$, and $\boldsymbol{C}^2$ and explain the meaning of your answers.

$$\boldsymbol{A}^2 = \begin{pmatrix} 1 & 0 \\ 0 & -1 \end{pmatrix}\begin{pmatrix} 1 & 0 \\ 0 & -1 \end{pmatrix} = \begin{pmatrix} 1 & 0 \\ 0 & 1 \end{pmatrix} = \boldsymbol{I}$$

$$\boldsymbol{B}^2 = \begin{pmatrix} -1 & 0 \\ 0 & 1 \end{pmatrix}\begin{pmatrix} -1 & 0 \\ 0 & 1 \end{pmatrix} = \begin{pmatrix} 1 & 0 \\ 0 & 1 \end{pmatrix} = \boldsymbol{I}$$

$$\boldsymbol{C}^2 = \begin{pmatrix} -1 & 0 \\ 0 & -1 \end{pmatrix}\begin{pmatrix} -1 & 0 \\ 0 & -1 \end{pmatrix} = \begin{pmatrix} 1 & 0 \\ 0 & 1 \end{pmatrix} = \boldsymbol{I}$$

Now $\boldsymbol{A}$, $\boldsymbol{B}$, $\boldsymbol{C}$ are the matrices which represent reflection in Ox, reflection in Oy and point-reflection in O, respectively; so it is clear that each of them, when repeated, leads to the '*identity*' transformation, which leaves every point unchanged and has matrix $\boldsymbol{I}$.

Exercise 0.7

1 $A = \begin{pmatrix} 1 & 0 \\ 0 & -1 \end{pmatrix}$, $B = \begin{pmatrix} -1 & 0 \\ 0 & 1 \end{pmatrix}$, $C = \begin{pmatrix} -1 & 0 \\ 0 & -1 \end{pmatrix}$.

Calculate (*i*) $\boldsymbol{AB}$ and $\boldsymbol{BA}$
(*ii*) $\boldsymbol{AC}$ and $\boldsymbol{CA}$
(*iii*) $\boldsymbol{BC}$ and $\boldsymbol{CB}$

and hence construct a multiplication table for the matrices $\boldsymbol{I}$, $\boldsymbol{A}$, $\boldsymbol{B}$, $\boldsymbol{C}$. Explain the meaning of your answers.

2 If $\boldsymbol{P} = \begin{pmatrix} 0 & 1 \\ -1 & 0 \end{pmatrix}$ and $\boldsymbol{Q} = \begin{pmatrix} 0 & -1 \\ 1 & 0 \end{pmatrix}$

calculate (*i*) $\boldsymbol{PQ}$ and $\boldsymbol{QP}$
(*ii*) $\boldsymbol{P}^2$ and $\boldsymbol{Q}^2$.

Explain the meaning of your answers.

3 If $\boldsymbol{U} = \begin{pmatrix} 0 & 1 \\ 1 & 0 \end{pmatrix}$ and $\boldsymbol{V} = \begin{pmatrix} -1 & 0 \\ 0 & 1 \end{pmatrix}$

calculate (*i*) $\boldsymbol{UV}$ and $\boldsymbol{VU}$
(*ii*) $\boldsymbol{U}^2$ and $\boldsymbol{V}^2$.

Explain the meaning of your answers.

4 If $\boldsymbol{L} = \begin{pmatrix} a & b \\ c & d \end{pmatrix}$, $\boldsymbol{M} = \begin{pmatrix} e & f \\ g & h \end{pmatrix}$, $\boldsymbol{N} = \begin{pmatrix} p & q \\ r & s \end{pmatrix}$

calculate $\boldsymbol{L}(\boldsymbol{MN})$ and $(\boldsymbol{LM})\boldsymbol{N}$, and so show that matrix multiplication is *associative*.

5 If $\boldsymbol{A} \neq \begin{pmatrix} 0 & 0 \\ 0 & 0 \end{pmatrix}$, is it true that $\boldsymbol{AB} = \boldsymbol{AC} \Rightarrow \boldsymbol{B} = \boldsymbol{C}$?

*0.8 The inverse of a matrix

Suppose that a linear transformation has matrix $\boldsymbol{A} = \begin{pmatrix} a & b \\ c & d \end{pmatrix}$ and transforms the point $P(x, y)$ into $P'(x', y')$.

Can we find the inverse transformation which moves P' into P?

In most cases this should be simple: the inverse of a stretch is a contraction and of a clockwise rotation is an anticlockwise rotation, whilst reflection is clearly its own inverse. We should expect, however, there to be some cases where the answer is not so simple. In Example 5 (*iii*) of p. 18, for instance, we saw that the whole of the plane was squashed on to the line $x = 2y$. So it appears that if P' is on this line there is any number of possible points P, whilst if P' is not on this line there is *no* point P. It was for this reason that we called it a 'singular' transformation and we see that such a transformation has no inverse.

In general, however

$$x' = ax + by$$

$$y' = cx + dy$$

$$\Rightarrow \quad \begin{aligned} dx' &= adx + bdy \\ by' &= bcx + bdy \end{aligned} \quad \Rightarrow \quad (ad - bc)x = dx' - by'$$

and $$\begin{aligned} cx' &= acx + bcy \\ ay' &= acx + ady \end{aligned} \quad \Rightarrow \quad (ad - bc)y = -cx' + ay'$$

For convenience, the value $ad - bc$ is called Δ and we shall examine separately the two cases (*i*) $\Delta \neq 0$, (*ii*) $\Delta = 0$.

(*i*) $\Delta \neq 0$. The above equations can then be written

$$x = \frac{d}{\Delta}x' - \frac{b}{\Delta}y'$$

$$y = -\frac{c}{\Delta}x' + \frac{a}{\Delta}y'$$

so that the inverse transformation has matrix

$$\begin{pmatrix} d/\Delta & -b/\Delta \\ -c/\Delta & a/\Delta \end{pmatrix}$$

We readily see that

$$\begin{pmatrix} a & b \\ c & d \end{pmatrix}\begin{pmatrix} d/\Delta & -b/\Delta \\ -c/\Delta & a/\Delta \end{pmatrix} = \begin{pmatrix} (ad - bc)/\Delta & 0 \\ 0 & (ad - bc)/\Delta \end{pmatrix}$$
$$= \begin{pmatrix} 1 & 0 \\ 0 & 1 \end{pmatrix} = \boldsymbol{I}$$

and $$\begin{pmatrix} d/\Delta & -b/\Delta \\ -c/\Delta & a/\Delta \end{pmatrix}\begin{pmatrix} a & b \\ c & d \end{pmatrix} = \begin{pmatrix} (ad - bc)/\Delta & 0 \\ 0 & (ad - bc)/\Delta \end{pmatrix}$$
$$= \begin{pmatrix} 1 & 0 \\ 0 & 1 \end{pmatrix} = \boldsymbol{I}$$

So $\begin{pmatrix} d/\Delta & -b/\Delta \\ -c/\Delta & a/\Delta \end{pmatrix}$ is called the *inverse* of $\begin{pmatrix} a & b \\ c & d \end{pmatrix}$ and we write

$$\boldsymbol{A}^{-1} = \begin{pmatrix} d/\Delta & -b/\Delta \\ -c/\Delta & a/\Delta \end{pmatrix}$$

Hence $\boldsymbol{AA}^{-1} = \boldsymbol{A}^{-1}\boldsymbol{A} = \boldsymbol{I}$.

For example, $\begin{aligned} x' &= 4x + 3y \\ y' &= x + 2y \end{aligned}$ has matrix $\begin{pmatrix} 4 & 3 \\ 1 & 2 \end{pmatrix}$

Now $\Delta = 4 \times 2 - 3 \times 1 = 5$

So $A^{-1} = \begin{pmatrix} \frac{2}{5} & -\frac{3}{5} \\ -\frac{1}{5} & \frac{4}{5} \end{pmatrix}$

and the inverse transformation is

$$x = \tfrac{2}{5}x' - \tfrac{3}{5}y'$$

$$y = -\tfrac{1}{5}x' + \tfrac{4}{5}y'$$

(*ii*) $\Delta = 0$. We have already considered the singular transformation

$$\begin{aligned} x' &= 2x - 2y \\ y' &= -x + y \end{aligned} \quad \text{with matrix} \quad \begin{pmatrix} 2 & -2 \\ -1 & 1 \end{pmatrix}$$

In this case $\Delta = (2 \times 1) - (-2 \times -1) = 0$, so that A^{-1} does not exist.

As the transformation is singular, this was only to be expected. But we now see that such a situation was determined by the fact that $\Delta = 0$. For this reason, the value of $\Delta = ad - bc$ is called

the *determinant* of $\begin{pmatrix} a & b \\ c & d \end{pmatrix}$

and we write $\Delta = \begin{vmatrix} a & b \\ c & d \end{vmatrix}$.

The fundamental difference between the matrix and the determinant should be quite clear:

The matrix $\begin{pmatrix} 1 & 2 \\ 3 & 4 \end{pmatrix}$ is simply a square array of numbers and one cannot speak of its value. By contrast the determinant $\begin{vmatrix} 1 & 2 \\ 3 & 4 \end{vmatrix}$ does possess a numerical value, namely $1 \times 4 - 2 \times 3 = -2$.

Example 1

Find the points of intersection of the lines:

(*i*) $x - y = 6$ and $2x + 3y = 4$

(*ii*) $x - y = 6$ and $-2x + 2y = 5$.

(*i*) $\begin{aligned} x - y &= 6 \\ 2x + 3y &= 4 \end{aligned}$ can be written $\begin{pmatrix} 1 & -1 \\ 2 & 3 \end{pmatrix}\begin{pmatrix} x \\ y \end{pmatrix} = \begin{pmatrix} 6 \\ 4 \end{pmatrix}$

The inverse of $\begin{pmatrix} 1 & -1 \\ 2 & 3 \end{pmatrix}$ is $\begin{pmatrix} 3/5 & 1/5 \\ -2/5 & 1/5 \end{pmatrix}$

So $\begin{pmatrix} x \\ y \end{pmatrix} = \begin{pmatrix} \frac{3}{5} & \frac{1}{5} \\ -\frac{2}{5} & \frac{1}{5} \end{pmatrix}\begin{pmatrix} 6 \\ 4 \end{pmatrix} = \begin{pmatrix} \frac{22}{5} \\ -\frac{8}{5} \end{pmatrix}$

and the required point is $(\frac{22}{5}, -\frac{8}{5})$.

(ii) $\begin{array}{r} x - y = 6 \\ -2x + 2y = 5 \end{array}$ can be written $\begin{pmatrix} 1 & -1 \\ -2 & 2 \end{pmatrix}\begin{pmatrix} x \\ y \end{pmatrix} = \begin{pmatrix} 6 \\ 5 \end{pmatrix}$

But $\begin{pmatrix} 1 & -1 \\ -2 & 2 \end{pmatrix}$ has no inverse (since $\Delta = 0$).

So it appears that the given lines have no point of intersection, which is immediately verified from the fact that they are parallel.

Exercise 0.8

1 Calculate, if possible, A^{-1} and check that $AA^{-1} = A^{-1}A = I$, when A is

(i) $\begin{pmatrix} 2 & 0 \\ 0 & 1 \end{pmatrix}$; (ii) $\begin{pmatrix} 3 & 0 \\ 0 & 3 \end{pmatrix}$; (iii) $\begin{pmatrix} -1 & 0 \\ 0 & 1 \end{pmatrix}$; (iv) $\begin{pmatrix} 0 & 1 \\ -1 & 0 \end{pmatrix}$;

(v) $\begin{pmatrix} 1 & -1 \\ -1 & 1 \end{pmatrix}$; (vi) $\begin{pmatrix} 0 & 1 \\ 1 & 0 \end{pmatrix}$.

2 Simplify: (i) I^{-1}; (ii) $(A^{-1})^{-1}$.

3 Evaluate the determinants

(i) $\begin{vmatrix} 4 & 3 \\ 1 & 2 \end{vmatrix}$; (ii) $\begin{vmatrix} -2 & 1 \\ -4 & 1 \end{vmatrix}$; (iii) $\begin{vmatrix} 8 & -6 \\ -4 & 3 \end{vmatrix}$.

4 Calculate the inverse of $\begin{pmatrix} 4 & 5 \\ 2 & 3 \end{pmatrix}$ and use your result to solve the equations

(i) $4x + 5y = 2,$ $2x + 3y = 4$ (ii) $4x + 5y = 3,$ $2x + 3y = 1$

(iii) $4x + 5y = -1,$ $2x + 3y = 2$ (iv) $4x + 5y = -4,$ $2x + 3y = -3$

5 Use inverse matrices where possible to find the points of intersection of the pairs of lines:

(i) $2x + 5y = 7,$ $x + 3y = 11$ (ii) $3x - y = -3,$ $x + 2y = 4$

(iii) $x + 2y = -6,$ $2x + 4y = 5$ (iv) $-7x - 3y = 2,$ $x + y = -3$

I

Functions and their graphs

1.1 Functions

Mathematics is largely concerned with relationships: how things depend on one another. In particular, when we are dealing with quantities or numbers, *that which depends* is called a *function* of *that* (*or those*) *on which it depends*. Because we usually have some freedom, however restricted, to change these latter quantities, they are usually called *independent variables*; the former is the *dependent variable*.

For example, if we are told the radius r of a circle, its area A is automatically determined.

So *A is a function of r.*

This function, which we call f, is simply the name of a piece of machinery for converting an input r into an output A.

We write $f: r \mapsto A$ and $A = f(r)$.

So $f(1)$ is shorthand for *the area of a circle of radius* 1 and $f(4.3)$ is shorthand for *the area of a circle of radius* 4.3.

We know, of course, that the particular machinery in this case is the formula πr^2, so we write

$$f(r) = \pi r^2,$$

and we can calculate $f(r)$ for any value of r.

$$\text{Hence } f(1) = \pi 1^2 = \pi \approx 3.142$$

$$f(4.3) = \pi(4.3)^2 \approx 58.1$$

The set of permitted values of the independent variable is called the *domain* of the function, and the corresponding set for the dependent variable is the *range* of the function. So, in this example, both domain and range of the function are the set of all positive numbers.

Exercise 1.1a

1 If $f(r) = \pi r^2$, where $\pi \approx 3.14$,
(*i*) Calculate $f(2), f(\frac{1}{2}), f(10), f(100)$
(*ii*) What is the value of $f(0)$, and what is its meaning?
(*iii*) Write in shorthand form the area of a circle of radius three metres. What is its value?
(*iv*) Repeat (*iii*) for a circle of radius one third of a metre.

2 When a stone is thrown vertically upwards with a velocity v m s^{-1}, its maximum height h m given approximately by the function

$$h = f(v) = \frac{1}{20}v^2$$

(*i*) Calculate its maximum height when thrown with velocity 10 m s^{-1}, 20 m s^{-1}, 30 m s^{-1}.
(*ii*) At what speed must it be projected in order to reach the height of a roof 125 m above its point of projection?

3 When a heavy mass swings at the end of a light string of length l m, its period of oscillation T s is given approximately by the function

$$T = f(l) = 2\sqrt{l}$$

(*i*) What is the period of a pendulum of length 1 m, 4 m?
(*ii*) What is the length of a pendulum whose period is 1 s, 3 s?

4 When you look out to sea on a clear day from a height h metre, the distance d kilometre of the horizon is given approximately by the function

$$d = f(h) = 3.6\sqrt{h}$$

(*i*) What is the distance of the horizon from a height 1 m, 4 m, 100 m respectively?
(*ii*) How high would you have to go in order to see 18 km, 36 km?

In these examples, the functions are expressible by an algebraic formula. But the only essential for a function is the existence of machinery for converting an *input* (the independent variable) into an *output* (the dependent variable), and this machinery could equally take the form of a set of rules, e.g., the Income Tax tables, whereby a man's gross income £I (and a welter of other information) is fed into the function-machine, called the Department of Inland Revenue, which all too certainly calculates his Income Tax £T.

So $f: I \mapsto T$ and $T = f(I)$

The way this machine works will vary from person to person, depending on other information about size of family, pension contributions and a host of other items, but for our particular man $f(1\ 000)$, $f(2\ 000)$ etc. represent his tax payments in the event of his earning £1 000, £2 000, etc.

For relatively low incomes no tax is payable; so for relatively small values of I, the value of T is zero. For these values of I, what we called the dependent *variable* (T) is in fact *constant*, and we usually say that the function itself is constant over this part of the domain.

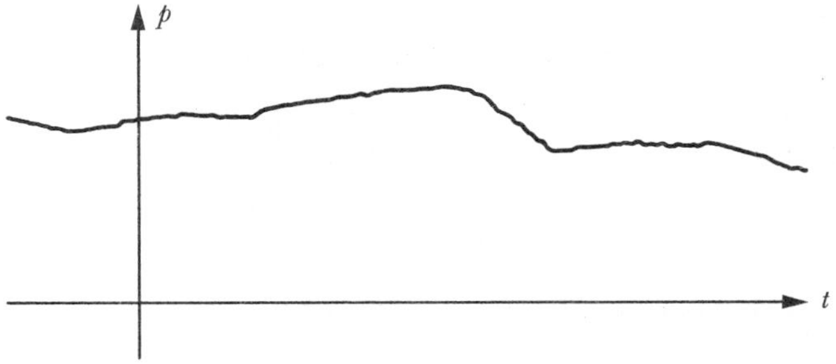

In another case, a function might exist in the form of a graph, e.g., the continuous trace of a barograph recording atmospheric pressure. Here the pressure p is a function of time t (measured, say, from midnight on a particular day). If we wish to know the pressure at a particular time, reference to the graph converts the input (t) into an output (p), and we write

$f: t \mapsto p$ and $p = f(t)$.

In the above examples there is no particular sanctity in the use of the letter f to describe the function machine, and if there is any danger of confusion between two functions different letters can be used.

So, for the brother of the man paying income tax, his function might be written

$T = g(I)$,

and for atmospheric pressures at a different weather-station we might use the Greek letter ϕ (phi), and write

$p = \phi(t)$

Inverse and product functions

If f is a function which converts an input x into an output $f(x)$, it can be represented by means of a *mapping* of the domain of f into its *codomain* (so that the range of a particular function is a subset of its codomain).

The only essential is that every element of the domain should be mapped on to a unique element (or *image*) in the codomain. Sometimes, as we have already seen, the reverse is not true, and a particular element of the co-domain can be the image of more than one element of the domain. In our illustration, for instance, $f(c) = f(d)$; and if $g(x) = x^2$, then $g(a) = g(-a)$ for all values of a.

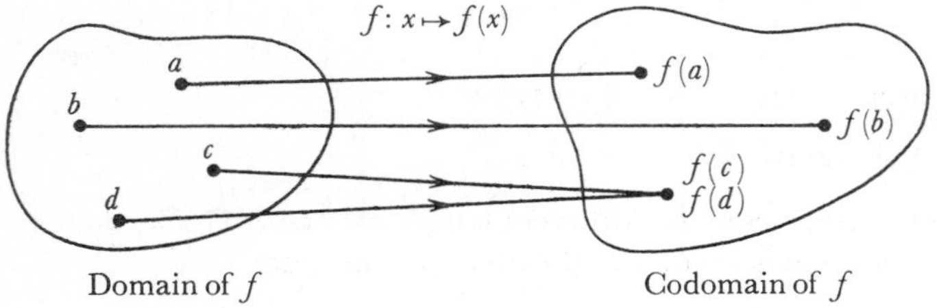

But if f is a function whose outputs all arise from unique inputs, we can regard the backwards mapping from output to input as also defining a function, which we call the inverse function f^{-1}.

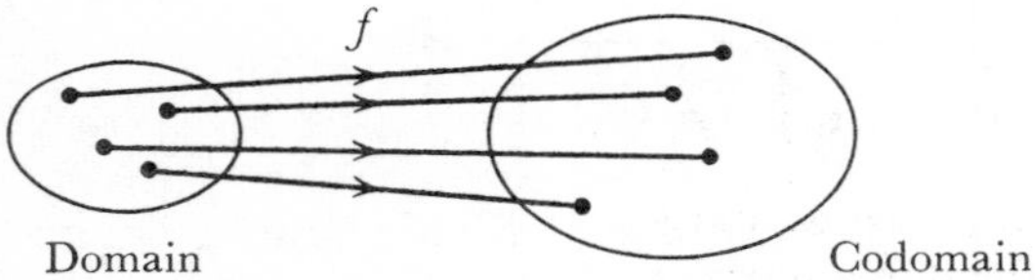

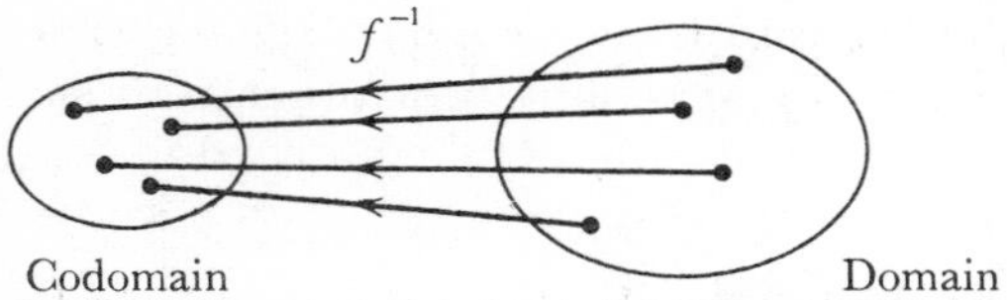

For example, if $f : x \mapsto 2x$, then $f^{-1} : x \mapsto \frac{1}{2}x$

$$g : x \mapsto x + 1, \text{ then } g^{-1} : x \mapsto x - 1$$

$$h : x \mapsto 3x + 2, \text{ then } h^{-1} : x \mapsto \tfrac{1}{3}(x - 2)$$

Sometimes we can define such inverse functions where they would not otherwise exist, by restricting the domain of the original function.

If, for instance, $f : x \mapsto x^2$ then $+4$ is the image of both $+2$ and -2, so there is no inverse function.

But if we restrict the domain of f by stating that $x \geqslant 0$, then

$$f : x \mapsto x^2$$

has the inverse function $f^{-1} : x \mapsto \sqrt{x}$.

Lastly, if f and g are two functions, we sometimes need to use the output of f as the input of g. We call this combination the *product function* gf:

For example, if $f : x \mapsto x^2$ and $g : x \mapsto x + 1$, then gf means 'square x, and then add one', and fg means 'add one to x, and then square'.

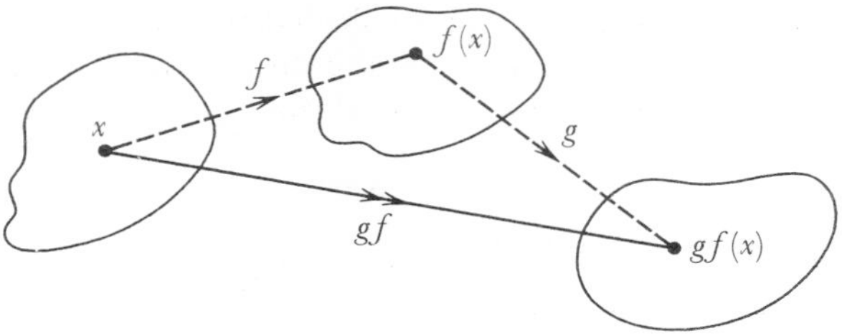

So $gf: x \mapsto x^2 + 1$ and $fg: x \mapsto (x + 1)^2$.

Hence gf and fg are, in general, different functions; and (to make matters worse!) neither of them is related to the algebraic product

$f(x)\,g(x) = x^2(x + 1)$

Exercise 1.1b

1 $f: x \mapsto x + 1$ and $g: x \mapsto x^3$.

Find the functions

(*i*) fg	(*ii*) gf	(*iii*) f^{-1}
(*iv*) g^{-1}	(*v*) $(fg)^{-1}$	(*vi*) $g^{-1}f^{-1}$

2 Repeat this question with $f: x \mapsto 3x$ and $g: x \mapsto x^2$.
Is it necessary to restrict the domain of f or g?

3 What do you notice about f^{-1}, g^{-1} and $(fg)^{-1}$?

1.2 Graphs of functions

We have seen, as in the case of the barograph, that a function could be defined by means of a graph. Certainly it is always possible to illustrate a function in this way, and this usually displays most fully its particular features. Using two perpendicular axes, we adopt the convention that the *independent* variable (or *input*) is plotted *horizontally*, and the *dependent* variable (or *output*) is plotted *vertically*.

Example 1

(Exercise 1.1a, No. 4),

$f(h) = 3.6\sqrt{h}$

This function can be calculated for different values of h:

h	0	1	2	3	4
$f(h)$	0	3.6	5.1	6.2	7.2

(Domain of f: all non-negative numbers
Range of f: all non-negative numbers)

and these values can be plotted as a graph:

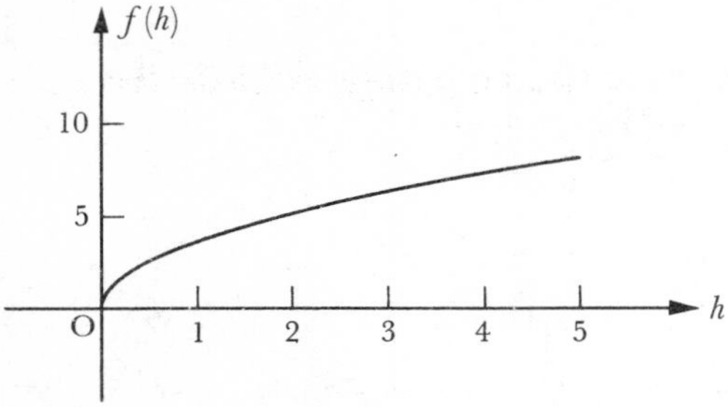

Example 2

$f(x) = |x|$

where $|x|$ means the *numerical value of* x

($|7| = 7$, $|-2.5| = 2.5$, etc.)

(Domain of f: all numbers
Range of f: all non-negative numbers)

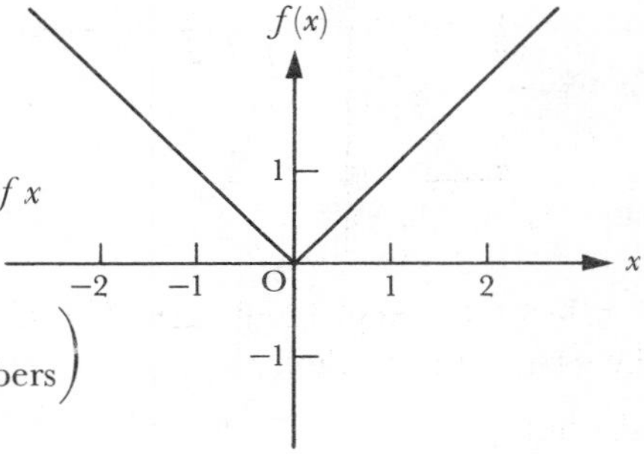

Example 3

$$g(x) = \begin{cases} x - 1 & \text{if } x < 0 \\ 0 & \text{if } x = 0 \\ x + 1 & \text{if } x > 0 \end{cases}$$

(In our illustrations we shall use blobs to make clear the values of the function.)

(Domain of g: all numbers
Range of g: all numbers greater than 1, less than -1, and zero)

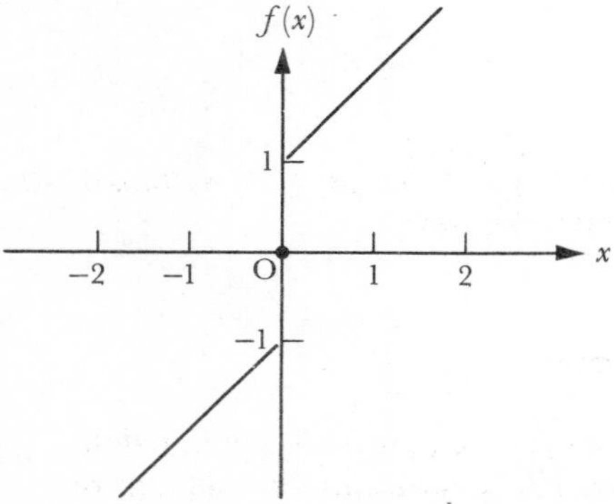

Example 4

$h(x) = [x]$

where [ ] means *the integral part of x* or *the greatest integer not greater than x*. (So that $[2.3] = 2$, $[3] = 3$, $[-2.7] = -3$, etc.)

(Domain of h: all numbers
Range of h: all integers)

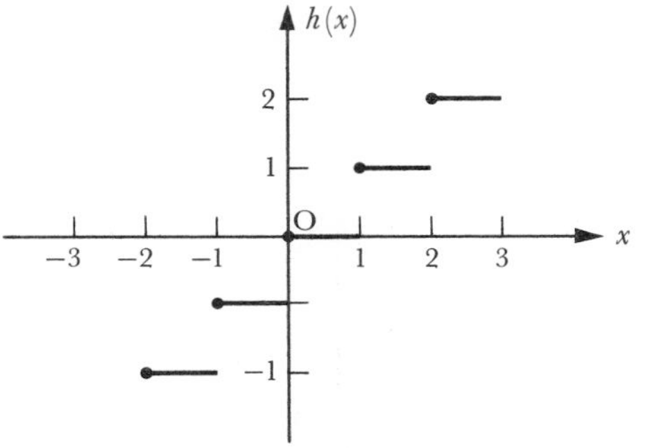

When the independent variable is an integer, we usually represent it by the letter n rather than x:

Example 5

$\phi(n) = (-1)^n$

(Domain of ϕ: all integers
Range of ϕ: ± 1)

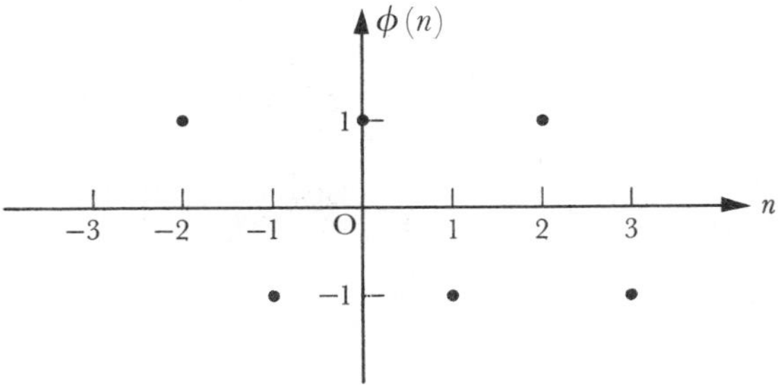

It will be noticed that in Examples 2 and 5 the graph is symmetrical about the vertical axis. This can also be seen since

$$f(-x) \equiv (-x)^2 \equiv x^2 \equiv f(x)$$

and $\phi(-n) \equiv (-1)^{-n} \equiv (-1)^n \equiv \phi(n)$.

Any such function, for which $f(-x) \equiv f(x)$, is called an *even* function. Similarly, in example 3, the graph has point-symmetry about the origin, i.e., is unaltered by reflection through the origin (or by rotation through two right angles or by successive reflections in the two axes).

Here $g(-x) \equiv -g(x)$ and any such function is called an *odd* function.

It is, of course, rather special for a graph to have either kind of symmetry, and functions (unlike integers) are generally neither even nor odd.

Finally it will be recalled that if a and b are constants, the graph of

$y = ax + b$ is a straight line.

The corresponding function

$$f(x) = ax + b$$

is therefore called a *linear function*.

So $f(x) = 2x + 1$, $7 - x$, -3, etc. are all linear functions.

Exercise 1.2

In the case of each of the following functions (Nos. 1–13),

(*i*) Calculate, if possible,

$$f(0),\quad f(3),\quad f(2.5),\quad f(-1);$$

(*ii*) State whether it is an even function, an odd function, or neither;

(*iii*) State its domain and range, and sketch its graph.

1 $f(x) = 3x - 2$

2 $f(x) = 7$

3 $f(x) = \dfrac{1}{x^2}$

4 $f(x) = x^3$

5 $f(x) = \sqrt{x}$

6 $f(x) = \dfrac{1}{x}$ if $x \neq 0$; 0 if $x = 0$

7 $f(x) = x - [x]$

8 $f(x) = x^2$ if $|x| < 1$; $\dfrac{1}{x^2}$ if $|x| \geqslant 1$

9 $f(x) = 1$ if x is rational; 0 if x is irrational

10 $f(x) = x$ if x is rational; 0 if x is irrational

11 $f(n) = \dfrac{(-1)^n}{n}$

12 $f(n)$ is the number of days in the nth month of a leap year.

13 $f(n)$ is the digit in the nth decimal place of π.

14 Can we say anything about the value of $f(0)$ if $f(x)$ is
(*i*) an even function?
(*ii*) an odd function?

15 For what values of k is $f(x) = x^k$
(*i*) an even function?
(*ii*) an odd function?

16 What can be said about the constants a and b if the linear function

$$f(x) = ax + b$$

is (*i*) an even function?
(*ii*) an odd function?

17 If $f(x) = x^3$, what are the following functions? Show in separate diagrams how their graphs are related to that of $f(x)$ and in each case state in words how the graph of $f(x)$ has been transformed.

(*i*) $f(x) + 2$; (*ii*) $f(x) - 3$; (*iii*) $f(x - 1)$; (*iv*) $f(x + 4)$; (*v*) $2f(x)$; (*vi*) $-3f(x)$; (*vii*) $f(2x)$; (*viii*) $f(-x)$.

1.3 Continuity

You will have noticed in the last section a marked difference between graphs which were unbroken, or *continuous*, and those which included sudden jumps. The concept of *continuity* is extremely important in mathematics and needs very careful definition. For the present we shall simply say that a function is *continuous at a particular point* if we can depart from the point along the graph both to the left and to the right without having to make a jump.

So in the examples of the last section,

1 $f(h) = 3.6\sqrt{h}$ is continuous if $h > 0$.

2 $f(x) = |x|$ is continuous everywhere.

3 $$g(x) = \begin{cases} x - 1, & x < 0 \\ 0, & x = 0 \\ x + 1, & x > 0 \end{cases}$$

is continuous except at $x = 0$ (since at this point we have to jump whether we move to left or to right).

4 $h(x) = [x]$

is continuous except when x is an integer.

(Since at these points we have to jump when we move to the left, though not if we move to the right.)

Exercise 1.3

For what values of x are the functions defined in Exercise 1.2, Nos 1–10 (*a*) continuous; (*b*) discontinuous?

1.4 Limits

We shall now reverse the process of the last section and look at the behaviour of a function not as we *leave* a particular point, but as we *approach* it.

This behaviour may be very erratic. Consider, for instance, the function

$$f(x) = \begin{cases} 1 \text{ when } x = \pm 1, \pm \frac{1}{2}, \pm \frac{1}{4}, \pm \frac{1}{8} \ldots \\ 0 \text{ otherwise} \end{cases}$$

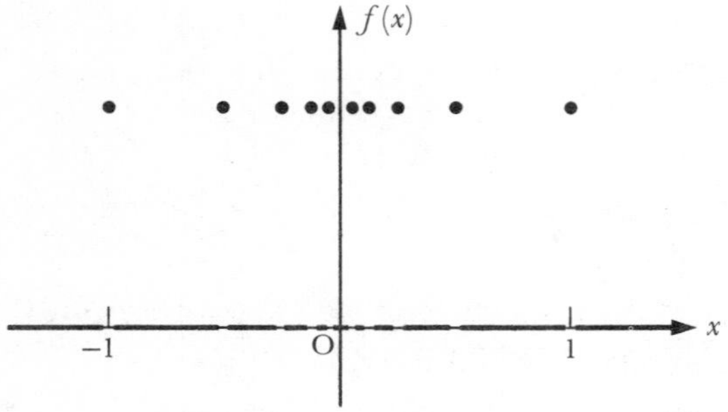

From the graph, we see that although the function is generally zero, it takes the value 1 with increasing frequency as x approaches 0, whether from above or below; so we cannot reasonably say that the function is tending towards any particular limit.

By contrast, the function

$$f(x) = \begin{cases} x^2 + 1 & \text{if } x > 0 \\ 0 & x = 0 \\ x^2 - 1 & x < 0 \end{cases}$$

approaches the value $+1$ as x tends to 0 through positive values, and -1 as x tends to 0 through negative values.

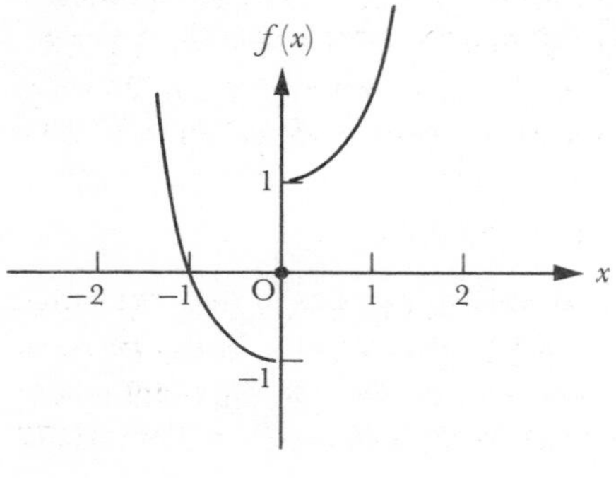

These two values, $+1$ and -1, are sometimes called the *right-hand limit* and *left-hand limit* of the function as x tends to 0, and we can write

$$\lim_{x\to 0+} f(x) = +1 \quad \text{and} \quad \lim_{x\to 0-} f(x) = -1$$

Now consider the function

$$f(x) = \begin{cases} x^2 + 1 & \text{if } x \neq 0 \\ 0 & \text{if } x = 0 \end{cases}$$

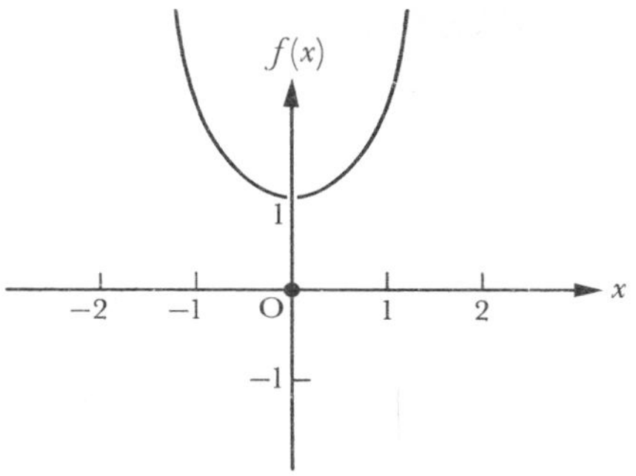

Here the right-hand limit and the left-hand limit have become identical, and when this happens we call their common value simply

the limit of $f(x)$ as x approaches 0,

and write:

$$\text{As } x \to 0, \quad f(x) \to 1$$
$$\text{or} \quad \lim_{x\to 0} f(x) = 1$$

It must be emphasised very strongly that this limit is an indication of the behaviour of the function *near* $x = 0$, not *at* $x = 0$. In this case,

$$\lim_{x\to 0} f(x) = 1, \quad \textit{but} \quad f(0) = 0.$$

So the limit (if it exists) of a function as we approach a point is not necessarily the same as the value of the function at the point.

As with continuity, so with limits, a very precise definition is necessary. For the present, however, we shall content ourselves with the following statement, even though the word *approach* has not been carefully defined:

If when x approaches a particular value $x = a$ (whether from above or below), the values of $f(x)$ approach a certain value l, we call this *the limit of $f(x)$ as x tends to a*, and write:

$$\text{As} \quad x \to a, \quad f(x) \to l \quad \text{or} \quad \lim_{x\to a} f(x) = l$$

The gap left in the curve overleaf is simply to emphasise that the value of the function *at* $x = a$ is irrelevant to the existence of the limit. Now, at last, we are able to come full circle and observe that the function is *continuous* at this point if (and only if) its value there is equal to its limit as the point is approached:

$f(x)$ is continuous at $x = a \quad \Leftrightarrow \quad \lim_{x\to a} f(x) = f(a)$

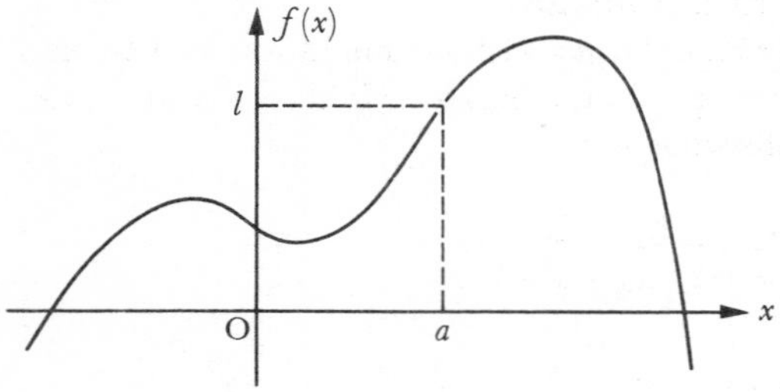

Exercise 1.4a

Investigate (*i*) the right-hand limit, (*ii*) the left-hand limit, (*iii*) the limit of the following functions as x approaches the specified points. In each case state whether the function is continuous at the point.

1 $\left.\begin{array}{ll} f(x) = x & \text{if } x \geqslant 0 \\ 0 & \text{if } x < 0 \end{array}\right\} \quad x = 0$

2 $\left.\begin{array}{ll} f(x) \doteq |x| & \text{if } x \neq 0 \\ 1 & \text{if } x = 0 \end{array}\right\} \quad x = 0$

3 $f(x) = [x]; \quad x = 1$

4 $f(x) = x - [x]; \quad x = 1$

5 $f(x) = [x]; \quad x = 1\frac{1}{2}$

There is another case where, in a rather different sense, we speak of the limit of $f(x)$ as $x \to a$, and that is when $f(x)$ becomes *unlimitedly* large (either positively or negatively).

For example:

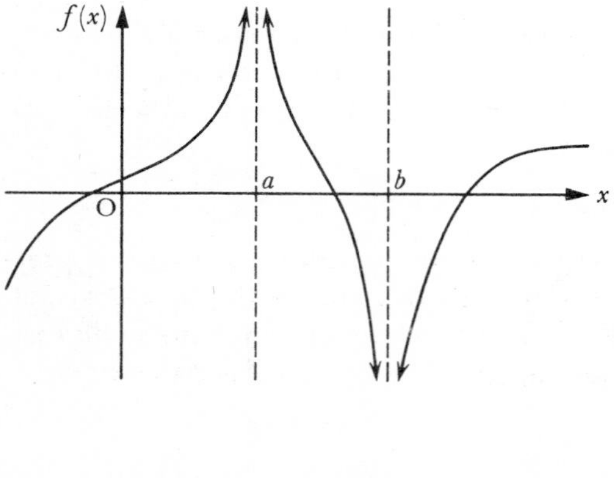

Here we write:

as $x \to a, f(x) \to +\infty$ or $\lim_{x \to a} f(x) = +\infty$

and as $x \to b, f(x) \to -\infty$ or $\lim_{x \to b} f(x) = -\infty$

and the two dotted lines are called *asymptotes*.

Lastly, when x becomes larger and larger without limitation, rather than approach a particular value, we say that x 'tends to infinity', and we use the notation of limits in the following way:

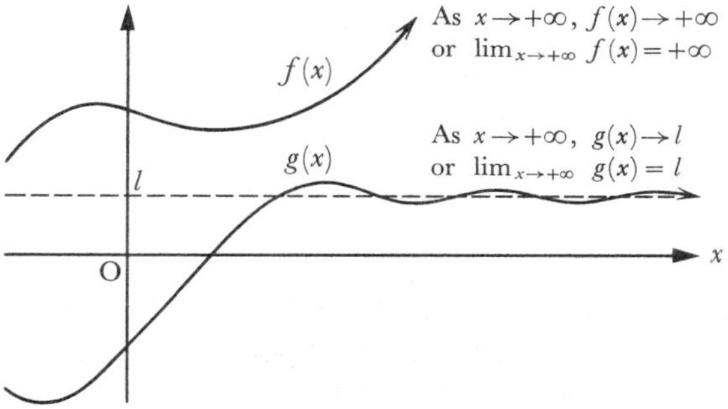

Similar notation is used when x becomes very large and negative $(x \to -\infty)$; and if we wish to speak of x becoming large without reference to whether it is positive or negative, we simply write $x \to \infty$, without a sign.

So, as $x \to \infty$, $x^2 + 3x \to +\infty$,
or $\lim_{x \to \infty} (x^2 + 3x) = +\infty$

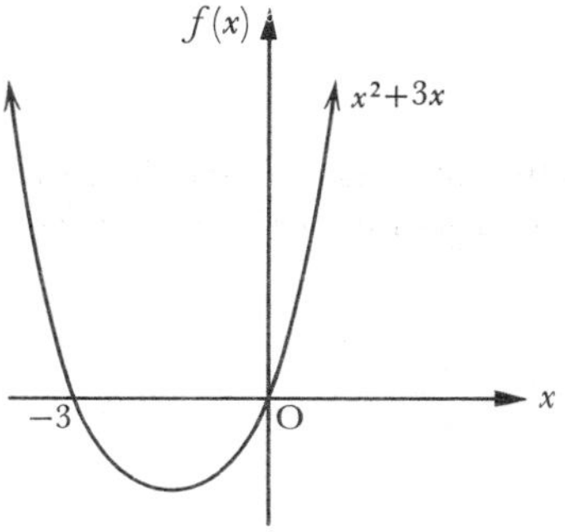

Example

If $f(x) = \dfrac{x+1}{x-1}$, investigate

(*i*) $\lim_{x \to 1} f(x)$, (*ii*) $\lim_{x \to \infty} f(x)$

(*i*) As $x \to 1$, $x - 1$ becomes very small.

So $f(x) = \dfrac{x+1}{x-1}$ becomes very large.

For example, $f(1.001) = \dfrac{2.001}{0.001} = 2\,001$

$$f(0.999) = \frac{1.999}{-0.001} = -1\,999$$

So the right-hand and left-hand limits are

$$\lim_{x\to 1+} f(x) = +\infty$$

and $\lim_{x\to 1-} f(x) = -\infty$;

and as these are different, $\lim_{x\to 1} f(x)$ does not exist.

(*ii*) As $x \to \infty$, $f(x) = \dfrac{x+1}{x-1} = \dfrac{1+\frac{1}{x}}{1-\frac{1}{x}} \to \dfrac{1}{1} = 1.$

So $\lim_{x\to\infty} f(x) = +1$.

These results are illustrated by the graph of $f(x)$:

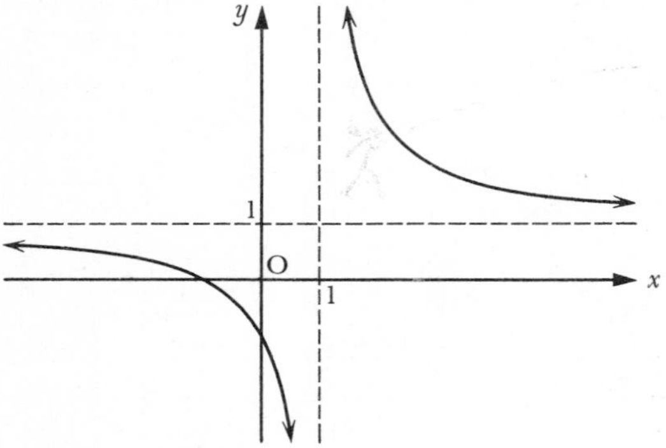

Exercise 1.4b

Investigate the following limits, and illustrate by sketch graphs:

1 $\lim_{x\to+\infty} (2x - x^2)$ $\lim_{x\to-\infty} (2x - x^2)$

2 $\lim_{x\to+\infty} \sqrt{x}$ $\lim_{x\to-\infty} \sqrt{x}$

3 $\lim_{x\to 0} \dfrac{1}{x^2}$ $\lim_{x\to\infty} \dfrac{1}{x^2}$

4 $\lim_{x\to 1} \dfrac{x}{x-1}$ $\lim_{x\to\infty} \dfrac{x}{x-1}$

5 $\lim_{x\to 0} \dfrac{x^2+1}{x}$ $\qquad \lim_{x\to \infty} \dfrac{x^2+1}{x}$

6 $\lim_{x\to 0} \dfrac{x+1}{x^2}$ $\qquad \lim_{x\to \infty} \dfrac{x+1}{x^2}$

7 $\lim_{x\to \infty} [x]$ $\qquad \lim_{x\to \infty} x - [x]$

1.5 Interlude $f: x \mapsto x^2$

This function is represented by a curve called a *parabola*, especially important in the study of the paths of projectiles and in the construction of reflecting telescopes for focusing a parallel beam of light.

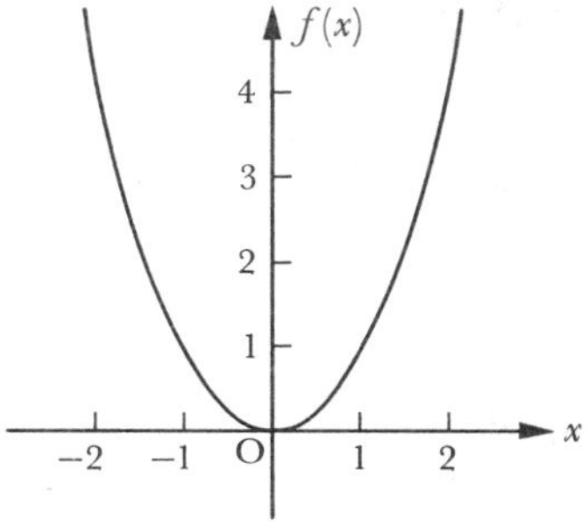

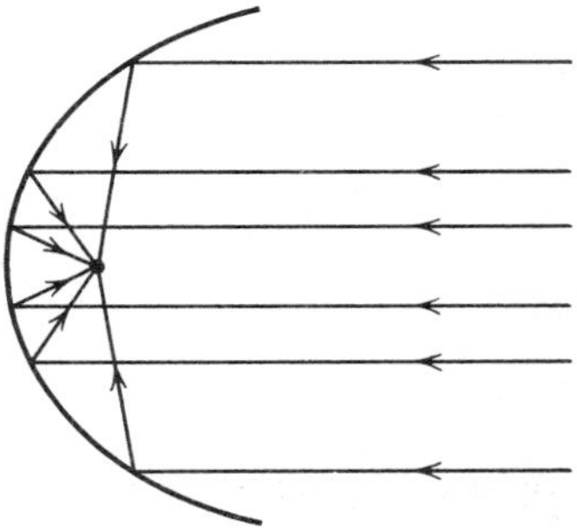

These cases will be discussed later, but the function $f: x \to x^2$ has also been one of the most important in the development of mathematics itself.

Rational and irrational numbers

Soon after men learnt to count, and to use what we call the *positive integers* 1, 2, 3, 4, 5, . . ., they must have realised that they could add them and multiply them and obtain answers which always remained safely within the same set of numbers.

As we know very well, however, inverse questions such as

'What do I need to add to 4 in order to obtain 3?'
and 'What number when multiplied by 4 gives 3?'

are *not* answerable in terms of positive integers and lead to the generation of whole new sets of numbers which we call, respectively, *negative* and *fractional*. This vastly increased field of numbers—positive and negative integers and fractions—is called the field of *rational numbers*. Within this field, the four processes of addition and subtraction, multiplication and division (except division by 0) can all be carried out with the answer again lying safely within the field. Here at last, together with geometry, Pythagoras and his followers felt they had the apparatus for explaining the universe; and this they set to do, ranging from the movement of planets to the harmonies of music.

But the function $f: x \mapsto x^2$ soon leads to the question,
'Which number has 2 as its square?'
or 'What is the side of a square which has area 2?'

Now $(1.4)^2 = 1.96$, so it is clear that the answer is roughly 1.4.

For a better approximation, we note that

$$2 \div 1.4 \approx 1.428$$

$$\Rightarrow \quad 2 \approx 1.4 \times 1.428$$

Taking the average of 1.4 and 1.428 leads us to look at the number 1.414. Now $2 \div 1.414\,00 \approx 1.414\,43$,

$$\Rightarrow \quad 2 \approx 1.414 \times 1.414\,43,$$

and we next look at their average, 1.414 215.

By this means we can approach as close as we like—1.4, 1.414, 1.414 215 etc.—to the number whose square is 2, but the number itself remains strangely elusive.

Let us therefore try another approach and suppose that the required number is the fraction p/q, which is expressed in its lowest terms (i.e., with p and q as integers having no common factor).

Then $(p/q)^2 = 2$

So $p^2 = 2q^2$, which is an even integer.

But the square of any odd integer is odd. So p cannot be odd, and must be even.

Let $p = 2m$ (where m is an integer).

Then $(2m)^2 = 2q^2$

$$\Rightarrow \quad q^2 = 2m^2$$

So q^2 is even, and therefore q must be even.

Hence p and q are both even, and so are both divisible by 2. But this is a blatant contradiction of our original statement that p and q have no common factor. There must, therefore, be a slip in our working, a flaw in our argument or an inconsistency in our original assumptions.

Close inspection fails to reveal either a slip or a flaw, so we are inevitably led to the conclusion that the contradiction which we have reached must spring from an internal inconsistency within our assumptions.

But there were only two such assumptions: that x is the quotient of two integers and that $x^2 = 2$.

We therefore see that there *cannot* be a number x which is the quotient of two integers and whose square is 2.

So $\sqrt{2}$ is not a rational number.

This proof, by the exceedingly powerful method of 'reductio ad absurdum', was known to Euclid, and its implications were immediately clear: that the apparent completeness of the rational numbers is an illusion and that others exist (in fact, as we now know, completely 'outnumbering' the rationals).

Worse than this, the stage in the argument which we called a 'contradiction' or an 'absurdity' must have seemed to the Greeks more like a catastrophe. For here were two integers p and q which both *did* and *did not* have 2 as a common factor. The whole rationality of mathematics seemed imperilled and numbers like $\sqrt{2}$ were said by some to be 'unspeakable'. Rather less dramatically, we simply call numbers *irrational* if they cannot be expressed as the ratio of two integers. Together with the rational numbers, they form the field of *real numbers*.

Exercise 1.5a

Prove that $\sqrt{3}$ and $\sqrt[3]{2}$ are both irrational numbers.

The number i

The function $f: x \mapsto x^2$ also leads us to ask which number has -1 as its square. For what value of x is $x^2 = -1$? What is $\sqrt{-1}$?

Clearly there is no answer within the field of numbers which we already know: their squares are never negative and the parabola $f(x) = x^2$ never goes below the x-axis.

But such impossibilities rarely deter mathematicians, and this particular question has been one of the most productive in the whole of mathematics. For the moment all we can say is that if we boldly use a symbol, usually chosen as i, with all the usual properties such as $i + 3i = 4i$ etc. *and the additional property that* i^2 *is replaceable by* -1, then this step will have a great unifying and creative power.

Since we can write $i^2 = -1$, it also follows that $(-i)^2 = -1$, so i and $-i$ are said to be the two square roots of -1.

But does $\sqrt{-1}$ actually exist? For that matter, does -1 exist? Or any other number?

Exercise 1.5b

1 Use the symbol i to express (*i*) $\sqrt{-4}$; (*ii*) $\sqrt{-3}$.

2 It might be thought that to ask further questions like $\sqrt[2]{i}$ or $\sqrt[3]{-1}$ might lead us to an even more extensive field of numbers. But is this so? Answer these questions by first calculating

(*i*) the squares of $\dfrac{1+i}{\sqrt{2}}$ and $\dfrac{-1-i}{\sqrt{2}}$

(*ii*) the cubes of -1, $\dfrac{1+\sqrt{3}\,i}{2}$, $\dfrac{1-\sqrt{3}\,i}{2}$

*1.6 Quadratic functions

These are especially important as the simplest type of non-linear function.

Example 1

$f(x) = x^2 - 2x - 3$

We can start by finding $f(x)$ for various values of x and plotting the graph of $f(x)$:

$x =$	-3	-2	-1	0	1	2	3	4	5
$x^2 =$	9	4	1	0	1	4	9	16	25
$-2x =$	6	4	2	0	-2	-4	-6	-8	-10
$-3 =$	-3	-3	-3	-3	-3	-3	-3	-3	-3
$f(x) = x^2 - 2x - 3$	12	5	0	-3	-4	-3	0	5	12

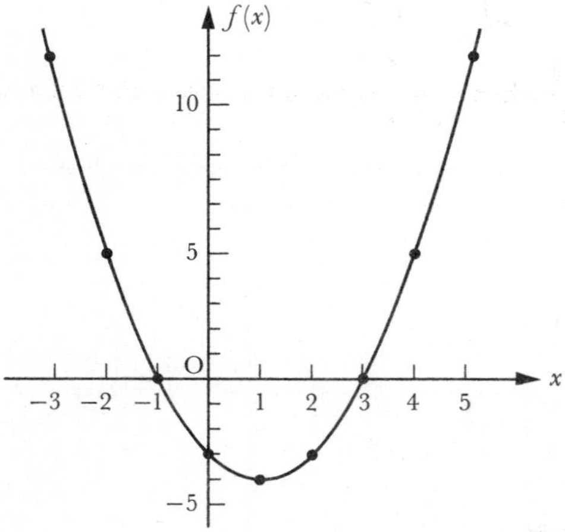

This curve clearly illustrates the important features:

(*i*) $f(0) = -3$
(*ii*) As $x \to \pm\infty, f(x) \to +\infty$
(*iii*) $f(x)$ has its *minimum* value -4 when $x = 1$
(*iv*) $f(x) = 0$ when $x = -1$ or $+3$.

The first two of these could have been seen without the labour of plotting the graph. What about (*iii*) and (*iv*)?

We can write

$$f(x) = x^2 - 2x - 3 = (x - 1)^2 - 4$$

Now $(x - 1)^2 \geqslant 0$. So $f(x) \geqslant -4$,
and we see that $f(x)$ has its minimum value -4 when $x - 1 = 0 \Rightarrow x = 1$.
Also $f(x) = 0 \Leftrightarrow (x - 1)^2 = 4 \Leftrightarrow x - 1 = \pm 2 \Leftrightarrow x = -1$ or $+3$.

Alternatively, we can write:

$$f(x) = x^2 - 2x - 3 = (x + 1)(x - 3).$$

So $f(x) = 0 \Leftrightarrow (x + 1)(x - 3) = 0 \Leftrightarrow x = -1$ or $x = 3$.
Hence $f(x) = 0$ when $x = -1$ or 3.

Example 2

Describe the important features of

$f(x) = x^2 - 3x + 5$, and then sketch its graph.

(*i*) $f(0) = 5$
(*ii*) As $x \to \pm\infty, f(x) \to +\infty$
(*iii*) $f(x) = x^2 - 3x + 5 = \left(x - \frac{3}{2}\right)^2 + \frac{11}{4} \geqslant \frac{11}{4}$

So $f(x)$ has a minimum value $\frac{11}{4}$ when $x = \frac{3}{2}$, and is therefore never zero.

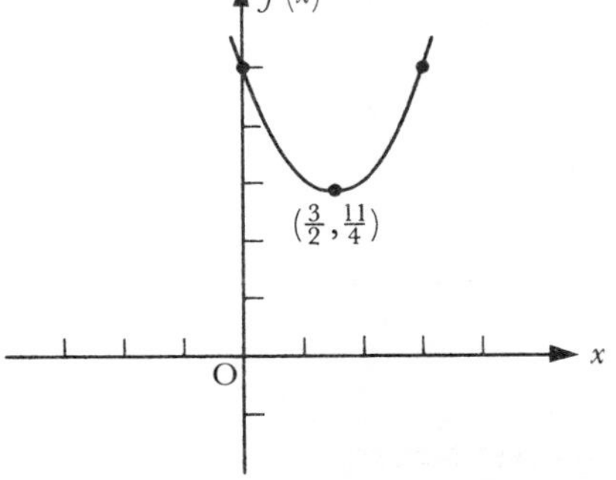

Example 3

Describe the important features of

$$f(x) = 1 + x - x^2$$

and then sketch its graph.

(*i*) $f(0) = 1$
(*ii*) As $x \to \pm\infty, f(x) \to -\infty$
(*iii*) $f(x) = 1 + x - x^2$

$$= \frac{5}{4} - \left(x - \frac{1}{2}\right)^2 \leqslant \frac{5}{4}$$

So $f(x)$ has a maximum value $\frac{5}{4}$ when $x = \frac{1}{2}$.

(*iv*) $f(x) = \frac{5}{4} - \left(x - \frac{1}{2}\right)^2$

$$\text{So } f(x) = 0 \Rightarrow \frac{5}{4} - \left(x - \frac{1}{2}\right)^2 = 0$$

$$\Rightarrow \left(x - \frac{1}{2}\right)^2 = \frac{5}{4}$$

$$\Rightarrow x - \frac{1}{2} = \pm\frac{\sqrt{5}}{2}$$

$$\Rightarrow x = \frac{1 \pm \sqrt{5}}{2}$$

$$\Rightarrow x = 1.62 \text{ or } -0.62$$

So the graph of $f(x)$ is

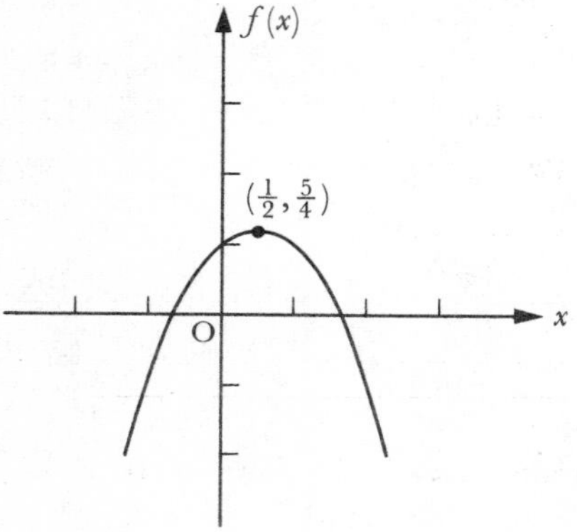

Example 4

$f(x) = 2x^2 + 3x - 5.$

Here we can say $f(x) = 2\left(x^2 + \frac{3}{2}x - \frac{5}{2}\right)$

and regard $f(x)$ as a magnification of the function $x^2 + \frac{3}{2}x - \frac{5}{2}$, which we can investigate in similar fashion.

The general quadratic function

$f(x) = ax^2 + bx + c \quad (a \neq 0)$

(*i*) $f(0) = c$

(*ii*) As $x \to \pm\infty$; if $a > 0, \quad f(x) \to +\infty$

$a < 0, \quad f(x) \to -\infty$

(*iii*)
$$\begin{aligned} f(x) &= ax^2 + bx + c \\ &= a\left\{x^2 + \frac{b}{a}x + \frac{c}{a}\right\} \\ &= a\left\{\left(x + \frac{b}{2a}\right)^2 + \frac{4ac - b^2}{4a^2}\right\} \\ &= a\left(x + \frac{b}{2a}\right)^2 + \frac{4ac - b^2}{4a} \end{aligned}$$

If $a > 0, f(x) \geqslant \frac{4ac - b^2}{4a}$, which is its minimum value $\left(\text{when } x = -\frac{b}{2a}\right)$

If $a < 0, f(x) \leqslant \frac{4ac - b^2}{4a}$, which is its maximum value $\left(\text{when } x = -\frac{b}{2a}\right)$

(*iv*)
$$\begin{aligned} f(x) = 0 \quad &\Rightarrow \quad a\left\{\left(x + \frac{b}{2a}\right)^2 + \frac{4ac - b^2}{4a^2}\right\} = 0 \\ &\Rightarrow \quad \left(x + \frac{b}{2a}\right)^2 = \frac{b^2 - 4ac}{4a^2} \\ &\Rightarrow \quad x + \frac{b}{2a} = \pm\frac{\sqrt{(b^2 - 4ac)}}{2a} \\ &\Rightarrow \quad \boxed{x = \frac{-b \pm \sqrt{(b^2 - 4ac)}}{2a}} \end{aligned}$$

	two distinct			$>$
So $f(x) = 0$ has	two equal	real roots	$\Leftrightarrow$	$b^2 - 4ac = 0.$
	no			$<$

Example 5

Solve the equation $3x^2 + 4x - 2 = 0$

Here $a = 3$, $b = 4$, $c = -2$.

$$\text{So} \quad x = \frac{-4 \pm \sqrt{(4^2 - 4\times 3\times -2)}}{2.3} = \frac{-4 \pm \sqrt{40}}{6} \approx \frac{-4 \pm 6.324}{6}$$

$$\approx \frac{+2.324}{6} \quad \text{or} \quad \frac{-10.324}{6}$$

$$\Rightarrow \quad x \approx 0.39 \quad \text{or} \quad -1.72$$

So $3x^2 + 4x - 2 = 0 \;\Rightarrow\; x \approx 0.39$ or -1.72.

Example 6

$4x^2 - 20x + 25 = 0$

Here $a = 4$, $b = -20$, $c = 25$.

$$x = \frac{20 \pm \sqrt{(400 - 400)}}{8} = \frac{20 \pm 0}{8} = 2\tfrac{1}{2} \quad \text{or} \quad 2\tfrac{1}{2}.$$

So $4x^2 - 20x + 25 = 0 \;\Rightarrow\; x = 2\frac{1}{2}$.

Example 7

$x^2 - x + 1 = 0$

Here $a = 1$, $b = -1$, $c = 1$

$$\Rightarrow \quad x = \frac{1 \pm \sqrt{(1 - 4)}}{2} = \frac{1 \pm \sqrt{-3}}{2}$$

So $x^2 - x + 1 = 0$ is true for no real value of x.

[But using the letter i of Section 1.5 we could say

$$x = \frac{1 \pm \sqrt{(3\times -1)}}{2} = \frac{1 \pm \sqrt{3}\sqrt{-1}}{2} = \frac{1 \pm \sqrt{3}\, i}{2}, \text{ etc.}$$

and we confirm that

$$f\left(\frac{1 + \sqrt{3}\, i}{2}\right) = \left(\frac{1 + \sqrt{3}\, i}{2}\right)^2 - \left(\frac{1 + \sqrt{3}\, i}{2}\right) + 1$$

$$= \frac{-2 + 2\sqrt{3}\, i}{4} - \frac{1 + \sqrt{3}\, i}{2} + 1 = 0$$

and similarly $f\left(\dfrac{1 - \sqrt{3}\, i}{2}\right) = 0$].

Exercise 1.6

1 Solve the following equations and illustrate your answers by sketching graphs of the corresponding functions:

(*i*) $x^2 - 4 = 0$
(*ii*) $x^2 - 4x = 0$
(*iii*) $x^2 - 4x + 3 = 0$
(*iv*) $x^2 - 4x + 4 = 0$
(*v*) $x^2 - 4x + 2 = 0.$

In each case check one of your answers and also calculate the sum of the roots. Do you notice anything special? What about the product of the roots?

2 Solve the following equations and again calculate the sum and product of the roots.

(*i*) $2x^2 + x = 1$
(*ii*) $2x^2 + x = 2$
(*iii*) $3x^2 = 2x + 5$
(*iv*) $x^2 = px + q.$

3 What is the range of the following functions, and for what values of x are the functions greatest or least? Illustrate by sketching their graphs.

(*i*) $x^2 - 6x$
(*ii*) $4 - 3x - x^2$
(*iii*) $(x - 2)(x - 4).$

4 For what values of x is

(*i*) $x^2 > x$
(*ii*) $x^2 \leqslant x - 2$
(*iii*) $x^2 \leqslant 1 - x$
(*iv*) $x^2 > x - 1$?

5 Find the points of intersection of:

(*i*) $x + y = 3$ and $4xy = 5$
(*ii*) $y = x + 2$ and $y = 2x^2.$

Illustrate by sketching graphs.

6 For what values of m does the line $y = mx$ meet the curve $y = x^2 - 4x + 3$ in (*i*) two points; (*ii*) one point; (*iii*) no points? Illustrate by a sketch.

7 The equation $ax^2 + bx + c = 0$ has roots α and β.

(*i*) By writing the equation with these two roots as

$$a(x - \alpha)(x - \beta) = 0,$$

show that $\alpha + \beta = -\dfrac{b}{a}$ and $\alpha\beta = +\dfrac{c}{a}$

Use these results to check your solutions of questions 1 and 2.

(*ii*) Use the formulae $\dfrac{-b \pm \sqrt{(b^2 - 4ac)}}{2a}$ to verify that

$$\alpha + \beta = -\frac{b}{a} \quad \text{and} \quad \alpha\beta = +\frac{c}{a}.$$

8 A stone thrown vertically upwards with velocity 20 m s^{-1} is at height y m after t s, where

$$y = 20t - 5t^2$$

For how long is it over 10 m above the ground?

9 A batsman hits a cricket ball so that its trajectory has the equation

$$y = x - \frac{1}{80}x^2,$$

where x and y are horizontal and vertical distances measured in metres. How far away does it land, and what is its greatest height?

10 The sum of n terms of

$1 + 4 + 7 + 10 + \cdots$ is $\frac{1}{2}n\,(3n - 1)$

How many terms have to be taken for this sum to exceed 1 000?

11 A car factory produces n cars per day and makes a profit of £p per day, where

$$p = 2n^2 - 300n$$

How many cars need to be produced per day
(*i*) for production to be profitable
(*ii*) for a daily profit of £10 800
(*iii*) for a daily loss of £11 200
(*iv*) for worst possible loss?
Would you expect the given function to be valid for all values of n?

*1.7 Polynomial functions

We have seen that the general linear function is $ax + b$ and the general quadratic function is

$$ax^2 + bx + c$$

So we now naturally proceed to

$ax^3 + bx^2 + cx + d$ (called a *cubic*)
and $ax^4 + bx^3 + cx^2 + dx + e$ (called a *quartic*).

More generally, any function

$$P(x) = a_0x^n + a_1x^{n-1} + a_2x^{n-2} + \ldots + a_{n-1}x + a_n$$

(where n is a positive integer and $a_0, a_1, \ldots, a_n$ are constants) is called a *polynomial of degree n*, and $a_0, \ldots a_n$ are its *coefficients*.

So linear, quadratic, cubic and quartic polynomials have degree 1, 2, 3, 4 respectively.

With more complicated polynomials, just as with linear and quadratic functions, it is useful to be able to recognise the main features and to be able to sketch their graphs. But first we must look at the structure of such polynomials and the way in which they can be multiplied and divided. Two examples will be sufficient to show similarities with long multiplication and division in ordinary arithmetic:

Example 1

Multiply $3x^2 - 2x - 1$ by $2x + 3$

$$\begin{array}{rrrr}
 & 3x^2 & -\ 2x & -\ 1 \\
 & & 2x & +\ 3 \\
\hline
6x^3 & -\ 4x^2 & -\ 2x & \\
 & +9x^2 & -\ 6x & -\ 3 \\
\hline
6x^3 & +\ 5x^2 & -\ 8x & -\ 3 \\
\hline
\end{array}$$

which can be written as:

$$\begin{array}{rrrr}
x^3 & x^2 & x & 1 \\
 & 3 & -2 & -1 \\
 & & 2 & 3 \\
\hline
6 & -4 & -2 & \\
 & 9 & -6 & -3 \\
\hline
6 & 5 & -8 & -3 \\
\hline
\end{array}$$

Example 2

Divide $x^3 - 5x^2 + 11x - 6$ by $x - 2$

$$\begin{array}{r|rrrr}
 & x^2 & -\ 3x & +\ 5 & \\
\hline
x - 2 & x^3 & -\ 5x^2 & +\ 11x & -\ 6 \\
 & x^3 & -\ 2x^2 & & \\
\hline
 & & -3x^2 & +\ 11x & \\
 & & -3x^2 & +\ 6x & \\
\hline
 & & & 5x & -\ 6 \\
 & & & 5x & -\ 10 \\
\hline
 & & & & +4 \\
\end{array}$$

or

$$\begin{array}{r|rrrr}
 & 1 & -3 & 5 & \\
\hline
1 - 2 & 1 & -5 & 11 & -6 \\
 & 1 & -2 & & \\
\hline
 & & -3 & 11 & \\
 & & -3 & 6 & \\
\hline
 & & & 5 & -6 \\
 & & & 5 & -10 \\
\hline
 & & & & +4 \\
\end{array}$$

So when $x^3 - 5x^2 + 11x - 6$ is divided by $x - 2$, the quotient is $x^2 - 3x + 5$ and the remainder $+4$.

Now when, for example, 25 is divided by 7,
quotient = 3 and remainder = 4
and 25 = 3×7 + 4
More generally,

number = divisor × quotient + remainder

So in our example, as can very easily be checked by multiplication:

$$x^3 - 5x^2 + 11x - 6 \equiv (x - 2)(x^2 - 3x + 5) + 4.$$

Exercise 1.7a

1 Find the product of
(*i*) $x^2 + 3x - 2$ and $2x - 1$
(*ii*) $x^2 - x + 1$ and $x + 1$
(*iii*) $x^3 + x^2 + x + 1$ and $x - 1$
(*iv*) $2x^3 + x^2 - 5x + 4$ and $2x - 3$.

2 Find the quotient and remainder when
(*i*) $2x^3 + 3x^2 - 4x + 5$ is divided by $x + 2$
(*ii*) $4x^3 - 6x^2 + 5$ is divided by $2x - 1$
(*iii*) x^5 is divided by $x + 1$
(*iv*) $x^4 - 1$ is divided by $x - 1$
(*v*) $x^4 + 2x^3 + 3x^2 + 4x + 5$ is divided by $x^2 + x + 1$.
Check your answers by multiplication.

The remainder theorem

It frequently happens that we wish to discover factors of a given polynomial, i.e., to find expressions which divide into the polynomial and leave zero remainder. Can this be done without going through the process of division? Is there any way, for instance, in Example 2 of finding the remainder when $x^3 - 5x^2 + 11x - 12$ is divided by $x - 2$, *without performing the division*?

Let us suppose that the quotient is $Q(x)$ and the remainder is R.

Then $x^3 - 5x^2 + 11x - 6 \equiv (x - 2)\,Q(x) + R.$

If we now take the value $x = 2$,

$$2^3 - 5\times2^2 + 11\times2 - 6 = 0 + R$$

$$\Rightarrow \quad R = 8 - 20 + 22 - 6$$

$$\Rightarrow \quad R = +4$$

and we have obtained the value of the remainder merely by calculating $P(2)$.

More generally, suppose that when $P(x)$ is divided by $x - \alpha$, the quotient is $Q(x)$ and the remainder is R.

Then $P(x) \equiv (x - \alpha)\,Q(x) + R.$

If we now put $x = \alpha$,

$$P(\alpha) = 0 + R$$

$$\Rightarrow \quad R = P(\alpha)$$

This is the *remainder theorem*, that when a polynomial $P(x)$ is divided by $x - \alpha$, the remainder is $P(\alpha)$. In particular,

$$P(\alpha) = 0 \quad \Leftrightarrow \quad \text{Remainder is zero.}$$

So $P(\alpha) = 0 \quad \Leftrightarrow \quad x - \alpha$ is a factor of $P(x)$,

and α is then called a *zero* of $P(x)$, or a *root* of the equation $P(x) = 0$.

So finding the root of $P(x) = 0$, or *solving* this equation, is the same problem as factorising $P(x)$.

Example 1

Find the zeros and factors of $x^3 - 2x^2 - 5x + 6$
and so sketch its graph.

We can see at a glance that one of the zeros is $x = 1$, since this value makes the given expression zero.

So one of its factors is $x - 1$.

By division we obtain

$$x^3 - 2x^3 - 5x + 6 \equiv (x - 1)(x^2 - x - 6)$$

$$\equiv (x - 1)(x - 3)(x + 2)$$

So there are three zeros, $x = 1, 3$ and -2.

We can also see that the sign of the polynomial for other values of x is given by

x	-3		-2		-1		0		1		2		3		4
$P(x)$	$- - -$	$- - -$	0	$+ + +$	$+ + +$	$+ + +$	$+ + +$	$+ + +$	0	$- - -$	$- - -$	$- - -$	0	$+ + +$	$+ + +$

Also, as $x \to +\infty$, $x^3 - 2x^2 - 5x + 6 \to +\infty$,

and as $x \to -\infty$, $x^3 - 2x^2 - 5x + 6 \to -\infty$.

So the graph of the function is

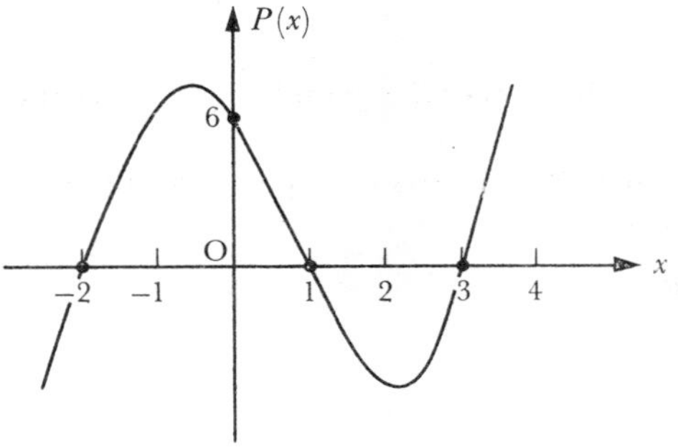

Example 2

Find the factors and zeros of $x^4 - 3x^2 - 2x$
and so sketch its graph.

Two of its zeros are clearly $x = 0$ and $x = -1$. So two factors are x and $x + 1$.

By division, we find that

$$x^4 - 3x^2 - 2x \equiv x(x+1)(x^2 - x - 2)$$

But $\quad x^2 - x - 2 \equiv (x+1)(x-2)$

So $\quad x^4 - 3x^2 - 2x \equiv x(x+1)^2(x-2)$

and has zeros 0, -1, and $+2$.

Also as $\quad x \to +\infty, x^4 - 3x^2 - 2x \to +\infty$

and as $\quad x \to -\infty, x^4 - 3x^2 - 2x \to +\infty$

Hence the values of $x^4 - 3x^2 - 2x$ are given by

x	-3	-2	-1	0	1	2	3	4
$P(x)$	+ + + + + + + + + + + 0 + + + 0 − − − − − − − − 0 + + + + + + + + + + +							

and its graph is

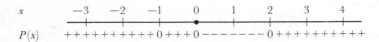

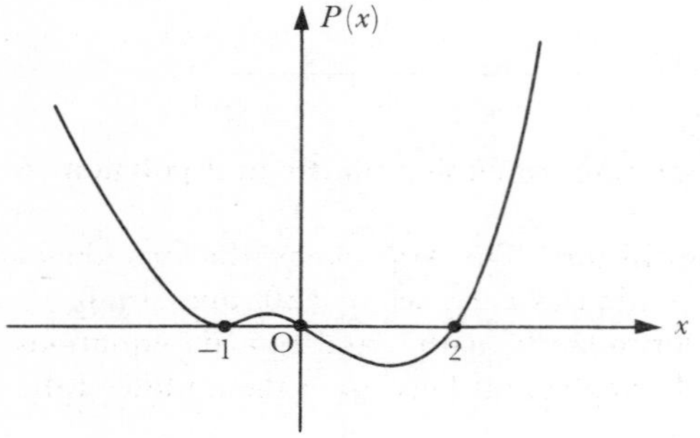

Exercise 1.7b

1 Find the remainder when

(*i*) $x^3 + 3x^2 - 4x + 5$	is divided by $x - 2$
(*ii*) $2x^3 + 8x^2 - 3$	is divided by $x - 1$
(*iii*) $x^3 + 8$	is divided by $x + 1$
(*iv*) $x^4 + 3x^3 + 2x - 3$	is divided by $x + 3$
(*v*) $x^3 + a^3$	is divided by $x + a$
(*vi*) $8x^3 - 12x^2 + 7$	is divided by $2x - 1$.

2 Factorise the following polynomials, find their zeros and sketch their graphs:

(*i*) $x^3 - 6x^2 + 11x - 6$

(*ii*) $x^3 - x^2$

(*iii*) $x^3 + x^2 - x - 1$

(*iv*) $2x^3 - x^2 - 8x + 4$

(*v*) $-x^4 + 4x^2$

(*vi*) $x^4 - 8x^3 + 14x^2 + 8x - 15$.

3 (*i*) Show that $3x + 1$ is a factor of the expression $3x^3 + 4x^2 + 7x + 2$.
(*ii*) Find the values of a and b if $x - 2$ and $x + 1$ are both factors of $ax^3 + 3x^2 - 9x + b$, and state the third factor of the expression.

(O.C.)

4 When $x^3 - x^2 + ax + b$ is divided by $x - 1$ the remainder is -8, and when the same expression is divided by $x + 1$ the remainder is 0.

Calculate the values of a and b, and factorise the expression completely.

(O.C.)

5 Factorise (*i*) $x^3 - a^3$; (*ii*) $x^3 + a^3$.

Polynomial equations

This exercise will have provided some practice in the solution of simple polynomial equations. Can we always obtain such solutions?

In the case of the quadratic we were successful in discovering the formula

$$x = \frac{-b \pm \sqrt{(b^2 - 4ac)}}{2a}.$$

Are there similar formulae for the roots of cubic, quartic and polynomial equations of higher degree?

This problem has had a strange history. The Arabs knew the formula for the solution of the quadratic equation as early as AD 830, and similar—but much more complicated—formulae for cubic and quartic equations were discovered by the Italians Tartaglia and Ferraro in the middle of the sixteenth century.

But for equations of higher degree, like the quintic (of degree 5), no such solution could be found. For centuries the problem remained unsolved and was only finally settled by the young Norwegian mathematician, Abel (1802–29). He discovered, astonishingly, not a solution of such equations, but a proof that a general solution giving the roots in terms of the coefficients is impossible. So their investigation (and in practice that of cubic and quartic equations too) has usually to be undertaken by graphical or numerical methods.

There remains the simpler question, whether we can decide on the number of roots of a polynomial equation. Firstly, is there a maximum number of roots? A little thought shows us that there is, and that a polynomial function of degree n cannot possibly have more than n zeros (unless it is identically zero: $0x^n + 0x^{n-1} + \cdots 0x + 0$).

Suppose for instance, that the cubic function

$$ax^3 + bx^2 + cx + d \ (a \neq 0)$$

has zeros α, β, and γ.

Then $x - \alpha$, $x - \beta$, and $x - \gamma$ are all factors.

$$\Rightarrow \quad a(x - \gamma)(x - \beta)(x - \gamma) \equiv ax^3 + bx^2 + cx + d$$

So if δ were also a zero,

$$a(\delta - \alpha)(\delta - \beta)(\delta - \gamma) = 0$$

$$\Rightarrow \quad \delta = \alpha, \quad \beta \quad \text{or} \quad \gamma$$

Hence a cubic polynomial cannot have more than three zeros, nor a cubic equation more than three roots. Similarly a polynomial equation of degree n cannot possibly have more than n roots.

To the other question, whether there is a *minimum* number of zeros of a polynomial, we have already seen the answer. For there are no real values of x for which

$$x^2 + 1 = 0 \quad \text{or} \quad x^4 + 3x^2 + 2 = 0$$

It is possible to show that, if we make use of the symbol i $(=\sqrt{-1})$, every polynomial has a zero; and thence that we can always factorise a polynomial of degree n into n linear factors. But for the present we confine ourselves to zeros and roots which are real.

Lastly, we can ask whether there are *any* connections between the zeros of a polynomial and its coefficients, and to this question we can provide some easy answers.

Suppose, for example, that the quadratic equation

$ax^2 + bx + c = 0$ has roots $x = \alpha$ and $x = \beta$.

Then $ax^2 + bx + c$ has factors $x - \alpha$ and $x - \beta$

So $\quad ax^2 + bx + c \equiv k(x - \alpha)(x - \beta)$

From the x^2 term, we see that

$$ax^2 \equiv kx^2 \Rightarrow a = k$$

Hence $\quad ax^2 + bx + c \equiv a(x - \alpha)(x - \beta)$

$$\Rightarrow \quad ax^2 + bx + c \equiv ax^2 - a(\alpha + \beta)x + a\alpha\beta$$

So $\quad -a(\alpha + \beta) = b$

and $\quad a\alpha\beta = c$

$$\Rightarrow \quad \boxed{\begin{aligned} \alpha + \beta &= -b/a \\ \alpha\beta &= +c/a \end{aligned}}$$

Similarly if $\quad ax^3 + bx^2 + cx + d \quad$ has zeros α, β, γ

then $\quad ax^3 + bx^2 + cx + d \equiv a(x - \alpha)(x - \beta)(x - \gamma)$

$$\equiv ax^3 - a(\alpha + \beta + \gamma)x^2 + a(\beta\gamma + \gamma\alpha + \alpha\beta)x - a\alpha\beta\gamma$$

$$\Rightarrow \quad -a(\alpha + \beta + \gamma) = b \quad \Rightarrow \quad \alpha + \beta + \gamma = -b/a$$

$$a(\beta\gamma + \gamma\alpha + \alpha\beta) = c \qquad \beta\gamma + \gamma\alpha + \alpha\beta = +c/a$$

$$-a\alpha\beta\gamma = d \qquad \alpha\beta\gamma = -d/a$$

These relationships are useful in a variety of ways. In example 1, for instance, we could have checked that $x^3 - 2x^2 - 5x + 6 = 0$ has roots 1, 3 and -2 by noting that

$$1 + 3 + (-2) = 2 = -(-2)$$

$$3\times -2 + -2\times 1 + 1\times 3 = -5 = +(-5)$$

and $$1 \times 3 \times -2 = -6 = -(6)$$

Exercise 1.7c

1 Find $\alpha + \beta + \gamma$, $\beta\gamma + \gamma\alpha + \alpha\beta$ and $\alpha\beta\gamma$ where α, β, γ are the roots of:

(*i*) $x^3 - 6x^2 + 11x - 6 = 0$

(*ii*) $x^3 - x = 0$

(*iii*) $x^3 + x^2 - x - 1 = 0$

(*iv*) $2x^3 - x^2 - 8x + 4 = 0$.

Check your results by using the answers to Exercise 1.7b, No. 2.

*1.8 Rational functions

If $P(x)$ and $Q(x)$ are polynomial functions of x, their quotient

$\dfrac{P(x)}{Q(x)}$ is called a *rational function* of x.

Example 1

Investigate the function $f(x) = \dfrac{2x}{x-1}$

We first note that

$$f(0) = 0$$

and $f(1) = \dfrac{2}{1-1} = \dfrac{2}{0}$, which is undefined.

The sign of $f(x)$ is indicated by

x		-2		-1		0		1		2		3	
$f(x)$	+ + +	+ + +	+ + +	+ + +	+	0	− − −	|	+ + +	+ + +	+ + +	+ + +	+

As $x \to 1, f(x)$ becomes very large, and by considering the sign of $f(x)$ we see that

as $\quad x \to 1-, \quad f(x) \to -\infty;$

as $\quad x \to 1+, \quad f(x) \to +\infty$

Finally, we can see how $f(x)$ behaves as $x \to \pm\infty$ by writing

$$f(x) = \frac{2x}{x-1} = \frac{2}{1-1/x} \to 2 \quad \text{as } x \to \pm\infty.$$

These can now all be assembled in the graph of the function, where it is seen that there are two asymptotes, one parallel to each of the axes, which the curve approaches but never meets:

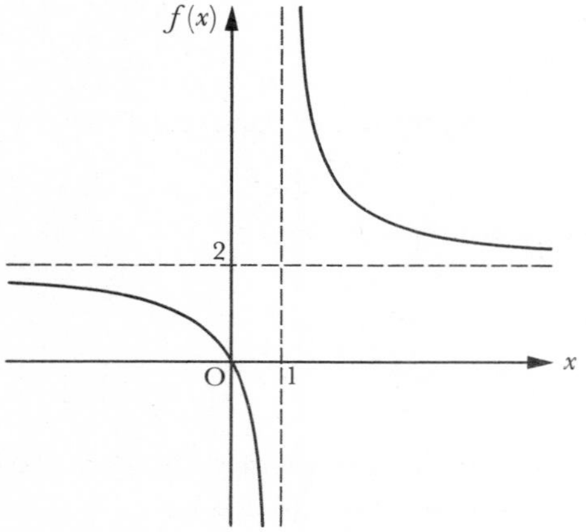

Example 2

Investigate the function $f(x) = \dfrac{x^2}{2-x}$

In this case $f(0) = 0$ and $f(2)$ is undefined.

The sign of $f(x)$ is given by

x	-2	-1	0	1	2	3
$f(x)$	+ + + + +	+ + + + +	0 + + +	+ + + +	|	− − − − − −

and we see that

as $\quad x \to 2-, \quad f(x) \to +\infty$

$\quad x \to 2+, \quad f(x) \to -\infty$

Finally, we can see how $f(x)$ behaves as $x \to \pm\infty$ by writing

$$f(x) = \frac{x^2}{2-x} = \frac{x}{2/x-1} \approx \frac{x}{-1} = -x$$

So the graph of $f(x)$ is

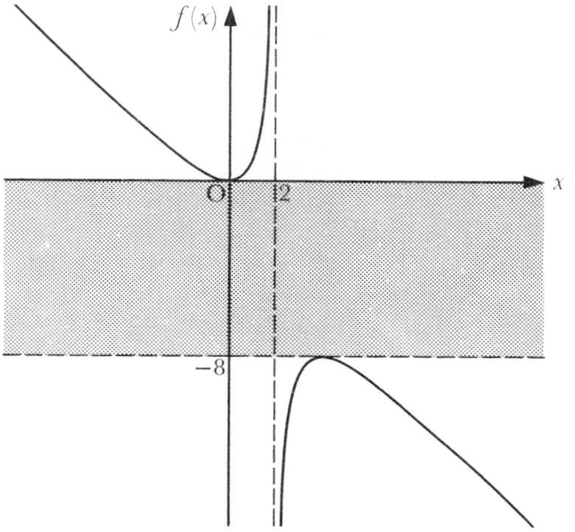

The range of values of $f(x)$ can be found by letting $f(x) = y$.

$$\Rightarrow \quad \frac{x^2}{2-x} = y$$

$$\Rightarrow \quad x^2 + yx - 2y = 0$$

So the values of x which produce a given value of y are the roots of this quadratic equation, and these are real if and only if

$$y^2 - 4(-2y) \geqslant 0$$

$$\Leftrightarrow \quad y(y+8) \geqslant 0$$

$$\Leftrightarrow \quad y \geqslant 0 \quad \text{or} \quad y \leqslant -8.$$

So $f(x)$ can take all values except those between 0 and -8, corresponding to the shaded area of the graph.

Exercise 1.8

Sketch graphs of the following functions, stating their range and clearly marking their asymptotes:

1 $\dfrac{1}{1-x}$ **2** $\dfrac{1}{x+1}$ **3** $\dfrac{x^2+1}{x}$ **4** $\dfrac{x^2-1}{x}$

5 $\dfrac{1}{(x-1)(x-2)}$ **6** $\dfrac{x^2}{x^2+4}$ **7** $\dfrac{x^2}{x^2-4}$ **8** $\dfrac{x}{(x+1)(x+2)}$

*1.9 Functions of several variables

At the start of this chapter it was mentioned that some quantities are functions of several variables, and in such cases a similar notation is used.

Example 1

The volume V of a solid circular cylinder is a function of both its base-radius r and its height h, so we write

$$V = f(r, h) = \pi r^2 h$$

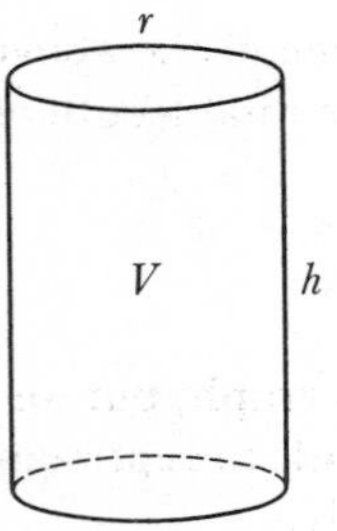

Hence the volume V of a cylinder of radius 1 m and height 2 m is described by

$$f(1, 2) = \pi \times 1^2 \times 2 \approx 6.28 \text{ m}^3.$$

Example 2

The outer surface area A of an open rectangular box of length l, breadth b, and height h is given by

$$A = \phi(l, b, h) = lb + 2lh + 2bh$$

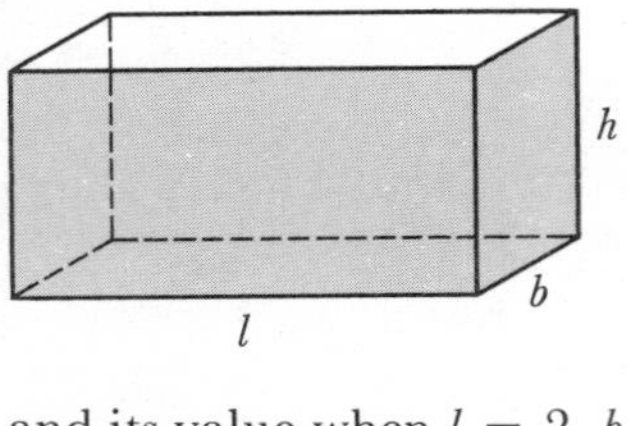

and its value when $l = 2$, $b = 3$, $h = 1$ is

$$\phi(2, 3, 1) = 2\times3 + 2\times2\times1 + 2\times3\times1 = 16.$$

Exercise 1.9a

Using these examples, find the value and explain the meaning of

1 (*i*) $f(2, 4)$; (*ii*) $f(4, 2)$; (*iii*) $\phi(1, 2, 3)$; (*iv*) $\phi(3, 2, 1)$.
Should $\phi(1, 2, 3)$ and $\phi(2, 1, 3)$ be equal?
What do you know about $\phi(a, b, c)$ and $\phi(b, a, c)$?
Explain your results.

2 Use the above functions f and ϕ to describe and to calculate:
(*i*) the volume of a cylinder of base-radius 10 m and height 5m,
(*ii*) the outer surface area of an open tank with a square base of side 1m and height 3m.

3 (*i*) Construct a function $g(r, h)$ for the total surface area of the solid cylinder.
(*ii*) Calculate g (2, 4) and explain what it represents.
(*iii*) Use your function to find another function which represents the total surface area S of a cylinder whose height and base-diameter are both equal to c.

4 Construct the function $V(l, b, h)$ to represent the volume of the above rectangular box, and the function $L(l, b, h)$ to represent the total length of its edges.

Illustration of functions of several variables

A function of one variable can be illustrated by means of a graph, but unfortunately a function of two variables is much more difficult to represent and functions of more than two variables defy us completely.

With two variables the most illuminating way is by means of a surface, as in a relief map showing height h above sea-level at a point (x, y):

$$h = f(x, y)$$

But construction of such models is a lengthy process and the most convenient method without using a third dimension is again that of the map-maker, by drawing contours connecting points (x, y) of equal h. These occur in a variety of other forms.

If, for instance, θ is the temperature at a point (x, y), then $\theta = f(x, y)$ and the contours are *isotherms*.

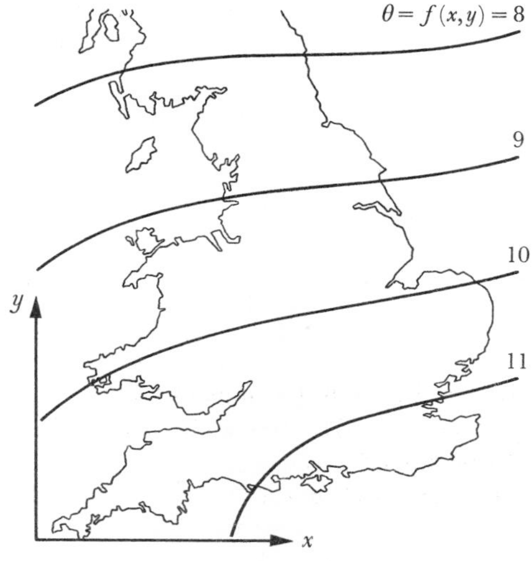

Similarly, if p is the atmospheric pressure, then $p = f(x, y)$ and the contours are *isobars*; and if V is the electric potential, then $V = f(x, y)$ and the contours are *equipotentials*.

Example 1

$f(x, y) = x^2 + y^2$ has contours

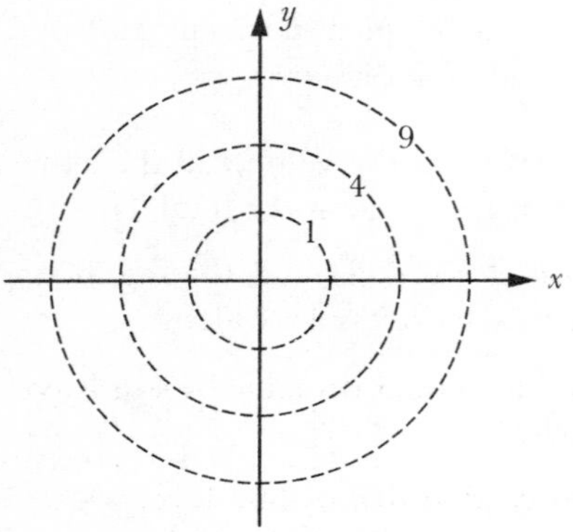

and so is represented by (a *paraboloid*)

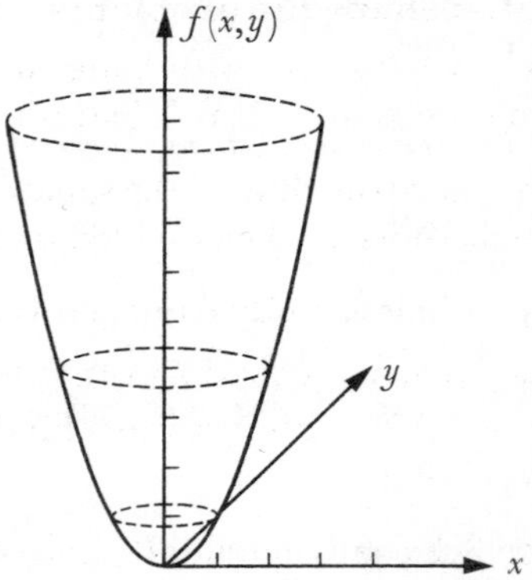

Example 2

$f(x, y) = xy$ has contours

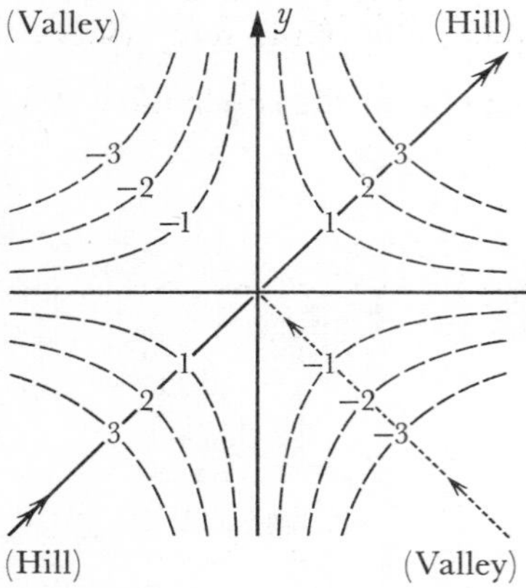

and so can be visualised as a saddle connecting two hills (and two valleys) (a *hyperboloid*)

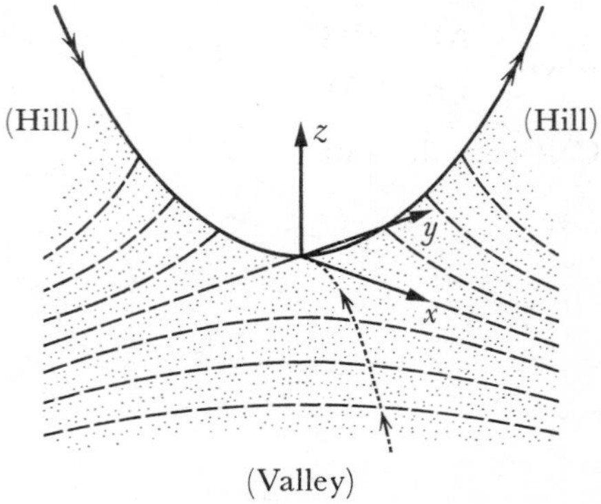

Exercise 1.9b

Describe, and illustrate:

1 $f(x, y) = x + y$

2 $f(x, y) = x - y$

3 $f(x, y) = 100 - x^2 - y^2$

4 $f(x, y) = |x - y|$

5 $f(x, y) = 10 - |x| - |y|$.

Miscellaneous problems 1

1 A telegraph wire hangs in the form of a parabola and the sag at its middle point is 1 m. What is the sag one quarter of the way along?

2 Is it true that all the squares of integers are either divisible by 4 or leave remainder 1 when divided by 8? If so, why?

3 If n is a positive integer, is $n^5 - n$ always divisible by 30? If so, why?

4 When a polynomial $P(x)$ is divided by $x - 1$, x, $x + 1$, the remainders are 1, 2, 3 respectively. Find the remainder when $P(x)$ is divided by $x(x^2 - 1)$.

5 The polynomial $P(x)$ is divided by $(x - a)(x - b)$, where $a \neq b$. Express the remainder in terms of $P(a)$ and $P(b)$.

6 The equation $4x^3 - 3x^2 - 10x - 49 = 0$ has a root which is approximately 3. By substituting $x = 3 + h$ and neglecting squares and higher powers of h, find a closer approximation to the value of this root. (M.E.I.)

7 Artists have sometimes paid special attention to the subdivision of a line AB by a point P, called its *golden section*,

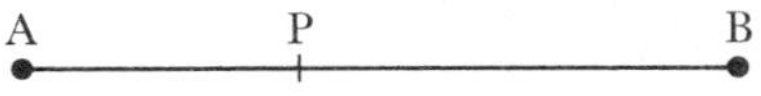

where $$\frac{AP}{PB} = \frac{PB}{AB}$$

Calculate this ratio.

2

Power functions and logarithmic functions

2.1 Power functions

We have so far concentrated our attention upon linear functions (like $2x + 1$, $6 - \frac{1}{2}x$), quadratic functions (like x^2, $3 - x + x^2$), higher degree polynomials and their sums, differences, products and quotients.

In this chapter we shall investigate functions where the independent variable occurs as an index or power, and we shall call them *power functions*.†

The simplest such functions are $f(x) = 2^x, \quad 10^x$, etc.
and we already know the values of these functions when x is a positive integer:

$2^5 = 32$	$10^5 = 100\,000$
$2^4 = 16$	$10^4 = 10\,000$
$2^3 = 8$	$10^3 = 1\,000$
$2^2 = 4$	$10^2 = 100$
$2^1 = 2$	$10^1 = 10$

As for such expressions as 2^0, 2^{-3}, and 10^{-4}, we are at liberty to define them however we like. But there is an obvious convenience if we maintain the pattern of this table, so we continue with

$2^0 = 1$	$10^0 = 1$
$2^{-1} = \frac{1}{2}$	$10^{-1} = \dfrac{1}{10}$

† Also commonly called exponential functions (see preface).

$$2^{-2} = \tfrac{1}{4} = \frac{1}{2^2} \qquad 10^{-2} = \frac{1}{100} = \frac{1}{10^2}$$

$$2^{-3} = \tfrac{1}{8} = \frac{1}{2^3} \qquad 10^{-3} = \frac{1}{1000} = \frac{1}{10^3}, \text{ etc.}$$

More generally, if n is a positive integer,

we *define* $2^{-n} = \dfrac{1}{2^n}$, $10^{-n} = \dfrac{1}{10^n}$

and (provided $a \neq 0$), $a^{-n} = \dfrac{1}{a^n}$ and $a^0 = 1$.

Exercise 2.1a

1 Calculate:

(*i*) 3^{-1}; (*ii*) 4^{-2}; (*iii*) $\left(\frac{3}{2}\right)^{-3}$; (*iv*) $(-5)^{-1}$; (*v*) $\left(\frac{1}{4}\right)^{-3}$; (*vi*) 2^{-6};

(*vii*) $\left(\frac{1}{3}\right)^{-2}$; (*viii*) 1^0; (*ix*) $\left(-\frac{1}{2}\right)^0$; (*x*) 0^1.

2 Calculate:

(*i*) $3^4 \times 3^2$ and 3^{4+2}
(*ii*) $2^3 \times 2^{-5}$ and 2^{3-5}
(*iii*) $4^{-2} \times 4^5$ and 4^{-2+5}
(*iv*) $10^{-3} \times 10^{-2}$ and 10^{-3-2}
(*v*) $5^0 \times 5^{-3}$ and 5^{0-3}

3 Calculate:

(*i*) $(3^4)^2$ and $3^{4\times 2}$
(*ii*) $(2^{-3})^2$ and $2^{-3\times 2}$
(*iii*) $(10^2)^3$ and $10^{2\times 3}$
(*iv*) $(4^{-2})^{-1}$ and $4^{-2\times -1}$
(*v*) $(5^0)^3$ and $5^{0\times 3}$.

From this exercise it appears that if $a \neq 0$ and m, n are integers, the above definition leads to

$$a^m \times a^n = a^{m+n}$$
and
$$(a^m)^n = a^{mn}$$
even if m or n is negative or zero.

We are now in a position to ask whether any useful meaning can be given to such expressions as $2^{1/2}$, $5^{2/3}$, and $4^{-3/2}$.

If, as before, we aim to maintain the existing pattern, it would be convenient if

$(2^{1/2})^2 = 2^{(1/2)\times 2} = 2^1 = 2$, so we *define* $2^{1/2}$ as $\sqrt{2}$

and $(5^{2/3})^3 = 5^{(2/3)\times 3} = 5^2$, so we *define* $5^{2/3}$ as $\sqrt[3]{5^2}$.

More generally, if $a \neq 0$ and p, q are positive integers,

$(a^{p/q})^q = a^{(p/q)\times q} = a^p$, so we define $a^{p/q}$ as $\sqrt[q]{a^p}$.

Lastly, the definition is extended to powers that are *negative* fractions, provided that

$$a^{-p/q} = \frac{1}{a^{p/q}} = \frac{1}{\sqrt[q]{a^p}}$$

So $$4^{-3/2} = \frac{1}{4^{3/2}} = \frac{1}{\sqrt[2]{4^3}} = \frac{1}{\sqrt[2]{64}} = \frac{1}{8}$$

The relationships

$$\boxed{a^m \times a^n = a^{m+n} \quad \text{and} \quad (a^m)^n = a^{mn}}$$

are now true for all rational values of m and n, whether positive or negative, integral or fractional.

Exercise 2.1b

1 Calculate:

(*i*) $25^{1/2}$	(*ii*) $16^{1/4}$	(*iii*) $27^{1/3}$	(*iv*) $8^{2/3}$
(*v*) $9^{3/2}$	(*vi*) $8^{5/3}$	(*vii*) 5^0	(*viii*) $4^{-1/2}$
(*ix*) $8^{-1/3}$	(*x*) $9^{-3/2}$	(*xi*) $16^{-5/4}$	(*xii*) $32^{-3/5}$.

2 If $2^{x+y} = 1$ and $10^{3x-y} = 100$, find the values of 5^{y-x} and x^y (M.E.I.)

2.2 The power function, 10^x

Suppose that a colony of insects is breeding so fast that each day its size is multiplied by 10. Then after x days it will have increased x times by a factor 10, i.e., by a factor 10^x.

In order to plot the growth of this *power function*, we first form a table of values of 10^x:

x	0	1	2	3
10^x	1	10	100	1 000

To obtain intermediate values, we first calculate $10^{1/2}$.

Now $10^{1/2} = \sqrt{10} \approx 3.162$

so that $10^{1\frac{1}{2}} \approx 31.62$, $10^{2\frac{1}{2}} \approx 316.2$,

and $10^{-\frac{1}{2}} \approx 0.316\,2$.

Similarly $10^{1/4} = (10^{1/2})^{1/2} \approx \sqrt{3.162} \approx 1.778$

$$\Rightarrow \quad 10^{3/4} = 10^{1/2} \times 10^{1/4} \approx 3.162 \times 1.778 \approx 5.623.$$

Using these results, we obtain the extended table:

x	−0.50	−0.25	0	0.25	0.50	0.75	1.00	1.25	1.50	1.75	2.00
10^x	0.3162	0.5623	1.0	1.778	3.162	5.623	10	17.78	31.62	56.23	100

and the graph:

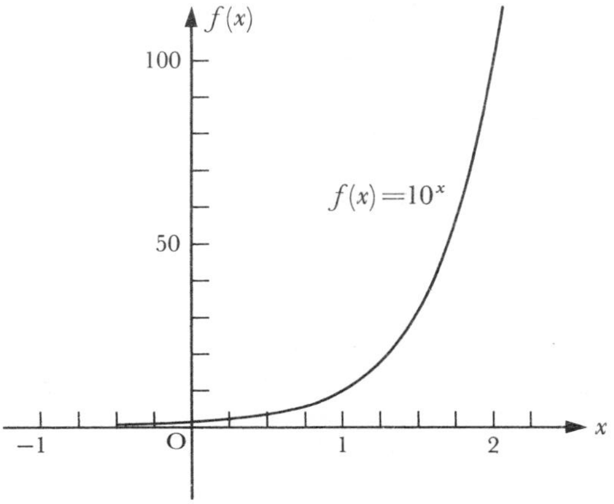

Exercise 2.2

(Start by plotting $f(x) = 10^x$ on graph paper, from $x = -1$ to $x = +2$.)

1 By what factor has the above population of insects increased after (*i*) 0.8 days; (*ii*) 1.3 days; (*iii*) 1.9 days?

2 How long does it take for the population to increase by a factor (*i*) 5; (*ii*) 27; (*iii*) 120?

3 Is there any meaning for the values of 10^x when x is negative?

4 By what factor does the population increase
(*i*) between $x = 1.0$ and $x = 1.2$;
(*ii*) between $x = 1.5$ and $x = 1.7$;
(*iii*) between $x = 1.8$ and $x = 2.0$?

5 Draw similar graphs for the functions 2^x and 3^x, taking values of x from $x = 0$ to 3 at intervals of 0.25, and using the same axes.

2.3 Increasing and decreasing power functions

The last section shows the importance of power functions whenever growth is being considered. Two familiar examples will provide further illustration:

Example 1

One of the commonest examples of growth is that of interest on a sum of money placed on deposit in a bank. If £1 is deposited and the interest rate is 5% per year, then usually one of two things happens:

Either (*i*) the annual interest £0.05 is withdrawn each year. In this case the total invested remains £1 and the amount earned in n years is the *simple interest*, £$0.05n$.

Or (*ii*) the interest is credited to the account each year, so that the future interest will be calculated upon the *new* amount on deposit. In this case the amount on deposit will be multiplied each year by a factor 1.05 and the total interest earned is called the *compound interest.*

The amount on deposit initially is £1,
so after 1 year is £(1.05),
after 2 years is £$(1.05)^2$,
and after n years is £$(1.05)^n$.

If we suppose that the sums invested are gaining interest continuously rather than at the end of each year, the difference between simple and compound interest is illustrated by the difference between the graphs of a linear function and a power function:

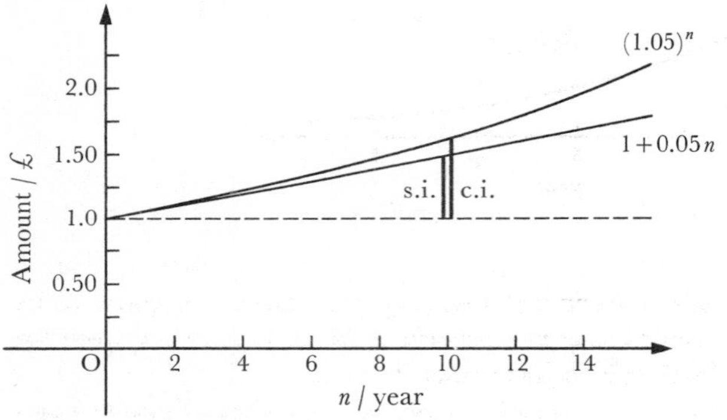

Example 2

Just as the growth of a quantity frequently involves a power function, so does decay; and the strength of these functions is just as marked with decay as with growth.

If, for instance, a car is bought for £1 000 and depreciates at 30% per year, then its value is changed each year by a factor $\frac{7}{10}$.

Initial value = £1 000,

so after 1 year, value $= £1\,000 \times \dfrac{7}{10} = £700$,

after 2 years, value $= £1\,000 \times \left(\dfrac{7}{10}\right)^2 = £490$,

and after n years, value $= £1\,000 \times \left(\dfrac{7}{10}\right)^n$.

This too is a power function, though here one of powerful decline. Such functions are extremely common, not only for measuring depreciations of value, but throughout nature, from the decay of radioactive elements to the cooling of one's bath-water. With an increasing power function, the factor of increase was always the same for a given time, whether the actual quantity was large or small: in just the same way, this factor remains constant for a decreasing power function, and the car in our example (for instance) is losing 30% of its value each year, no matter whether it is new and valuable or old and relatively worthless.

If, as with compound interest, we regard the value of the car as depreciating *continuously*, we can represent it by the curve

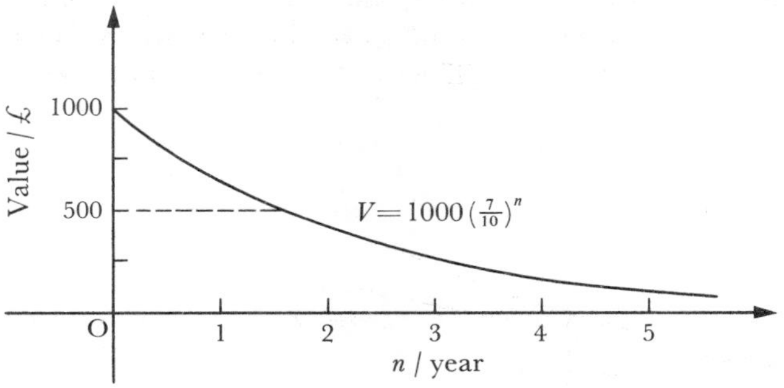

When a quantity is decreasing in this way, we sometimes speak of its 'half life', the time it would take to diminish by 50%. We can see from the graph that our car's 'half-life' is just under 2 years.

We are now in a position to sketch graphs of a number of power functions:

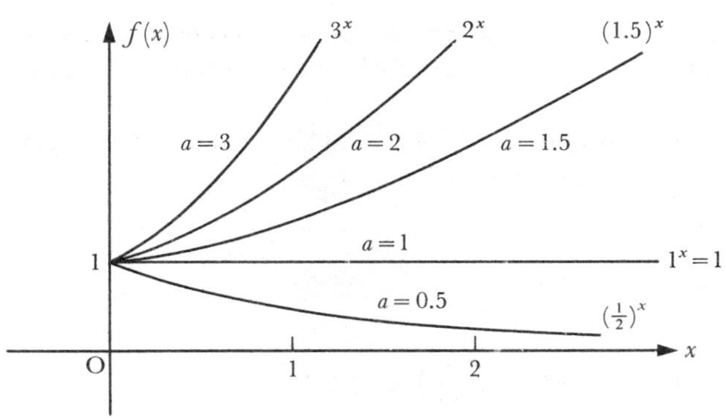

The reader will probably wonder how the increasing power functions compare with, say, x^2 and x^{10}. A little investigation should produce values of x for which 2^x overtakes both of these, and the reader can then ask himself whether a more impressive function like x^{1000} would be similarly overtaken.

Exercise 2.3

1 The cost of living in a certain country is increasing by 5% each year. What will be the total percentage increase after:
(*i*) 2 years; (*ii*) 3 years; (*iii*) 4 years?

2 In the early stages of a measles epidemic there were 100 people infected and each day the number was rising by 10%. How many would be infected: (*i*) after 2 days; (*ii*) after 3 days; (*iii*) after 4 days; (*iv*) after 1 week?

3 An absent-minded mathematician makes a pot of tea and immediately puts it in his refrigerator at 0° C. If each minute it loses a quarter of its excess temperature, how hot is it when discovered 5 minutes later?

4 If each year the purchasing value of £1 falls by 4%, what is the total percentage fall in the course of (*i*) 2 years; (*ii*) 3 years?

2.4 The logarithmic function, $\log_{10}x$

It was noticed in section 2.1 that indices can be used for the multiplication of two numbers.
For instance, 1 000 × 100 can be written as

$$10^3 \times 10^2 = 10^{3+2} = 10^5 = 100\,000$$

At first this seems a very unpractical method, but the graph of the power function can be used to extend it a little further, to perform the sum 4 × 5.

From our graph we see that

$$4 \approx 10^{0.6} \quad \text{and} \quad 5 \approx 10^{0.7}$$

So $4 \times 5 \approx 10^{0.6} \times 10^{0.7} = 10^{0.6+0.7} = 10^{1.3} \approx 20$

This still seems a very perverse and inaccurate way of calculating 4 × 5, but it will quickly be seen that the replacement of multiplication by addition can have enormous advantages, once there is a quick means of expressing any number as a power of 10.

The power itself is called the *logarithm* (to the base 10) of the number. From our graph, we found that

$$4 \approx 10^{0.6} \quad \text{and} \quad 5 \approx 10^{0.7}$$

So we write

$$\log_{10} 4 \approx 0.6 \qquad \text{and} \quad \log_{10} 5 \approx 0.7$$

More generally, the logarithm (to the base 10) of any number is the power to which 10 has to be raised to give the number:

$$x = 10^y$$
$$\Leftrightarrow \quad y = \log_{10}x$$

So we see that $\log_{10} x$ is the inverse function of 10^x (and vice-versa), and that as 10^x is always positive, we can only speak of the logarithms of positive numbers.

Rather than use a graph, we can of course obtain logarithms more quickly and more accurately by means of published tables.

Example 1

Calculate $4.163 \times 516.4 \times 0.7296$,

we can express this as

$$
\begin{aligned}
& 4.163 \times 10^2 \times 5.164 \times 10^{-1} \times 7.296 \\
&= 10^{0.6194} \times 10^2 \times 10^{0.7129} \times 10^{-1} \times 10^{0.8631} \\
&= 10^{0.6194} \times 10^{2.7129} \times 10^{\bar{1}.8631} \\
&= 10^{0.6194+2.7129+\bar{1}.8631} \\
&= 10^{3.1954} \\
&= 10^3 \times 10^{0.1954} \\
&= 10^3 \times 1.568 \\
&= 1568 \\
&= 1570 \text{ to 3 s.f.}
\end{aligned}
$$

This is, of course, more usually set down in the form:

	no.	log
$4.163 \times 516.4 \times 0.7296$	4.163	0.6194
$= 1568$	516.4	2.7129
$= 1570$ to 3 s.f.	0.7296	$\bar{1}.8631$
	1568	3.1954

Example 2

$(2.639)^6$	2.639	0.4215
$= 338.1$		6
$= 338$, to 3 s.f.	338.1	2.5290

Example 3

$\sqrt[3]{0.02}$	0.02	$\bar{2}.3010$ or
$= 0.2714$		$3 \mid \bar{3} + 1.3010$
$= 0.271$ to 3 s.f.	0.2714	$\bar{1}.4337$

Exercise 2.4

1 Express the following as powers of 10. Hence, without using tables, find their logarithms to the base 10.

(*i*) 100 000; (*ii*) $\sqrt[2]{1\,000}$; (*iii*) 1; (*iv*) $\dfrac{1}{100}$; (*v*) $\sqrt[3]{10}$; (*vi*) $\dfrac{1}{\sqrt{10}}$

2 It is known that, to five places of decimals,
$2 = 10^{0.30103}$ and $3 = 10^{0.47712}$.

Hence express as powers of 10:
(*i*) 20; (*ii*) 300; (*iii*) 0.03; (*iv*) 0.002; (*v*) 6; (*vi*) 4; (*vii*) $\frac{1}{2}$; (*viii*) 1.5.

3 Using logarithm tables to express numbers as powers of 10, calculate the following:

(*i*) 5.314×1.297	(*ii*) $6.243 \div 2.109$
(*iii*) 31.26×29.47	(*iv*) $(3.296)^4$
(*v*) $\sqrt{41.26}$	(*vi*) $3^{1.5}$
(*vii*) 0.026×0.4791	(*viii*) $0.2971 \div 4.26$
(*ix*) $1 \div 0.02671$	(*x*) $\sqrt[3]{0.0145}$

*2.5 Properties of logarithmic functions

To investigate the properties of logarithms,
let $\log_{10} x = p$ and $\log_{10} y = q$.

Then $x = 10^p$ and $y = 10^q$

So $xy = 10^{p+q}$, $\dfrac{x}{y} = 10^{p-q}$, $x^n = 10^{np}$

$\Rightarrow \log_{10} xy = p + q$, $\log_{10} \dfrac{x}{y} = p - q$, $\log_{10} x^n = np$

So

$$\begin{aligned} \log_{10} xy &= \log_{10} x + \log_{10} y \\ \log_{10} \frac{x}{y} &= \log_{10} x - \log_{10} y \\ \log_{10} x^n &= n \log_{10} x \end{aligned}$$

These, of course, are the properties which underlie the use of logarithms in calculations. By adding the logarithms of two numbers we obtain the logarithm of their product, and so are able to find the product itself; and similarly for quotients and powers.

The construction of a slide rule is based on the same principle, the length to a certain number being proportional to the logarithm of the number. So two numbers can be multiplied simply by placing these lengths—or logarithms—end to end.

So far we have considered logarithms to the base 10. But it is clear that we can also define logarithms to other bases. The logarithm to any base of a number is simply the power to which the base must be raised in order to give the number.

For example,

$9 = 3^2$, so $\log_3 9 = 2$; and $8 = 2^3$, so $\log_2 8 = 3$

More generally,

If $a > 0$, $\boxed{x = a^p \Leftrightarrow \log_a x = p}$ †

To investigate the properties of such logarithms,

let $\log_a x = p \Leftrightarrow x = a^p$

and $\log_a y = q \Leftrightarrow y = a^q$

Hence $xy = a^{p+q}$, $\frac{x}{y} = a^{p-q}$, $x^n = a^{np}$

and $\log_a xy = p + q$, $\log_a \frac{x}{y} = p - q$, $\log_a x^n = np$.

So
$$\boxed{\begin{aligned} \log_a xy &= \log_a x + \log_a y \\ \log_a \frac{x}{y} &= \log_a x - \log_a y \\ \log_a x^n &= n \log_a x \end{aligned}}$$

and the three basic laws of logarithms are therefore true whatever base is being used.

In order to calculate logarithms to a base other than 10 it is usually most convenient to proceed as follows:

To find $\log_3 7$

Let $\log_3 7 = x$

$\Rightarrow \quad 3^x = 7$

$\Rightarrow \quad \log_{10} 3^x = \log_{10} 7$

$\Rightarrow \quad x \log_{10} 3 = \log_{10} 7$

$\Rightarrow \quad x = \dfrac{\log_{10} 7}{\log_{10} 3}$

Hence $\log_3 7 = \dfrac{\log_{10} 7}{\log_{10} 3} = \dfrac{0.8451}{0.4771}$

which can then be calculated (by logarithms!), so that we finally obtain

$\log_3 7 = 1.77$ (to 3 s.f.)

† And $x = a^p = \text{antilog}\, p$.

Exercise 2.5

1 Given that

$\log_{10} 2 = 0.301\,030$ and $\log_{10} 3 = 0.477\,121$, find:

(*i*) $\log_{10} 4$	(*ii*) $\log_{10} 5$	(*iii*) $\log_{10} 8$
(*iv*) $\log_{10} 9$	(*v*) $\log_{10} 10$	(*vi*) $\log_{10} 12$
(*vii*) $\log_{10} 15$	(*viii*) $\log_{10} 30$	(*ix*) $\log_{10} 180$
(*x*) $\log_{10} 0.25$	(*xi*) $\log_{10} \frac{3}{2}$	(*xii*) $\log_{10} \frac{2}{3}$
(*xiii*) $\log_{10} \sqrt{24}$	(*xiv*) $\log_{10} \sqrt[3]{16}$	(*xv*) $\log_{10} \frac{1}{\sqrt{3}}$

2 Use logarithm tables to solve the equations:
(*i*) $3^x = 100$ (*ii*) $2^x = 1\,000\,000$
(*iii*) $(1.293)^x = 10$ (*iv*) $10^x = 3$

3 Use logarithm tables to evaluate:
(*i*) $\log_4 7$ (*ii*) $\log_5 10$
(*iii*) $\log_2 40$ (*iv*) $\log_3 2$.

4 Find, correct to three significant figures, the values of x if
(*i*) $3^{x+1} = 4^{2x-1}$ (*ii*) $5^x \times 2^{-2x} = 4$ (O.C.)

5 A slide rule is graduated so that the distance between the graduations '1' and 'n' is proportional to $\log_{10} n$. If the distance between '1' and '100' is 40 cm, find the distance between '25' and '45', correct to three significant figures. (S.M.P.)

6 What is the value after 10 years of £100 invested at compound interest if the rate of interest is (*i*) 5%; (*ii*) 10%?

7 How long does a sum of money take to double itself when invested at compound interest of (*i*) 5%; (*ii*) 10%?

8 A country's population at the end of each year is 2% greater than at the start of the year. How long will it take for its population to increase by 50%?

9 A car bought for £1 000 depreciates in value by 15% each year.
(*i*) What is its value after 6 years?
(*ii*) What is the half-life of its value?

10 Archaeologists can sometimes estimate the age of their discoveries by finding the concentration of the radioactive isotope Carbon 14 in them. The half-life of this isotope is 5 600 years. How long does it take for 30% to be transmuted?

11 2% of light is absorbed when passing through a glass screen of thickness 1 mm. How much would be absorbed by a screen of thickness 1 cm?

12 Atmospheric pressure diminishes approximately as a power function of height. If at a height of 5 000 m it is roughly halved, by what percentage does it diminish for each 1 000 m?

*2.6 Graphs of logarithmic functions

The graph of $y = \log_{10} x$ is the same as that of $x = 10^y$.
But $x = 10^y$ is obtained
from $y = 10^x$ simply by interchanging x and y.

Such an interchange of x and y corresponds geometrically to a reflection in the line $y = x$.

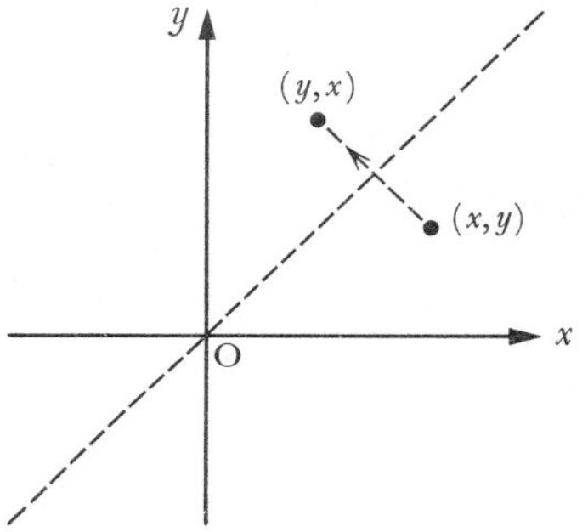

So $y = \log_{10} x$ and $y = 10^x$ are reflections of each other in the line $y = x$.

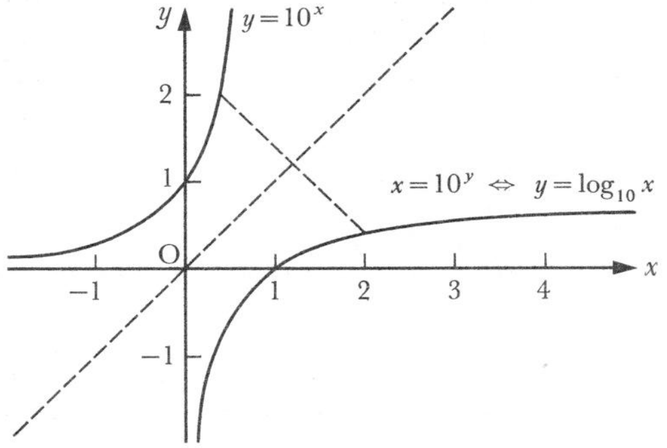

Exercise 2.6

1 What is the domain and the range of the logarithmic function $\log_{10} x$? Use its graph to illustrate your answer.

2 Draw accurately, in one diagram, graphs of the power functions 2^x, 3^x, 5^x, 10^x and of the logarithmic functions $\log_2 x$, $\log_3 x$, $\log_5 x$, $\log_{10} x$.

*2.7 Logarithmic scales

Logarithms are frequently used in graphical work when indices are involved and often enable us to replace curved graphs by straight lines.

For example

if $y = 3\times2^x$,

we can take logarithms (base 10) and obtain:

$$\log_{10} y = \log_{10} 3\times2^x$$

$$\Rightarrow \quad \log_{10} y = \log_{10} 3 + x \log_{10} 2.$$

So if we plot $\log_{10} y$ against x, we obtain a straight line graph with gradient $\log_{10} 2 \approx 0.3$ and

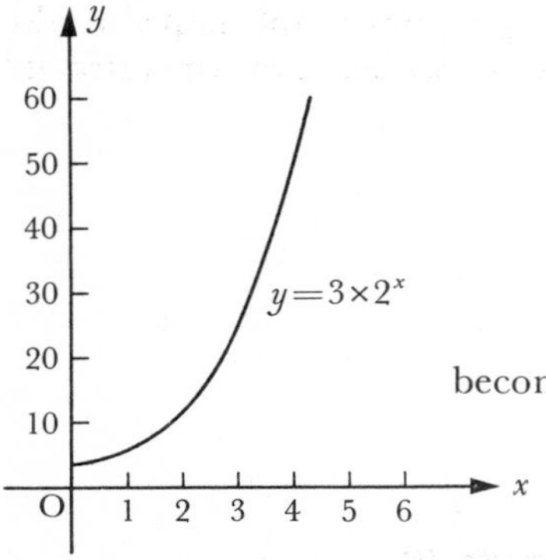

becomes

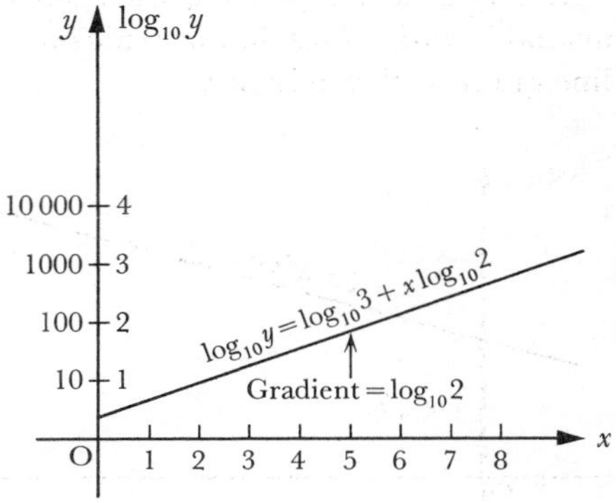

Such representation of power functions by a straight line graph is so common that logarithmic graph paper is available to make the taking of logarithms automatic:

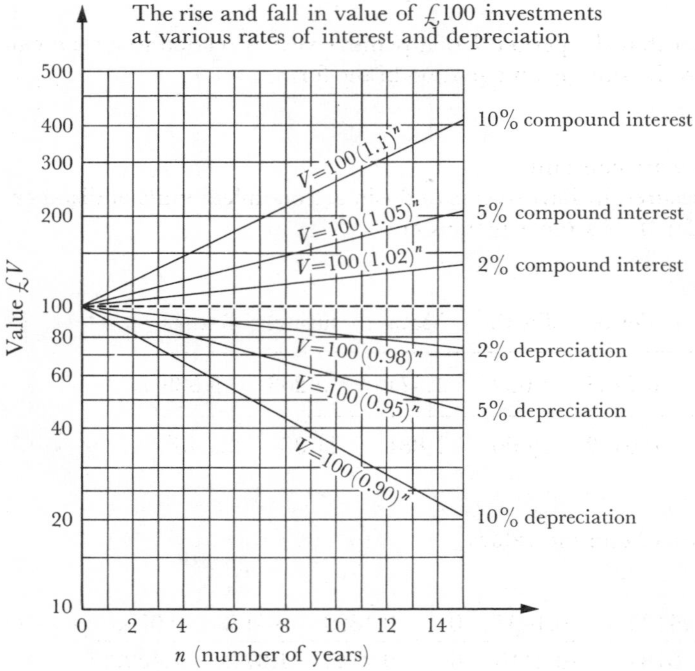

It sometimes happens that y is a function of x of the form

$$y = bx^n$$

Again taking logarithms:

$$\log_{10} y = \log_{10} bx^n$$

$$\Rightarrow \quad \log_{10} y = \log_{10} b + n \log_{10} x$$

If this time we plot $\log_{10} y$ against $\log_{10} x$ (and graph paper is also available with a logarithmic scale on *each* axis), we again obtain a straight line graph with gradient n.

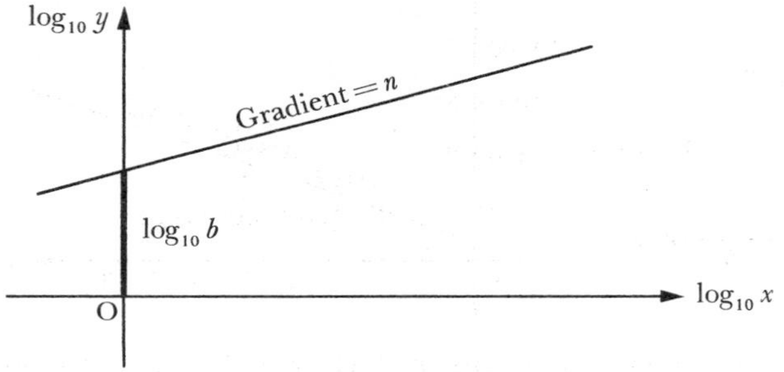

Example

It is suspected that the period T of a planet's year is related to r, its mean distance from the sun, by an equation of the form

$$T = kr^n,$$

where k and n are constants.

If T is measured in Earth-years and r in astronomical units of distance, their values for the six inner planets are

	Mercury	Venus	Earth	Mars	Jupiter	Saturn
r	0.3871	0.7233	1.000	1.524	5.203	9.539
T	0.2408	0.6152	1.000	1.881	11.86	29.46

From these we obtain the values

$\log r$	−0.4122	−0.1407	0	0.1829	0.7162	0.9795
$\log T$	−0.6184	−0.2110	0	0.2744	1.0742	1.4692

These can be plotted on a graph:

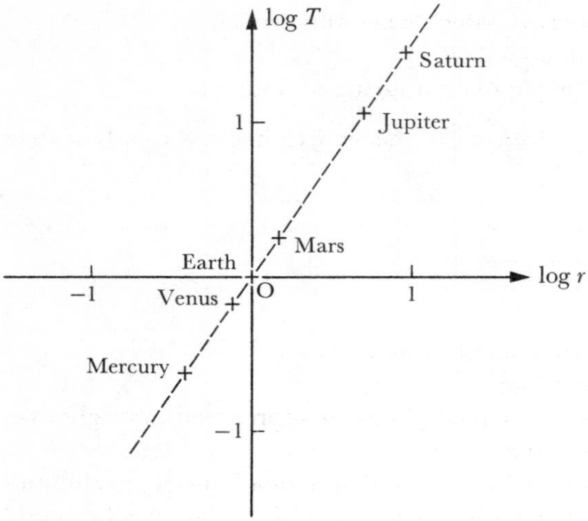

The gradient of the line is 1.5

So $\log T = 1.5 \log r = \log r^{1.5}$

$\Rightarrow \quad T = r^{1.5}$

and our suspicions are confirmed.

Hence $T^2 = r^3$,

a result known as Kepler's Third Law of Planetary Motion, after the remarkable German mathematician, John Kepler, who discovered it in 1618.

Exercise 2.7

[To be worked accurately on graph paper.]

1 The population of a rapidly developing country at successive ten-yearly intervals from 1940 was estimated to be:

Year	1940	1950	1960	1970
No. of years x	0	10	20	30
Population P (millions)	2.72	4.14	6.29	9.54

It is suspected that this growth can be expressed by a power function

$P = ab^x$

where a and b are constants.

(*i*) Use logarithms to plot these values as an approximate straight line.
(*ii*) Estimate values for a and b.
(*iii*) What is the percentage growth rate per annum?
(*iv*) Estimate the population in 1980.
(*v*) How many years is the population taking to double?

2 The times of oscillation, T second, of heavy weights on the ends of wires of length l metre are given by

l	2	4	6	8
T	2.81	4.01	4.98	5.63

It is suspected that these are related by an equation $T = kl^n$, where k and n are constants.
(*i*) Use logarithms to plot these readings as an approximate straight line.
(*ii*) Estimate values for k and n.
(*iii*) From your graph find the time of oscillation of a 5-metre pendulum.
(*iv*) What is the length of a pendulum whose time of oscillation is 1 second?

3 A car costing £2 000 depreciates at a rate of 20% each year.
(*i*) Express its value £P as a function of its age n years.
(*ii*) Use a logarithmic scale to plot its depreciation as a straight-line graph.
(*iii*) With the same axes plot the depreciation lines when the annual rate is 10%. 30%. 40%.

4 Find y in terms of x if it is known to be proportional to a power of x, with $y = 7.27$ when $x = 2$ and $y = 13.92$ when $x = 3$

As further examples, exercise 2.5 Nos. 6–10 may be answered graphically, and preferably by use of logarithmic graph paper.

Miscellaneous problems 2

1 $f(x) = x^x, \quad g(x) = (x^x)^x, \quad h(x) = x^{(x^x)}$
Calculate these functions when $x = 1, 2$ and 3.

2 A fine drizzle is falling steadily and after 1 minute 10% of a pavement is wet. What proportion will be wet after 2 minutes; 3 minutes; 4 minutes?

How long will it take for half of the pavement to become wet?

3 The musical notes Middle C and Top C are taken to have frequencies 256 and 512 respectively. In an equal-tempered scale the octave between them is divided into 12 intervals by the notes

C#, D, D#, E, F, F#, G, G#, A, A#, B

and the frequencies of successive notes are in constant ratio.

Find this ratio and the frequencies of the notes E and G.

4 Prove that $\log_a b = \dfrac{\log_c b}{\log_c a}$

5 If £1 gains 100% interest in 1 year, it becomes £2.

If, however, 50% interest is added each half-year, it becomes firstly £1.5 and finally £2.25.

What would it become

(*i*) if 25% were added quarterly
(*ii*) if 10% were added at the end of 10 equal intervals
(*iii*) if 1% were added 100 times
(*iv*) if 0.1% were added 1 000 times?

To what do these amounts seem to be tending?

[Take $\log_{10} 1.1 = 0.041\,393$
$\log_{10} 1.01 = 0.004\,321$
$\log_{10} 1.001 = 0.000\,434$]

3

Rates of change: differentiation

'Everything is in movement', wrote Heraclitus in about 500 B.C. Yet there was little attention to the mathematics of movement and variation until the sixteenth and seventeenth centuries. Two thousand years earlier, Archimedes had calculated areas by the 'method of exhaustion' (4.1), which is at the root of the *integral calculus*. But the discovery (or invention?) of the *differential calculus* had to wait for Newton and Leibniz in the second half of the seventeenth century.

Taken together, the *differential* and *integral calculus* form the two edges of a tool which, first used in astronomy, has developed outstanding versatility and power. We shall begin to shape it by a simple example.

3.1 Average velocity and instantaneous velocity

If a stone is dropped over the edge of a cliff, the distance y m in which it has fallen after t s, is given approximately by

$$y = 5t^2.$$

This can be represented by the table

t	0	1	2	3	4	5	6
y	0	5	20	45	80	125	180

or by the graph

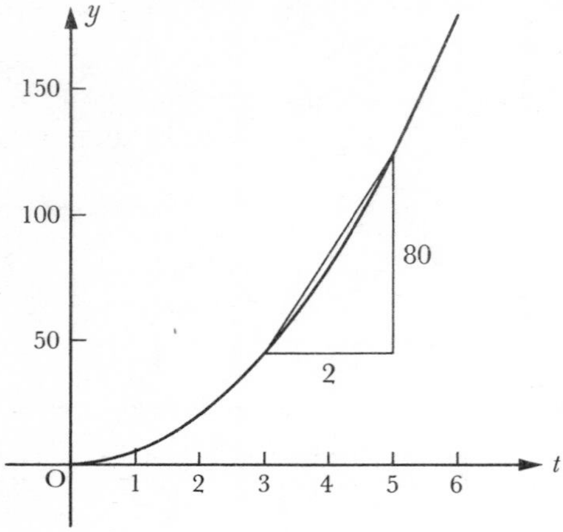

The *average velocity*† of this stone during an interval of time is defined as

$$\frac{\text{Displacement during the interval}}{\text{Duration of the interval}}$$

So in the 2 second interval between $t = 3$ and $t = 5$,

$$\text{average velocity} = \frac{125 - 45 \text{ m}}{2 \text{ s}} = 40 \text{ m s}^{-1}$$

This is, of course, also the gradient‡ of the line joining corresponding points of the graph.

Exercise 3.1a

When a stone is catapulted with velocity 40 m s^{-1} vertically upwards, its height y m after t s is given by

$y = 40t - 5t^2$

1 Construct a table showing the height of the stone at second intervals from $t = 0$ to $t = 10$.

2 Draw accurately the corresponding graph, with clear labels on the axes.

3 What is the maximum height reached by the stone? After how many seconds?

4 How long does it take for the stone to return to ground level?

† We shall use the word velocity rather than speed whenever we are concerned about the direction of motion.

‡ Throughout this and the next section we shall regard all gradients as being expressed in the appropriate units, usually m s^{-1}. The actual slope of lines drawn on the graph paper will, of course, depend on the scales chosen for distance and time.

5 Could there possibly be any meaning for values of y calculated from the above formula when (*i*) $t > 8$, (*ii*) $t < 0$?

6 What is the average velocity of the stone during the following intervals (velocities being counted positive if upwards and negative if downwards)?
(*i*) $t = 0$ to $t = 2$
(*ii*) $t = 1$ to $t = 4$
(*iii*) $t = 4$ to $t = 7$
(*iv*) $t = 1$ to $t = 7$.

Instantaneous velocity

Although we often wish to know the average velocity over a given interval, we more frequently need the velocity *at a certain instant*. We might, for example, want to know how fast the stone dropped over the cliff was falling after 3 seconds, i.e., at the instant when $t = 3$.

For this purpose, we first examine its behaviour in some short intervals near $t = 3$. A few calculations will show that

t	3	3.1	3.01	3.001
y	45	48.05	45.3005	45.030 005

So, between $t = 3$ and $t = 3.1$,

$$\text{average velocity} = \frac{3.05\text{ m}}{0.1\text{ s}} = 30.5\text{ m s}^{-1}$$

$t = 3$ and $t = 3.01$,

$$\text{average velocity} = \frac{0.300\,5\text{ m}}{0.01\text{ s}} = 30.05\text{ m s}^{-1}$$

$t = 3$ and $t = 3.001$,

$$\text{average velocity} = \frac{0.030\,005\text{ m}}{0.001\text{ s}} = 30.005\text{ m s}^{-1}$$

It appears that the smaller the interval we take, the nearer the average velocity approaches to 30 m s^{-1}, so we are tempted to say that this is the velocity *at the instant* when $t = 3$.

This becomes still more evident if we consider the interval between

$$t = 3, \quad \text{when } y = 45$$
$$\text{and } t = 3 + h, \text{ when } y = 5(3 + h)^2 = 45 + 30h + 5h^2$$

The average velocity over this interval

$$= \frac{30h + 5h^2\text{ m}}{h\quad\text{s}}$$

$$= 30 + 5h\text{ m s}^{-1}$$

So, as $h \to 0$, average velocity $\to 30$ m s^{-1}.

We therefore say that the velocity *at the instant when* $t = 3$ is 30 m s^{-1}.

Looking at this on the graph, we see that as $h \to 0$, the line whose gradient we are calculating is gradually approaching the tangent to the curve at the point where $t = 3$:

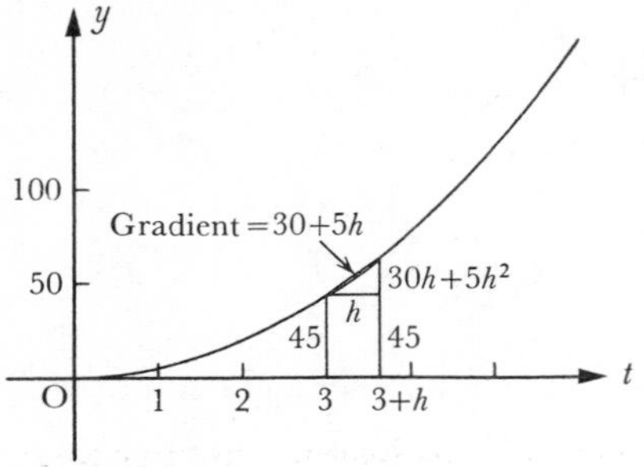

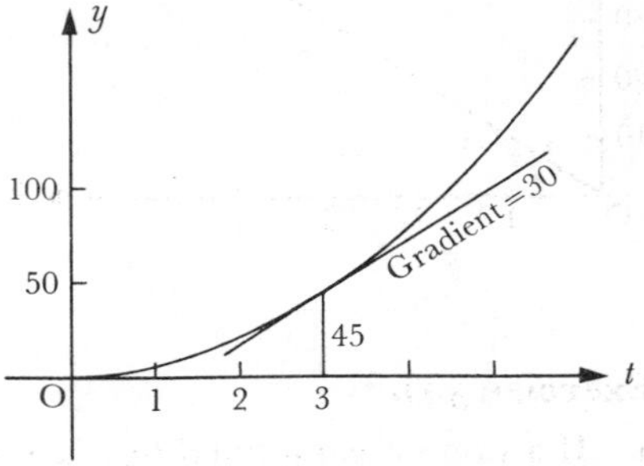

The velocity is therefore represented by the *gradient of the tangent* to the distance-time graph.

Velocity at any instant

Suppose that we need to know the velocity v of the falling stone at any instant.

After time t the stone has fallen a distance $5t^2$

After time $t + h$ the stone has fallen a distance $5(t + h)^2 = 5t^2 + 10th + 5h^2$

Average velocity during this interval

$$= \frac{10ht + 5h^2}{h} = 10t + 5h$$

As $h \to 0$, average velocity $\to 10t$

So $v = 10t$

This enables us to calculate the velocity at *any* instant:

t/s	0	1	2	3	4	5
$v = 10t$/m s^{-1}	0	10	20	30	40	50

So we can represent v, as well as y, graphically as a function of t:

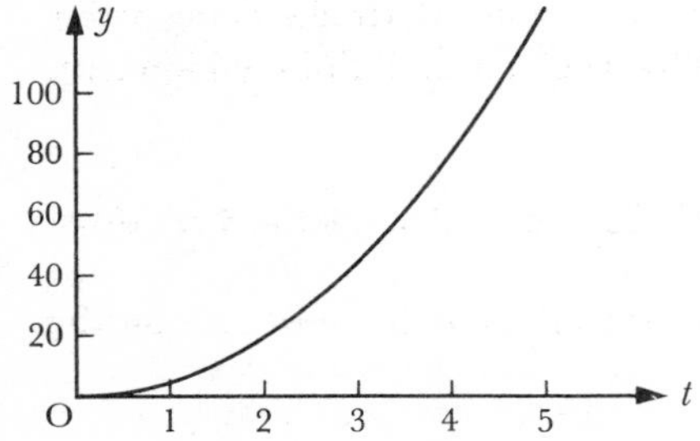

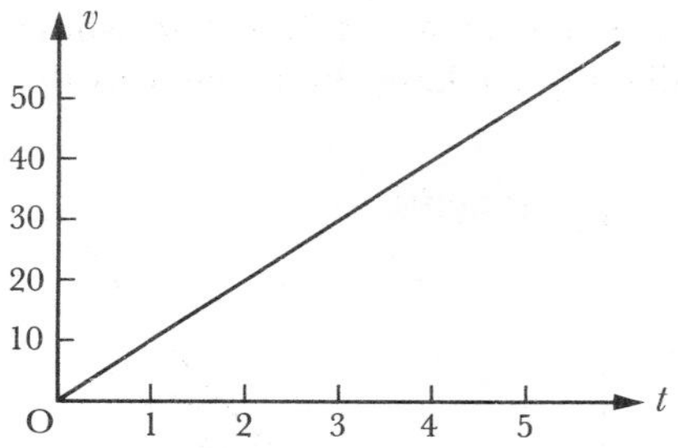

Exercise 3.1b

1 If a stone is catapulted upwards at 40 m s^{-1}, its height y m after time t s is given by $y = 40t - 5t^2$

(*i*) Calculate its average velocity during the intervals
(*a*) from $t = 2$ to $t = 2.1$,
(*b*) from $t = 2$ to $t = 2.01$,
and so estimate its velocity when $t = 2$.
(*ii*) Calculate its average velocity over the interval from $t = 3$ to $t = 3 + h$, and so find its velocity when $t = 3$.
(*iii*) Use the same method to calculate its velocity when $t = 7$.
(*iv*) More generally, find its velocity after time t, checking your results for $t = 2, 3, 7$.
(*v*) Sketch the curves of y and v against t.

2 (*i*) Repeat 1 (*i*) by considering the intervals
(*a*) from $t = 1.9$ to $t = 2$, and (*b*) from $t = 1.99$ to $t = 2$.
(*ii*) Could you obtain a closer approximation by averaging the values obtained from the intervals (1.99, 2) and (2, 2.01)?
(*iii*) How does this compare with the average velocity over the interval (1.99, 2.01)?

3 If, instead of a stone, a ball is dropped over the edge of a cliff, the distance y m through which it has fallen in t s is found to be

$$y = 5t^2 - \tfrac{1}{300}t^3,$$

the term $\frac{1}{300}t^3$ arising from air-resistance and the formula being valid only for the first four seconds of fall. (Why, even for an infinitely high cliff, can this formula not always be valid?)

Calculate:
(*i*) the effect of air-resistance on the distance fallen in the first three seconds;
(*ii*) the velocity when $t = 3$ and hence the effect of air-resistance on the velocity at this instant;
(*iii*) the velocity v at time $t (t < 4)$.

3.2 Differentiation

It is clear from these examples that the process of finding the rate of change of a function, or the gradient of a graph, is going to be very important. Sometimes the task is impossible at a particular point or set of points, perhaps because the function has a discontinuity or has different gradients on the two sides of the point; sometimes, and frequently in practical work, it can only be tackled numerically, or by drawing a tangent and estimating its gradient.

But if a function $f(t)$ has a definite rate of change (i.e., its graph has a gradient) this rate of change will itself be a function, which we call the *derivative*, of $f(t)$, and write as $f'(t)$.

For example, in the last section

$$f(t) = 5t^2 \Rightarrow f'(t) = 10t$$

The process of finding such derivatives is called *differentiation*.

Lastly, we see that

when $f'(t) > 0$; $f(t)$ is increasing
$f'(t) < 0$; $f(t)$ is decreasing
if $f'(t) = 0$; $f(t)$ is said to be *stationary*, or to have a *stationary value*

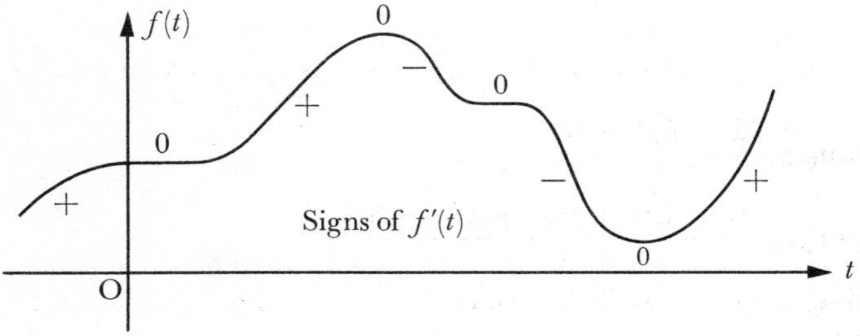

Example 1

If $f(x) = x^3$, find $f'(2)$

Gradient of chord

$$= \frac{(2 + h)^3 - 8}{h}$$

$$= \frac{(8 + 12h + 6h^2 + h^3) - 8}{h}$$

$$= 12 + 6h + h^2$$

As $h \to 0$, $12 + 6h + h^2 \to 12$

So $f'(2) = 12$

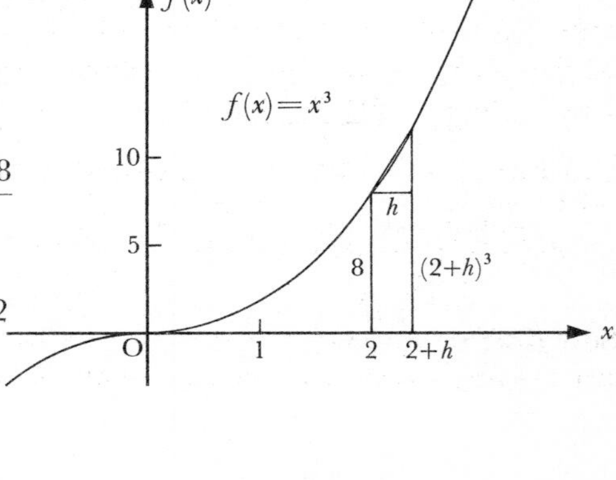

Example 2

Differentiate $f(x) = \frac{1}{x}$

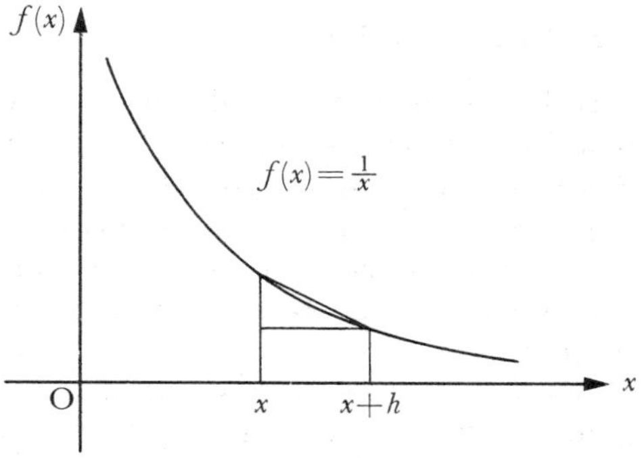

Gradient of chord

$$= \frac{f(x+h) - f(x)}{h}$$

$$= \frac{\frac{1}{x+h} - \frac{1}{x}}{h} = \frac{\frac{-h}{x(x+h)}}{h}$$

$$= -\frac{1}{x(x+h)}$$

As $h \to 0$, $x + h \to x$

So gradient $\to -\frac{1}{x^2}$

and $f'(x) = -\frac{1}{x^2}$

Alternatively, we can use the notation of limits:

Example 1

$$f'(2) = \lim_{h\to 0} \frac{(2+h)^3 - 8}{h}$$
$$= \lim_{h\to 0} \frac{(8 + 12h + 6h^2 + h^3) - 8}{h}$$
$$= \lim_{h\to 0} (12 + 6h + h^2) = 12$$

So $f'(2) = 12$

Example 2

$$f'(x) = \lim_{h\to 0} \frac{\frac{1}{x+h} - \frac{1}{x}}{h}$$

$$= \lim_{h\to 0} \frac{\frac{-h}{x(x+h)}}{h}$$

$$= \lim_{h\to 0} -\frac{1}{x(x+h)} = -\frac{1}{x^2}$$

So $f'(x) = -\frac{1}{x^2}$

More generally, given a function $f(x)$, we define its derivative as

$$f'(x) = \lim_{h\to 0}\frac{f(x+h) - f(x)}{h}$$

wherever this limit exists and is finite.

Two further examples will show the force of the words in italics:

Example 3

$f(x) = |x|$

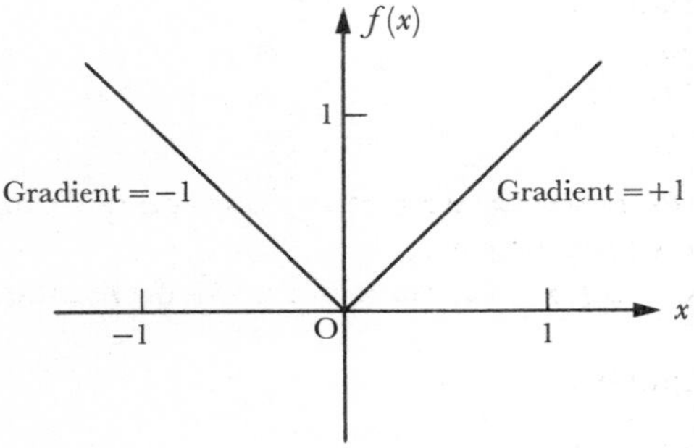

At $x = 0$, we see that, although there is a 'right-hand gradient' $+1$ and a 'left-hand gradient' -1, these are different and the graph has no genuine slope when $x = 0$. So $f'(0)$ does not exist.

Example 4

$$f(x) = \begin{cases} +1 & \text{if } x > 0 \\ 0 & \text{if } x = 0 \\ -1 & \text{if } x < 0 \end{cases}$$

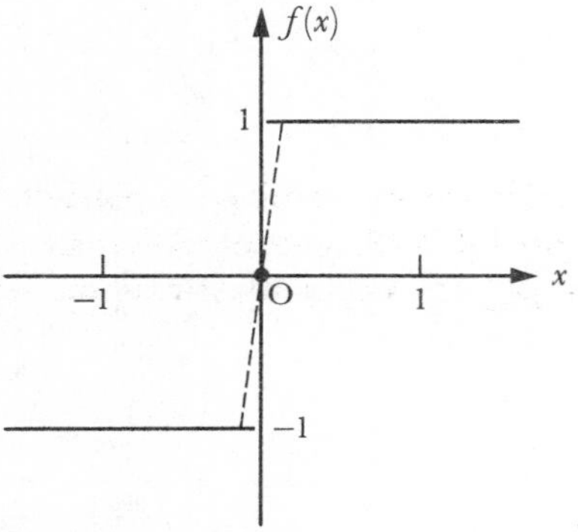

Here it is clear that at $x = 0$ the right-hand gradient and the left-hand gradient both tend to $+\infty$ as the origin is approached.

So here again the graph has no genuine slope, and $f'(0)$ does not exist.

It is readily seen that $f'(x)$ cannot exist at any point where $f(x)$ has a discontinuity.

Exercise 3.2

1 Using the definition

$$f'(x) = \lim_{h\to 0}\frac{f(x+h)-f(x)}{h}$$

(*i*) If $f(x) = x^2$, calculate $f'(3)$.
(*ii*) If $f(x) = 2x^3$, calculate $f'(0)$.
(*iii*) If $f(x) = 4x^2 + x + 7$, calculate $f'(-1)$.
(*iv*) If $f(x) = 3x^4 + 12x$, calculate $f'(1)$.
(*v*) If $f(x) = 1$, calculate $f'(-2)$.
(*vi*) If $f(x) = 3/x$, calculate $f'(2)$.
(*vii*) If $f(x) = 1/x^2$, calculate $f'(1)$.

2 For each of the functions in no. 1, find the derivative $f'(x)$ and mention particularly any point at which it does not exist.

For what range of values of x is $f(x)$: (*a*) increasing, (*b*) decreasing, (*c*) stationary?

Illustrate your answers with graphs.

3.3 Standard methods of differentiation

Throughout the above sections, we have discovered each derivative separately 'from first principles'. But it is easily seen that the results of some of the examples can be put in the form of a table:

$f(x)$	x^4	x^3	x^2	$x = x^1$	$1 = x^0$	$\frac{1}{x} = x^{-1}$	$\frac{1}{x^2} = x^{-2}$
$f'(x)$	$4x^3$	$3x^2$	$2x$	$1 = 1x^0$	$0 = 0x^{-1}$	$-\frac{1}{x^2} = -1x^{-2}$	$-\frac{2}{x^3} = -2x^{-3}$

So it appears that

$$\begin{array}{|ll|}\hline & f(x) = x^n \\ \Rightarrow & f'(x) = nx^{n-1} \\ \hline\end{array}$$

This has, of course, not been proved, and in any case n has been restricted to integral values. We shall prove it later (pp. 118, 223), and for more general values of n (e.g., $f(x) = x^{1/2} \Rightarrow f'(x) = \frac{1}{2}x^{-1/2}$), but in the meantime we shall assume it true for all values of n.

Exercise 3.3a

Find $f'(x)$ if $f(x)$ is

1 x^5 2 x^8 3 $x^{4/3}$ 4 $\sqrt[3]{x}$ 5 $x\sqrt{x}$

6 $\frac{1}{x^3}$ 7 $\frac{1}{x^4}$ 8 $x^{-1/2}$ 9 $\frac{1}{x\sqrt{x}}$ 10 $\frac{1}{\sqrt[3]{x}}$

Multiples, sums and differences

Whilst we are now able to differentiate at sight any function of the form x^n, we shall frequently need to find the derivatives of multiples and combinations of such functions.

Here again, the rules of procedure should be apparent from the above exercises. For instance, in exercise 3.2 no. 2 we found (or should have found!) that

(*i*) $f(x) = 2x^3 \Rightarrow f'(x) = 6x^2$, (*ii*) $f(x) = 3x^4 + 12x \Rightarrow f'(x) = 12x^3 + 12$.

So it appears that
(*i*) the derivative of a constant multiple of a function is this multiple of its derivative; and,
(*ii*) the derivative of the sum of two functions is this sum of its derivatives; or, more briefly, that
(*i*) $(kf)' = kf'$ (if k is a constant)
(*ii*) $(f + g)' = f' + g'$.

These can very easily be shown as follows:

(*i*)
$$\{kf(x)\}' = \lim_{h\to 0}\frac{kf(x+h) - kf(x)}{h}$$
$$= k\lim_{h\to 0}\frac{f(x+h) - f(x)}{h}$$
$$= kf'(x)$$

So $$(kf)' = kf'$$

(*ii*)
$$\{f(x) + g(x)\}' = \lim_{h\to 0}\frac{\{f(x+h) + g(x+h)\} - \{f(x) + g(x)\}}{h}$$
$$= \lim_{h\to 0}\frac{\{f(x+h) - f(x)\} + \{g(x+h) - g(x)\}}{h}$$
$$= \lim_{h\to 0}\frac{f(x+h) - f(x)}{h} + \lim_{h\to 0}\frac{g(x+h) - g(x)}{h}$$
$$= f'(x) + g'(x)$$

So $$(f + g)' = f' + g'$$

These may seem to the reader very laborious proof of results which are self-evident. If so, what about products and quotients? Is it always true, only sometimes true, or never true,

that $\{f(x)g(x)\}' = f'(x)g'(x)$ and $\left\{\dfrac{f(x)}{g(x)}\right\}' = \dfrac{f'(x)}{g'(x)}$?

You could start by investigating particular cases, e.g., $f(x) = x^3$, $g(x) = x^2$.

But perhaps you have chosen an exceptional case. Try some more.

The results $(kf)' = kf'$ (i)

$(f + g)' = f' + g'$ (ii)

can now be used. When, for instance, we differentiate

$f(x) = 2x^3 - 3x$

we use both the above results (and the first one twice). Though the notation is not recommended for general use, we can write

$$\begin{aligned} & f(x) = 2x^3 - 3x \\ \Rightarrow\ & f'(x) = (2x^3 - 3x)' \\ & = \{2x^3 + (-3x)\}' \\ & = (2x^3)' + (-3x)' && \text{using (ii)} \\ & = 2(x^3)' - 3(x)' && \text{using (i) twice} \\ & = 2\times 3x^2 - 3\times 1 \\ & = 6x^2 - 3; \end{aligned}$$

which we usually abbreviate to

$$\begin{aligned} & f(x) = 2x^3 - 3x \\ \Rightarrow\ & f'(x) = 6x^2 - 3 \end{aligned}$$

Exercise 3.3b

Using the above rules, find $f'(x)$ where $f(x)$ is

1	$x^4 + x^3$	2	$x^5 - x^2$
3	$3x^4 + 2x$	4	$4x^5 - 5x^3$
5	$2\sqrt{x}$	6	$x + \dfrac{1}{x}$
7	$\dfrac{1}{3x^3}$	8	$(x^2 + 1)^2$
9	$x^2 - \dfrac{1}{x^2}$	10	$3x\sqrt[3]{x}$.

3.4 Higher derivatives

Just as the derivative of f is called f', so the derivative of f' is called f'' (the *second derivative* of f); and so on for higher derivatives.

For example,

$$\begin{aligned} & f(x) = 2x^3 - 3x \\ \Rightarrow\ & f'(x) = 6x^2 - 3 \\ \Rightarrow\ & f''(x) = 12x \\ \Rightarrow\ & f'''(x) = 12 \\ \Rightarrow\ & f''''(x) = 0, \quad \text{etc.} \end{aligned}$$

and the original curve, with its first, second, third and fourth *derived curves* are:

$f(x) = 2x^3 - 3x,$

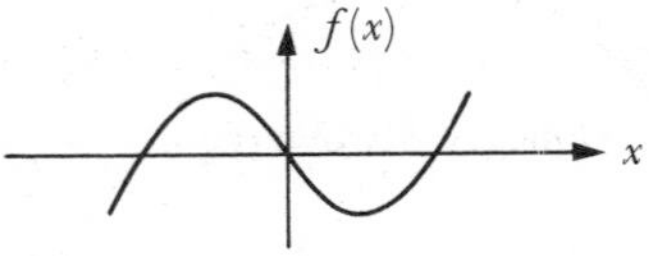

whose gradient is

$f'(x) = 6x^2 - 3,$

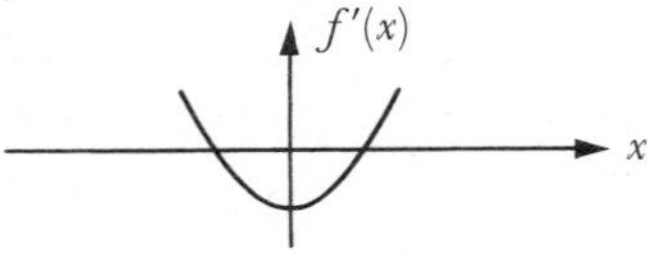

whose gradient is

$f''(x) = 12x,$

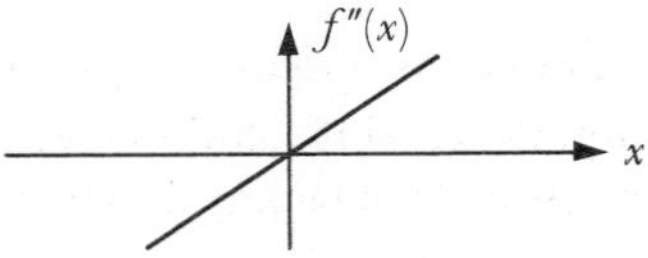

whose gradient is

$f'''(x) = 12,$

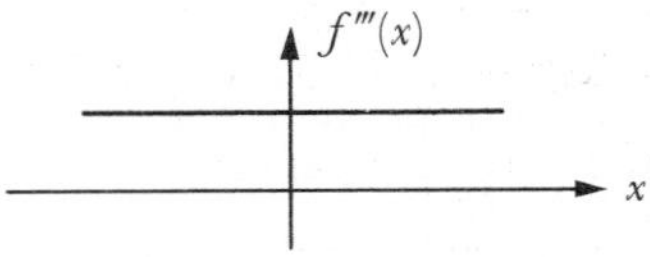

whose gradient is

$f''''(x) = 0.$

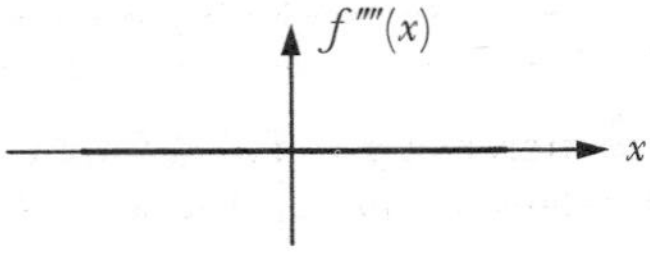

Exercise 3.4a

Find $f'(x)$ and $f''(x)$ where $f(x)$ is:

1 (*i*) $5x^3$; (*ii*) $x^4 + x$; (*iii*) $x^4 - x^2 + 2$; (*iv*) $\frac{1}{2}x^{10}$.

2 (*i*) x^{-6}; (*ii*) $14x^{-3}$; (*iii*) $\dfrac{1}{x^2}$; (*iv*) $\dfrac{5}{x^3}$.

3 (*i*) $x^{1/3}$; (*ii*) $\sqrt{x}$; (*iii*) $2x\sqrt{x}$; (*iv*) $\dfrac{1}{\sqrt{x}}$.

4 (*i*) $(x^3 + 1)^2$; (*ii*) $\dfrac{2x^2 + 3}{x}$; (*iii*) $\dfrac{x + 1}{2\sqrt{x}}$; (*iv*) $[\sqrt{x} + 1]^2$.

Concavity and convexity: points of inflection

It is readily seen that the sign of $f''(x)$ determines whether $f'(x)$ is increasing or decreasing, i.e., whether the curve of $f(x)$ is concave upwards or downwards (i.e., convex downwards or upwards).

Alternatively, we can think of x increasing and the first stretch as a bend taken with a 'right-hand lock', followed by the second stretch as a bend taken with a left-hand lock. The point where the lock changes (and the curve crosses its tangent, and $f''(x)$ changes sign) is called a *point of inflection*.

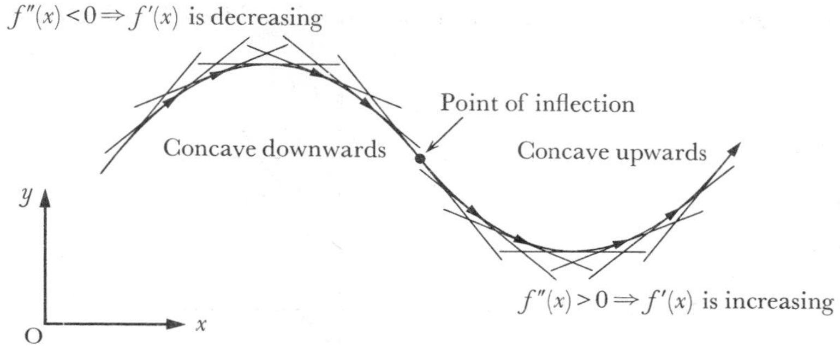

Example

For what values of x is the function $f(x) = x^3 - 3x^2 + 4$

(*i*) increasing, decreasing, stationary;

(*ii*) concave upwards, concave downwards, at a point of inflection?

$$f(x) = x^3 - 3x^2 + 4$$
$$\Rightarrow f'(x) = 3x^2 - 6x = 3x(x-2)$$

and the sign of $f'(x)$ is given by

x	−3	−2	−1	0	1	2	3	4	5
$f'(x)$	++++	++++	++++	0	−−−−	0	++++	++++	++++

Hence $f(x)$ is increasing if $x < 0$ or $x > 2$
decreasing if $0 < x < 2$
stationary if $x = 0$ or 2.

Also $f''(x) = 6x - 6 = 6(x-1)$ and the sign of $f''(x)$ is given by

x	−3	−2	−1	0	1	2	3	4	5
$f''(x)$	−−−−	−−−−	−−−−	−−−−	0	++++	++++	++++	++++

Hence $f(x)$ is concave upwards when $x > 1$
concave downwards when $x < 1$
at a point of inflection when $x = 1$

These can all be illustrated in the graph of $f(x)$:

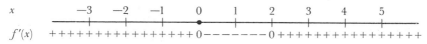

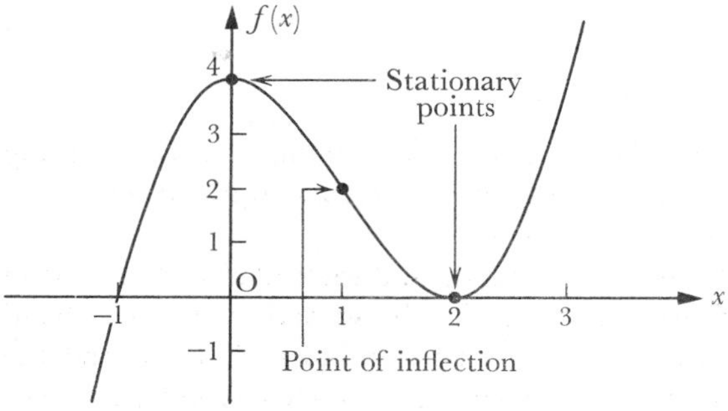

Exercise 3.4b

In each of the following cases
(*i*) find $f'(x)$ and $f''(x)$;
and state the values of x for which the function is
(*ii*) increasing, decreasing, stationary;
(*iii*) concave up, concave down, at a point of inflection;
(*iv*) sketch the graph of $f(x)$.

1 $x^2 - 2x$, **2** $x^3 - 3x$, **3** $3x^2 - 2x^3$,
4 $x^4 - 8x^2$, **5** $x^3 - 3x^2 - 9x + 11$.

3.5 Alternative notation

Newton and Leibniz, when they made their independent discoveries of the Calculus in the 1660's, used different notations. It could hardly have been otherwise for men who were saying something for the first time, but ever since their day the history of the subject has been one of varied notation. Sometimes this has been unfortunate, as in the eighteenth century when English mathematics stagnated because Newton's followers became hide-bound to his language and symbols. But more usually it has been a strength: different situations call for different ways of saying the same thing. It was perhaps felt in section 3.3 when we were proving $(f + g)' = f' + g'$ that our language was becoming rather laborious, and we shall soon see how, with a different notation, such proofs can be expressed more simply. But first we shall return to our previous example:

Suppose that we wish to find the gradient at any point of the curve $y = x^3$.

As before, we must find the effect of a small increase in x. We write δx (pronounced 'delta x' and simply meaning 'a bit more x') to indicate this small increase in x, and δy ('a bit more y') to indicate the corresponding increase in y.

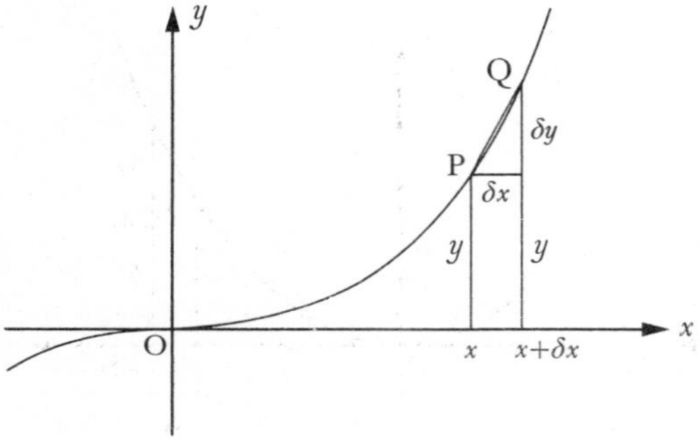

To find the gradient of the tangent at P, we first notice that

at P $\qquad y = x^3;$

and at Q $\quad y + \delta y = (x + \delta x)^3$

$$= x^3 + 3x^2\,\delta x + 3x(\delta x)^2 + (\delta x)^3$$

$$\Rightarrow \quad \delta y = 3x^2\delta x + 3x(\delta x)^2 + (\delta x)^3$$

$$\Rightarrow \quad \text{Gradient of PQ} = \frac{\delta y}{\delta x} = 3x^2 + 3x\,\delta x + (\delta x)^2$$

So as $\delta x \to 0$, $\dfrac{\delta y}{\delta x} \to 3x^2$, which is clearly the derivative of y and the gradient of the tangent. As this derivative is the limit of $\dfrac{\delta y}{\delta x}$, it is convenient to denote it by the symbol $\dfrac{\mathrm{d}y}{\mathrm{d}x}$.

So $\dfrac{\mathrm{d}y}{\mathrm{d}x} = \lim_{\delta x \to 0} \dfrac{\delta y}{\delta x} = 3x^2$

Two points must be emphasised:

(*i*) that δ is *not* a quantity, but an adjective meaning 'a bit more', and $\dfrac{\delta y}{\delta x}$ is simply the ratio $\dfrac{\text{'a bit more } y\text{'}}{\text{'a bit more } x\text{'}}$.

(*ii*) that, unlike δx and δy, $\mathrm{d}x$ and $\mathrm{d}y$ do not (yet) have any meaning by themselves, any more than the two dots in the symbol $\div$.

> $\dfrac{\mathrm{d}y}{\mathrm{d}x}$ is simply a shorthand way of writing $\lim_{\delta x \to 0} \dfrac{\delta y}{\delta x}$

Comparison of notations

To make the position quite clear we shall carry out a differentiation from first principles using the two notations in parallel.

Given $f(x) = 3x^2$, find $f'(x)$	Given $y = 3x^2$, find $\dfrac{\mathrm{d}y}{\mathrm{d}x}$

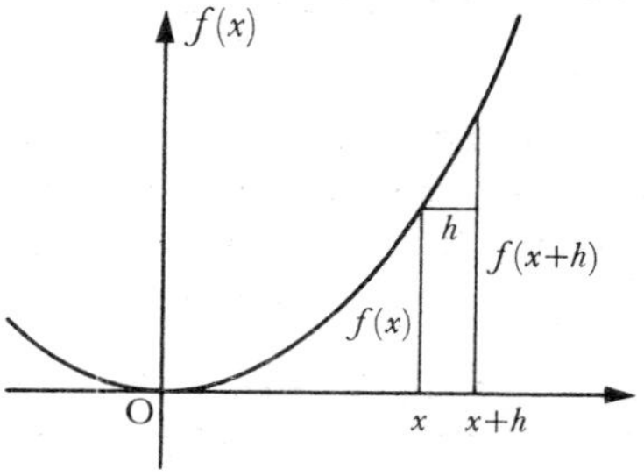

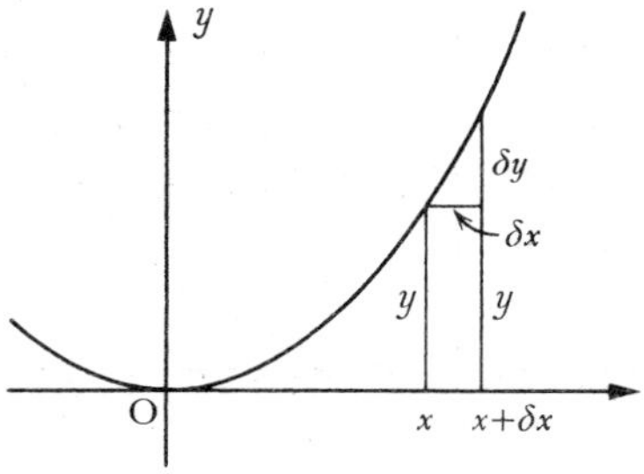

$f(x) = 3x^2$	$y = 3x^2$
$\Rightarrow \quad f(x+h) = 3(x+h)^2$	$\Rightarrow \quad y + \delta y = 3(x + \delta x)^2$
$\Rightarrow \quad f(x+h) - f(x) = 6hx + 3h^2$	$\Rightarrow \quad \delta y = 6x\delta x + 3(\delta x)^2$

$$\Rightarrow \quad \frac{f(x+h)-f(x)}{h} = 6x + 3h \qquad\qquad \Rightarrow \quad \frac{\delta y}{\delta x} = 6x + 3\delta x$$

$$\text{But } f'(x) = \lim_{h\to 0}\frac{f(x+h)-f(x)}{h} \qquad\qquad \text{But} \quad \frac{dy}{dx} = \lim_{\delta x\to 0}\frac{\delta y}{\delta x}$$

$$\Rightarrow \quad f'(x) = 6x \qquad\qquad \Rightarrow \quad \frac{dy}{dx} = 6x$$

It is also convenient to have a symbol to denote the *operation of differentiation*. Just as we use $\sqrt{}$ for the operation of taking the square root of a number, so we use the symbol $\frac{d}{dx}$ to denote differentiation, and write

$$\frac{d}{dx}(x^4 + 3x^2) = 4x^3 + 6x,$$

$$\text{and} \quad \frac{d}{dx}(y) = \frac{dy}{dx}$$

Further, we can use this notation to denote higher derivatives:

$$\frac{d}{dx}\left(\frac{dy}{dx}\right) = \frac{d^2}{dx^2}(y) \quad \text{or} \quad \frac{d^2y}{dx^2}$$

$$\text{and} \quad \frac{d}{dx}\left(\frac{d^2y}{dx^2}\right) = \frac{d^3}{dx^3}(y) \quad \text{or} \quad \frac{d^3y}{dx^3}$$

So $\quad y = x^4 + 3x^2$

$$\Rightarrow \quad \frac{dy}{dx} = \frac{d}{dx}(x^4 + 3x^2) = 4x^3 + 6x$$

$$\frac{d^2y}{dx^2} = \frac{d^2}{dx^2}(x^4 + 3x^2) = 12x^2 + 6$$

$$\frac{d^3y}{dx^3} = \frac{d^3}{dx^3}(x^4 + 3x^2) = 24x$$

Example

Given the curve $y = x^3 - x$,

find the equation of its tangent and normal (i.e., line perpendicular to tangent) at the point (2, 6).

$$y = x^3 - x$$

$$\Rightarrow \quad \frac{dy}{dx} = 3x^2 - 1$$

At (2, 6), $\frac{dy}{dx} = 3\times4 - 1 = 11$ and the equation of the tangent at (2, 6) is

$$y - 6 = 11(x - 2)$$

$$\Rightarrow \quad y = 11x - 16$$

Also the gradient of the normal is $-\frac{1}{11}$. So the equation of the normal is

$$y - 6 = -\frac{1}{11}(x - 2)$$

$$\Rightarrow \quad 11y - 66 = -x + 2$$

$$\Rightarrow \quad x + 11y - 68 = 0$$

Exercise 3.5

1 Use the notation δx, δy to differentiate from first principles:

(*i*) $y = 3x^2$; (*ii*) $y = \frac{2}{x}$.

2 Use standard methods to find $\frac{dy}{dx}$ and $\frac{d^2y}{dx^2}$ if

(*i*) $y = x^3 + 3x^2$; (*ii*) $y = 4x$; (*iii*) $y = \frac{1}{x^2}$; (*iv*) $y = \frac{1}{\sqrt{x}}$.

3 Use the operators $\frac{d}{dx}$ and $\frac{d^2}{dx^2}$ to state the first and second derivatives of

(*i*) x^4; (*ii*) $3x^2$; (*iii*) $2\sqrt{x}$; (*iv*) $\frac{1}{x}$.

4 Prove that the following curves go through the given points and in each case find:

(*a*) the gradient of the tangent at the point;
(*b*) the angle it makes with the positive x-axis;
(*c*) the equation of the tangent;
(*d*) the equation of the normal.

(*i*) $y = x^2$, (2, 4);
(*ii*) $y = x^3 - 2x$, (1, −1);
(*iii*) $y = x - x^2$, (1, 0);
(*iv*) $y = \frac{2}{x}$, (−2, −1).

5 Find the equation of the tangent to the curve $y = x^3 - 11x$ at the point (1, −10) and also the equations of the tangents which are parallel to the line $x - y = 0$. (O.C.)

6 Find the equation of the tangent at the point (2, 2) on the curve $y = x^3 - 3x$ and the coordinates of the point at which this tangent meets the curve again. (O.C.)

7 A man throws a stone horizontally with velocity 25 m s^{-1} from a point on top of a cliff 500 m above sea-level.

The equation of the path of the stone, referred to axes through its point of projection, is $y = -\frac{1}{125}x^2$.

How far out to sea does it reach and at what angle does it enter the water?

8 The muzzle-velocity of a shell is 300 m s^{-1} and when fired at an angle 45° with the horizontal its path is given by the equation

$$y = x - \frac{x^2}{9\,000}$$

Find:
(*i*) the range of the shell,
(*ii*) the angles which its direction makes with the horizontal when it has travelled horizontal distances 3 000 m, 4 500 m, and 6 000 m.

9 Sketch together the curves $y = x^2$ and $y = \frac{8}{x}$.

Find:
(*i*) the coordinates of their point of intersection,
(*ii*) their angle of intersection.

3.6 Kinematics

In section 3.1 we introduced the process of differentiation by considering the velocity of a falling stone. The study of the position of a moving body, and how it changes, is called *kinematics*. This forms part of *dynamics*, which also includes study of the forces acting upon the body and their relationship with its movements.

For the present, we shall ignore the *causes* of motion and confine ourselves to its *description* in the case of a small body, or *particle*, which is moving in a straight line.

Displacement, velocity, and acceleration

Suppose that O is a fixed point of a straight line and that a point P is moving so that at time t it has a displacement x (positive or negative) from O

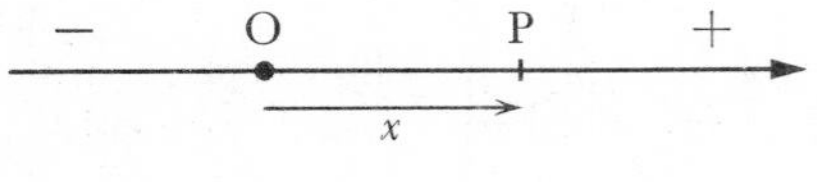

Its *velocity* v is the rate of change of its displacement, and its *acceleration* a is the rate of change of its velocity.

So $\quad v = \dfrac{dx}{dt}$

and $\quad a = \dfrac{dv}{dt} \left(= \dfrac{d^2x}{dt^2} \right)$

At a particular time t, x tells us where the particle is,
v how it is moving,
and a how its movement is varying.

Units

As the unit of distance is 1 m and the unit of time is 1 s, the unit of velocity is a rate of change of displacement of 1 m in 1 s = 1 $\mathrm{m\,s^{-1}}$, and the unit of acceleration is a rate of change of velocity of 1 $\mathrm{m\,s^{-1}}$ in 1 s = 1 $\mathrm{m\,s^{-2}}$.

At the surface of the Earth, and ignoring air resistance, the acceleration of an object falling under gravity is constant and roughly 9.81 $\mathrm{m\,s^{-2}}$; on the Moon it is 1.62 $\mathrm{m\,s^{-2}}$. For simplicity, however, the acceleration due to gravity at the Earth's surface will normally be taken as 10 $\mathrm{m\,s^{-2}}$.

Example

In the simplest case of a stone dropped over the edge of a cliff, let us indicate displacement, velocity and acceleration (in a downwards direction) after time t by the letters y, v and a.

$$y = 5t^2$$

$$\Rightarrow \quad v = \frac{dy}{dt} = 10t$$

$$\Rightarrow \quad a = \frac{dv}{dt} = 10$$

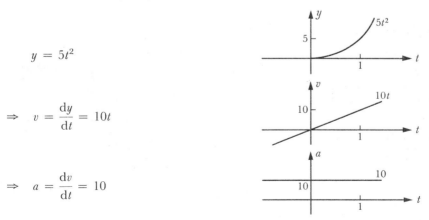

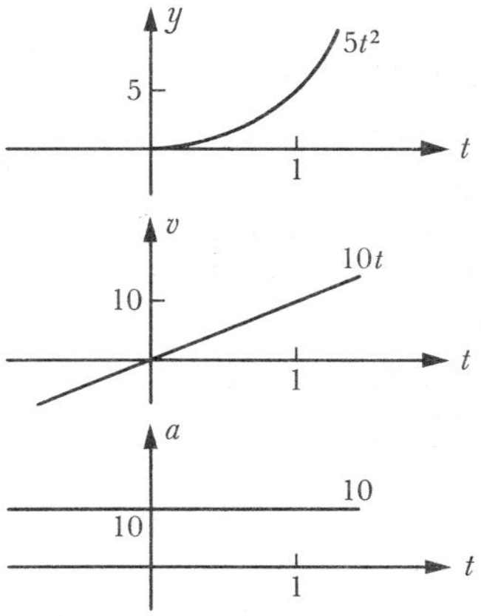

Exercise 3.6

1 The coordinate x m of a moving point A at time t s is given by the formula

$$x = t^3 - 2t^2 + 3t - 4$$

Find the velocity and the acceleration of A at the instant $t = 4$. (O.C.)

2 A particle P is travelling along a straight line so that its distance from the starting point, x m, after time t s is given by the equation $x = 8t - \frac{1}{2}t^2$. Calculate the velocity and acceleration of P after 3 s, and find the distance travelled by P when it first comes to rest.

3 A particle is moving in a straight line and its distance s m from a fixed point in the line after t s is given by the equation $s = 12t - 15t^2 + 4t^3$. Find:
(*i*) the velocity and acceleration of the particle after 3 s;
(*ii*) the distance travelled between the two times when the velocity is instantaneously zero. (O.C.)

4 The velocity of a particle travelling in a straight line is given by $v = t^2 - t - 12$, when v is measured in m s^{-1} and t is the time in seconds measured from a definite instant of the motion.

Find the time that elapses before the particle is instantaneously at rest, and find its acceleration when $t = 6$. (S.M.P.)

5 A particle moves along the x-axis, its position when t s have elapsed being given by $x = 27 - 36t + 12t^2 - t^3$. Show that for the first 2 s the particle moves in the negative direction, and for the next 4 s it moves in the positive direction.

Find also the accelerations at the instants when the particle is at rest. (O.C.)

6 A point moves on the x-axis and at time t its position is given by

$$x = t(t^2 - 6t + 12)$$

Show that its velocity at the origin is 12 and that the velocity decreases to zero at the points where $x = 8$ and thereafter continues to increase. (O.C.)

7 If an astronaut on the Moon catapults a stone vertically upwards at a speed of 24 m s^{-1}, its height y m after t s is given by

$$y = 24t - \frac{4}{5}t^2$$

(*i*) How high does it go?
(*ii*) How long does it take to reach its highest point?
(*iii*) What are its velocity v and acceleration a after time t?
(*iv*) Sketch the graphs of y, v, and a plotted against t.
(*v*) Is there any meaning to the sections of (*a*) the v-t curve, and (*b*) the y-t curve, when they drop below the t-axis?

3·7 Reverse processes

We now ask the reverse question: given $f'(x)$, can we find $f(x)$? If so, we call $f(x)$ a *primitive*† of $f'(x)$.

Example

$$f'(x) = 12x - 3x^2$$

In this case it is immediately clear that

$$f(x) = 6x^2 - x^3$$

is a primitive of $12x - 3x^2$.

But a moment's reflection reminds us that, as the derivative of any constant is zero,

$$6x^2 - x^3 + 15,\ 6x^2 - x^3 - 10, \text{ etc.}$$

are all equally possible primitives.

Indeed $6x^2 - x^3 + c$, where c is any constant (and therefore sometimes called an 'arbitrary constant') is also such a function. Looking at this graphically, we readily see the significance of the arbitrary constant c: the original instruction was simply to find a curve whose gradient is $12x - 3x^2$. The first such curve to be found was $6x^2 - x^3$ and the relationship between these two curves can be represented:

The gradient of $6x^2 - x^3$ is $12x - 3x^2$

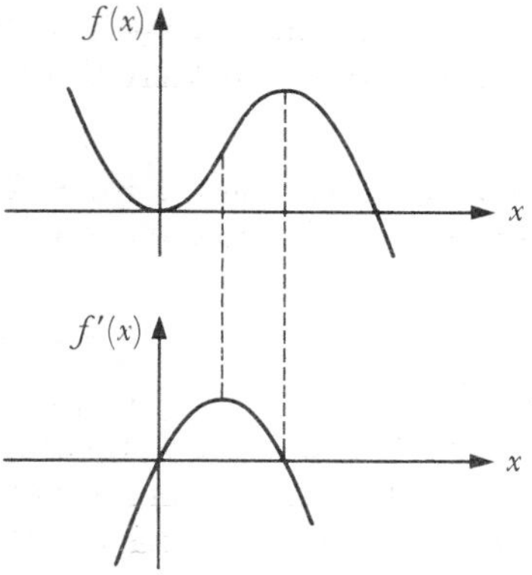

$6x^2 - x^3$ is a primitive of $f(x) = 12x - 3x^2$

But the gradient of $6x^2 - x^3$ is unaltered by any vertical displacement of the curves, so *any* function

$$f(x) = 6x^2 - x^3 + c$$

is equally a primitive of $f'(x) = 12x - 3x^2$.

† Primitives presumably being those from which others are derived, with functions just as with animals!

The gradient of $6x^2 - x^3 + c$ is always $12x - 3x^2$

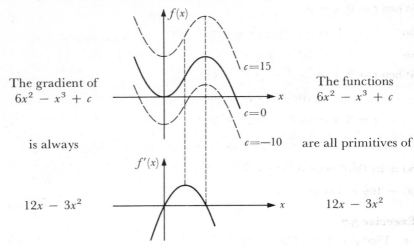

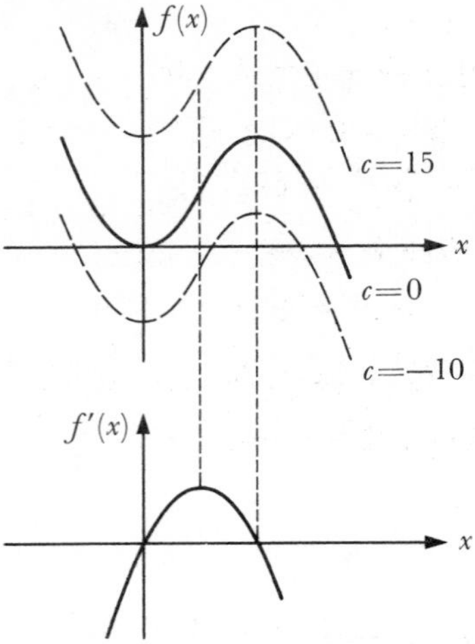

The functions $6x^2 - x^3 + c$ are all primitives of $12x - 3x^2$

In the last section we learnt to discover the behaviour of velocity and acceleration from that of displacement. We can now see how to reverse this process by use of primitives.

Example

A particle leaves the point O with a velocity $4\ \mathrm{m\,s^{-1}}$ and its acceleration a after time t is given by

$a = 5 - t$

Find:

(*i*) its velocity v after time t;
(*ii*) its displacement x after time t; and,
(*iii*) the distance travelled in its third second of motion.

$$\frac{\mathrm{d}v}{\mathrm{d}t} = a = 5 - t$$

$$v = 5t - \tfrac{1}{2}t^2 + c, \quad \text{where } c \text{ is a constant.}$$

When $t = 0$, $v = 4$.

So $\quad 4 = 0 - 0 + c \Rightarrow c = 4$

$\Rightarrow \quad v = 5t - \tfrac{1}{2}t^2 + 4$

So $\quad \dfrac{\mathrm{d}x}{\mathrm{d}t} = 5t - \tfrac{1}{2}t^2 + 4$

$\Rightarrow \quad x = \tfrac{5}{2}t^2 - \tfrac{1}{6}t^3 + 4t + d$, where d is a constant.

When $t = 0, x = 0$.

So $$0 = 0 - 0 + 0 + d \Rightarrow d = 0$$

$$\Rightarrow \quad x = \tfrac{5}{2}t^2 - \tfrac{1}{6}t^3 + 4t$$

When $t = 2$, $x = \tfrac{5}{2}\times 2^2 - \tfrac{1}{6}\times 2^3 + 4\times 2$

$= 10 - \tfrac{4}{3} + 8 = 16\tfrac{2}{3}$

$t = 3$, $x = \tfrac{5}{2}\times 3^2 - \tfrac{1}{6}\times 3^3 + 4\times 3$

$= 22\tfrac{1}{2} - 4\tfrac{1}{2} + 12 = 30$

So in its third second the particle travels

$30 - 16\tfrac{2}{3} = 13\tfrac{1}{3}$ m

Exercise 3·7

1 Find $f(x)$ if it is known that

(*i*) $f'(x) = 2x^3$; (*ii*) $f'(x) = 1 - x$;

(*iii*) $f'(x) = 3x^2 + 2$ and $f(0) = 4$; (*iv*) $f'(x) = \dfrac{1}{x^2}$ and $f(2) = 1$.

2 Find the equation of a curve

(*i*) whose gradient is $4x - 5$;

(*ii*) whose gradient is $4x - 5$, and which goes through the point (0, 4);

(*iii*) whose gradient is 3 and which goes through the point (1, 2);

(*iv*) whose gradient is $-x$ and which goes through the point (1, 1).

3 Find the displacement x m of a particle at time t s if we know that its velocity v m s^{-1} is given by

(*i*) $v = 10t$;

(*ii*) $v = 8t^3 - 6t^2$;

(*iii*) $v = 3t^2 + 4t$, and $x = 2$ when $t = 0$;

(*iv*) $v = 1 - 2t$, and $x = 3$ when $t = 1$.

4 Find the velocity v m s^{-1} and displacement x m of a particle at time t s if we know that its acceleration a m s^{-2} is given by

(*i*) $a = 6t - 8$, and when $t = 0$, $s = 4$ and $v = 6$;

(*ii*) $a = -10$, and when $t = 1$, $s = 5$ and $v = 2$.

5 A particle starts from rest at O and moves along a straight line. After t s its velocity is v m s^{-1}, where $v = 3t^2 - t^3$. Show that the particle is momentarily at rest after 3 s and find its distance from O at this time.

Find also the maximum velocity of the particle during the first 3 s of the motion. (O.C.)

6 The velocity of a particle travelling in a straight line is given by $v = t^2 - t - 6$, where v is measured in m s^{-1} and t is the time in s, reckoned from a definite instant of the motion. Find the time that elapses before the particle is instantaneously at rest, and find the acceleration when $t = 5$.

Find also the distance between its position when $t = 3$ and its position when $t = 9$. (O.C.)

7 A train, while travelling from its start at a point A to its stop at a point B, is moving, t hours after it leaves A, with a speed of $(240t - 150t^2)$ km/h. Find:
(*i*) the time taken in going from A to B;
(*ii*) the distance from A to B;
(*iii*) the greatest speed reached during the journey. (O.C.)

8 A particle starts from rest at a point 6 m from O, and moves in a straight line away from O with a velocity v m s^{-1} at time t s given by $v = t - \frac{1}{18}t^2$. Find:
(*i*) its acceleration and distance from O, each in terms of t;
(*ii*) the time at which it begins to return, and the time at which it again reaches its starting-point. (O.C.)

9 A particle travels in a straight line, being at a point A at zero time. Its velocity in m s^{-1} at time t s is given by the expression $6t^2 - 48t + 42$.
Find expressions at time t s for:
(*i*) its acceleration;
(*ii*) its distance from A.

Find the times at which its velocity is zero. Show also that it passes through A twice after zero time, and find (to one place of decimals) the length of time before its first return. (O.C.)

10 In the first 20 seconds of its motion down a runway, an aircraft is subject to an acceleration given by $(10 - \frac{1}{2}t)$ m s^{-2}. Find its velocity at the end of this time, and how far it has travelled, to the nearest metre. (M.E.I.)

11 The acceleration of a particle t s after starting from rest is $(2t - 1)$ m s^{-2}. Prove that the particle returns to the starting-point after $1\frac{1}{2}$ s, and find the distance of the particle from the starting-point after a further $1\frac{1}{2}$ s.
Find also at what time after starting the particle attains a velocity of 20 m s^{-1}. (O.C.)

12 The acceleration of a car t s after starting from rest is $\dfrac{75 + 10t - t^2}{20}$ m s^{-2} until the instant when this expression vanishes. After this instant the speed of the car remains constant. Find:
(*i*) the maximum acceleration;
(*ii*) the time taken to attain the greatest speed;
(*iii*) the greatest speed attained. (O.C.)

*3.8 Stationary values: maxima and minima

In section 3.2 we said that $f(x)$ was stationary whenever $f'(x) = 0$. We now investigate the various types of stationary values.

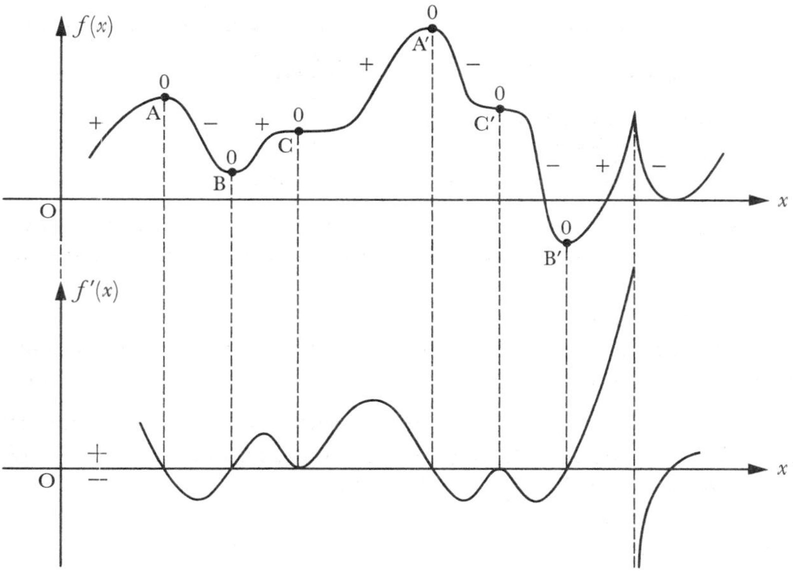

It is seen from the graph of $f(x)$ that stationary values are of three types:

(*i*) when the function is *less on both* sides of its stationary value. This is called a *maximum* (as at A, A′) and it is clear that $f'(x)$ must be decreasing from positive values through zero to negative values;
(*ii*) when the function is *greater* on *both* sides of its stationary value. This is called a *minimum* (as at B, B′) and it is clear that $f'(x)$ must be increasing from negative values through zero to positive values;
(*iii*) when the function is greater on one side of its stationary value, but less on the other side. This is a *point of inflection* (as at C, C′) and it is clear that $f'(x)$ must have the same sign (whether positive or negative) on each side of the point where it is zero.

Example 1

Investigate the stationary values of $f(x) = x^3 - 3x$

$f(x) = x^3 - 3x$
$\Rightarrow \quad f'(x) = 3x^2 - 3 = 3(x^2 - 1)$

At stationary values, $f'(x) = 0 \quad \Rightarrow \quad x^2 - 1 = 0 \Rightarrow x = \pm 1$

So $f(x)$ is stationary when $x = \pm 1$ and its two stationary values are

$f(+1) = 1 - 3 = -2$ and $f(-1) = -1 + 3 = +2$

So the graph of $f(x)$ is stationary at $(+1, -2)$ and $(-1, +2)$

We can now investigate in more detail by finding how $f'(x)$ varies as x

increases through the values $+1$ and -1.

As x increases through $+1$,

$f'(x) = 3(x^2 - 1)$ changes from $-$ to 0 to $+$

$\Rightarrow \quad f(+1) = -2$ is a *minimum.*

As x increases through -1,

$f'(x) = 3(x^2 - 1)$ changes from $+$ to 0 to $-$

$\Rightarrow \quad f(-1) = +2$ is a *maximum.*

So the graph of $f(x) = x^3 - 3x$ is:

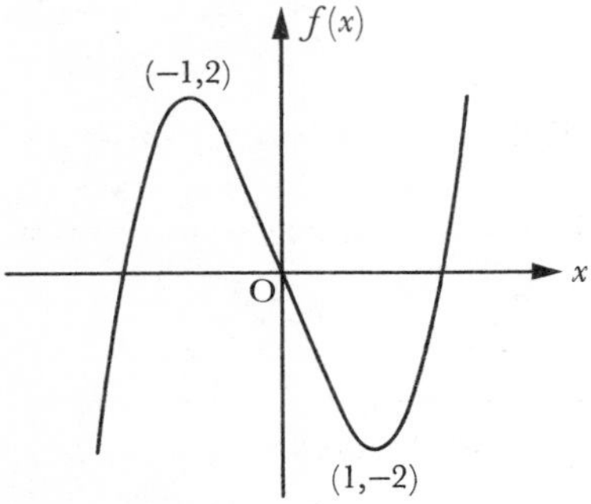

Use of $f''(x)$ at stationary values

If $f'(a) = 0$, i.e., $f(a)$ is a stationary value of $f(x)$, we can often decide whether it is a maximum or a minimum value simply by finding the value of $f''(a)$.

Case 1: $f''(a) < 0$.

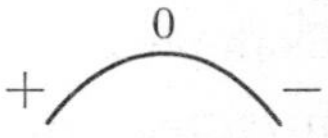

$f''(a) < 0 \Rightarrow f'(x)$ is *decreasing* at $x = a$.

But $f'(a) = 0$.

So, as x increases through a, $f'(x)$ must change from $+$ to 0 to $-$. So $f''(a) < 0 \Rightarrow f(a)$ is a *maximum.*

Case 2: $f''(a) > 0$.

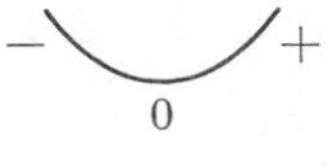

$f''(a) > 0 \Rightarrow f'(x)$ is increasing at $x = a$.

But $f'(a) = 0$.

So, as x increases through a, $f'(x)$ must change from $-$ to 0 to $+$. So $f''(a) > 0 \Rightarrow f(a)$ is a *minimum.*

In the above example,

$f(x) = x^3 - 3x$

$f'(x) = 3x^2 - 3$

$f''(x) = 6x \Rightarrow f''(+1) = +6$ and $f''(-1) = -6$.

Now $f(x)$ is stationary when $x = \pm 1$,

$$f''(+1) = +6 \Rightarrow f(+1) = -2 \text{ is a minimum,}$$
$$\text{and } f''(-1) = -6 \Rightarrow f(-1) = +2 \text{ is a maximum,}$$

which confirm what we previously discovered.
Summarising,

$$\text{when } f'(a) = 0, f''(a) < 0 \Rightarrow f(a) \text{ is a maximum}$$
$$f''(a) > 0 \Rightarrow f(a) \text{ is a minimum.}$$

Case 3: $f''(a) = 0$.

What happens if $f''(a) = 0$?

Sketch the graphs of $f(x)$ and calculate $f(0)$ and $f''(0)$ when (*i*) $f(x) = x^3$, (*ii*) $f(x) = x^4$, (*iii*) $f(x) = -x^4$.

What conclusions can be drawn about the nature of a stationary point if $f''(a) = 0$?

Summary:

Maximum	Minimum	Points of inflection

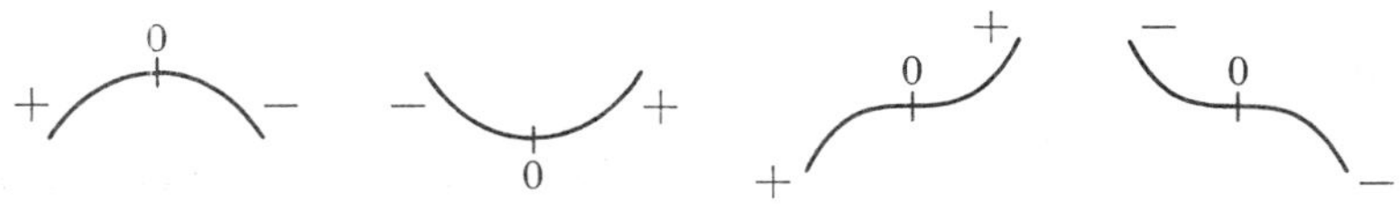

$f''(a) \leqslant 0$	$f''(a) \geqslant 0$	$f''(a) = 0$

Example 2

Investigate the stationary values of

$$f(x) = x + \frac{4}{x}$$

It follows that $f'(x) = 1 - \dfrac{4}{x^2}$

$$\text{and } f''(x) = \frac{8}{x^3}$$

At stationary values
$$f'(x) = 0$$
$$\Rightarrow 1 - \frac{4}{x^2} = 0$$
$$\Rightarrow x = \pm 2$$

$$\text{Now } f''(-2) = \frac{8}{(-2)^3} = -1$$

$$\Rightarrow f(-2) = -4 \text{ is a maximum value}$$

And $\quad f''(2) = \dfrac{8}{2^3} = +1$

$\Rightarrow \quad f(2) = +4$ is a minimum value.

At first it seems surprising that the maximum value should be less than the minimum value. But it is clear that $f(-2) = -4$ is a maximum because it is greater than other values *in the immediate neighbourhood* of $x = -2$; and similarly $f(2) = +4$ is less than other values *in the immediate neighbourhood* of $x = 2$. Their relationship is apparent from the graph of $f(x)$:

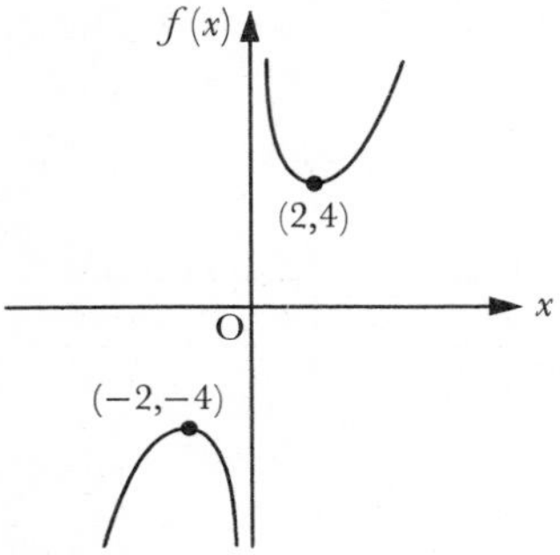

Example 3

Investigate the stationary points of

$y = 3x^4 - 4x^3$

and sketch its graph.

$$\frac{dy}{dx} = 12x^3 - 12x^2 = 12x^2(x-1)$$

and $\quad \dfrac{d^2y}{dx^2} = 36x^2 - 24x = 12x(3x-2)$

Now $\quad \dfrac{dy}{dx} = 0 \Rightarrow x = 0$ or 1.

So y is stationary at $x = 0,\ y = 0$
and at $x = 1,\ y = -1$

(*i*) At $x = 0,\ \dfrac{d^2y}{dx^2} = 0$, which is indecisive.

But as x increases through $x = 0$,

$\dfrac{dy}{dx} = 3x^2(4x-1)$ and changes from $-$ to 0 to $-$.

So at $(0, 0)$ the curve has a point of inflection.

(*ii*) At $x = 1$, $\frac{d^2y}{dx^2} = 12\times1\,(3\times1 - 2) = 12$

So at $(1, -1)$ the curve has a minimum.

Also $y = 0 \Rightarrow x = 0$ or $\frac{4}{3}$, and as $x \to \pm\infty$, $y \to +\infty$.

Summarising these, we obtain the graph

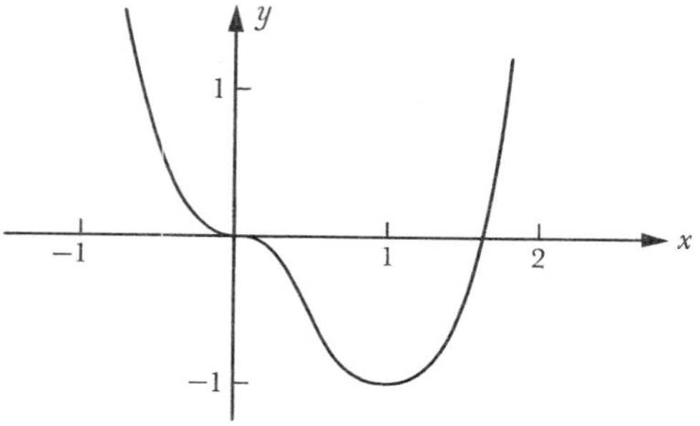

Example 4

When a ship is travelling at a speed of v km h^{-1} its rate of consumption of fuel in tonne h^{-1} is approximately $20 + 0.001\,v^3$.

Find an expression for the total fuel used on a voyage of 5 000 km at speed v, and proceed to find the speed for greatest fuel economy and the amount of fuel then used.

Time taken over journey $= \frac{5\,000}{v}$ h.

So if total amount of fuel used is A tonne,

$$A = (20 + 0.001\,v^3) \times \frac{5\,000}{v}$$

$$= \frac{100\,000}{v} + 5v^2$$

$$\Rightarrow \quad \frac{dA}{dv} = -\frac{100\,000}{v^2} + 10\,v$$

and $$\frac{d^2A}{dv^2} = +\frac{200\,000}{v^3} + 10$$

For A to be a minimum, $\frac{dA}{dv} = 0$

$$\Rightarrow \quad 10v = \frac{100\,000}{v^2}$$

$$\Rightarrow \quad v^3 = 10\,000$$

$$\Rightarrow \quad v \approx 21.54$$

Also $\dfrac{d^2A}{dv^2} > 0$, so A is definitely a minimum.

When $v = 21.54$, $A \approx \dfrac{100\,000}{21.54} + 5\,(21.54)^2$

$$\approx 4\,640 + 2\,320 = 6\,960$$

So the most economical speed for the voyage is approximately 21.5 km h^{-1} when the total fuel consumption will be 6 960 tonne.

Example 5

A cylindrical block of ice is melting at a rate proportional to its surface area. If its length, the radius of its ends and its volume are l, r, V respectively, prove that its surface area is

$$S = \frac{2V}{r} + 2\pi r^2$$

Hence find the ratio of length to radius if a cylindrical block of given volume is to melt as slowly as possible.

$$\text{Area of curved surface} = 2\pi rl$$
$$\text{Area of two ends} = 2\pi r^2$$
$$\Rightarrow S = 2\pi rl + 2\pi r^2$$

This expression for S is a function of two variables r and l, so we cannot yet find its derivative.

But r and l are connected by the fact that $\pi r^2 l = V$, which is a constant.

So $S = \dfrac{2V}{r} + 2\pi r^2$, which is a function of a single variable r.

Hence $\dfrac{dS}{dr} = -\dfrac{2V}{r^2} + 4\pi r$

and $\dfrac{d^2S}{dr^2} = +\dfrac{4V}{r^3} + 4\pi$

Now S is stationary $\Rightarrow \dfrac{dS}{dr} = 0$

$$\Rightarrow 4\pi r = \frac{2V}{r^2}$$

$$\Rightarrow 2\pi r^3 = V$$

Also $\dfrac{d^2S}{dr^2} > 0$

So S is a minimum when $2\pi r^3 = V$

$$\Rightarrow 2\pi r^3 = \pi r^2 l$$

$$\Rightarrow \frac{r}{l} = \tfrac{1}{2}$$

Hence the ice-block melts most slowly if its length is equal to its diameter.

Exercise 3.8

1 Find the stationary values of the following functions. In each case find whether the function has a maximum value, a minimum value or a point of inflection, and sketch the graph of the function.

(*i*) $f(x) = 4x^2 - 8x - 5$; (*ii*) $f(x) = 3 + 2x - x^2$;
(*iii*) $f(x) = x^3 - 3x$; (*iv*) $f(x) = x^3 + 3x$.

2 Investigate the stationary points of the following curves, deciding in each case whether the point is a maximum, a minimum or a point of inflection. Find also any points of oblique inflection and sketch the graph of the function.

(*i*) $y = x + \frac{4}{x}$; (*ii*) $y = x^2 + \frac{2}{x}$;
(*iii*) $y = 4x^5 - 5x^4$; (*iv*) $y = 6x^4 - 4x^6$.

3 (*i*) Find the coordinates of points on the curve

$$y = 2x^3 + 3x^2 - 12x + 6$$

at which y has a stationary value and state in each case whether the value of y is a maximum or a minimum.
(*ii*) Find the maximum and minimum values of

$$f(x) = 2x^3 - 3x^2 - 12x + 21.$$

Hence, or otherwise, show that the equation $f(x) = 0$ has only one real root. Sketch the curve $y = f(x)$.
(*iii*) Given that the function $y = ax^3 + bx^2$, where a and b are constants, has a stationary value -4 when $x = 2$, find the values of a and b. (O.C.)

4 At all points on a certain curve it is known that $\frac{d^2y}{dx^2} = 6x - 2$. If $\frac{dy}{dx} = -1$ and $y = 1$ when $x = 0$,

(*i*) find the equation of the curve;
(*ii*) find its stationary points, stating whether y has a maximum or a minimum value. (O.C.)

5 A farmer erects a fence along three sides of a rectangle in order to make a sheepfold; the fourth side of the rectangle is provided by a hedge already in existence. Find the maximum area of the enclosure thus made if the total length of the fence is to be 80 m. (O.C.)

6 A square sheet of metal of side 12 cm has four equal square portions removed at the corners and the sides are then turned up to form an open rectangular box. Prove that, when the box has maximum volume, its depth is 2 cm. (O.C.)

7 The bottom of a rectangular tank is a square of side x m and the tank is open at the top. It is designed to hold 4 m^3 of liquid. Express in terms of x

the total area of the bottom and the four sides of the tank. Find the value of x for which this area is a minimum. (O.C.)

8 A closed rectangular box is made of very thin sheet metal, and its length is three times its width. If the volume of the box is 288 cm^3, show that its surface area is equal to

$$\frac{768}{x} + 6x^2 \text{ cm}^2.$$

where x cm is the width of the box.

Find by differentiation the dimensions of the box of least surface area. (O.C.)

9 A rectangular box without a lid is made of cardboard of negligible thickness. The sides of the base are $2x$ cm and $3x$ cm, and the height is y cm. If the total area of the cardboard is 200 cm^2, prove that $y = \frac{20}{x} - \frac{3x}{5}$.

Find the dimensions of the box when its volume is a maximum. (O.C.)

10 The despatch department of a certain firm limits parcels to those whose length and girth are together less than 2 m. If the length of a parcel is x m, show that its volume V m^3, is given by

$$16V = x(2-x)^2$$

Find the maximum volume of a parcel of this shape, and prove that your result is, in fact, a maximum.

(*The girth of the parcel is the perimeter of its square cross-section.*)

11 An open tank is to be constructed with a square horizontal base and vertical sides. The capacity of the tank is to be 2 000 m^3. The cost of the material for the sides is £4 per square metre and for the base is £2 per square metre. Find the minimum cost of the material, and give the corresponding dimensions of the tank. (O.C.)

12 A piece of wire, 150 cm long, is cut into two parts; one is bent to form a square and the other to form a circle. Find the lengths of the two parts when the sum of the two areas is least.

13 A beam of rectangular cross-section is to be cut from a cylindrical log of radius a. The stiffness of the beam is proportional to xy^3 where x is the breadth and y the depth of the section. Find the cross-sectional area of the beam

(*i*) of greatest volume;
(*ii*) of greatest stiffness;
that can be cut from the log. (O.C.)

14 The cost £S and the time t min of manufacture of a certain article are connected by the formula

$$S = \frac{16}{t^3} + \frac{3t^2}{4}$$

Find:
(*i*) the rate of charge of cost when $t = 4$;
(*ii*) the minimum cost. (J.M.B.)

15 ABCD is a square field and the length of a diagonal is $2a$ km. A man starts to walk from A straight across to the opposite corner C at a speed of 4 km/h. At the same instant a second man starts to walk from B straight across to D at a speed of 3 km/h. Show that after t h ($t \leqslant \frac{1}{2}a$) their distance apart is

$$\{(a - 3t)^2 + (a - 4t)^2\}^{1/2}$$

Prove that their least distance apart is $\frac{1}{5}a$. Find how far each man is from the centre of the square at this moment. (O.C.)

16 A cylinder of height h fits exactly into a sphere of radius a. Show that its volume V is given by

$$V = \tfrac{1}{4}\pi h(4a^2 - h^2)$$

and hence find the maximum fraction of the sphere's volume which can be occupied by the cylinder.

*3.9 Further differentiation

Function of a function: the Chain Rule

So far we have confined our attention to functions which are powers of x, and to multiples, sums and differences of such functions. But a function is fundamentally a process which converts an input into an output; and, as with industrial processes, many functions are the succession of a series of such processes. Where the output of one function machine is used as the input of the next, the combined process is known as a 'function of a function'.

Example

If $y = \sqrt{(x^2 + 1)}$

we can calculate y when $x = 2$ by three successive processes, each different:

(*i*) using 2 as input, we square it to obtain 4;
then (*ii*) using 4 as input, we add 1 to obtain 5;
and finally (*iii*) using 5 as input, we take its square root to obtain $\sqrt{5} \approx 2.236$.

If we regard x as passing through intermediate stages u and v in the course of production of y, we can write

Input		Output
x	$\mapsto$	$x^2 = u$
u	$\mapsto$	$u + 1 = v$
v	$\mapsto$	$\sqrt{v} = y$

So $y = \sqrt{v} = \sqrt{(u + 1)} = \sqrt{(x^2 + 1)}$

We now meet the obvious question, whether we can differentiate y when it is known as the final output of a number of functions.

For convenience, let us suppose that we have a two-stage process:

Input	Intermediate	Output
$x \quad \mapsto$	$u \quad \mapsto$	y

Now if x is increased by δx,
u will increase by δu,
and so y will increase by δy.

Moreover, $\dfrac{\delta y}{\delta x} = \dfrac{\delta y}{\delta u} \times \dfrac{\delta u}{\delta x}$

Now suppose that $\delta x \to 0$ and $\delta u \to 0$, $\delta v \to 0$;

and also that $\dfrac{\delta u}{\delta x}, \dfrac{\delta y}{\delta u}, \dfrac{\delta y}{\delta x} \to \dfrac{du}{dx}, \dfrac{dy}{du}, \dfrac{dy}{dx}$ respectively.

Then $$\boxed{\frac{dy}{dx} = \frac{dy}{du} \times \frac{du}{dx}}$$

If there are more intermediate stages, say $x \mapsto u \mapsto v \mapsto w \mapsto y$, this result is easily extended to

$$\frac{dy}{dx} = \frac{dy}{dw} \times \frac{dw}{dv} \times \frac{dv}{du} \times \frac{du}{dx}$$

and for this reason it is often called the *Chain rule*.

Though it is one of the advantages of the notation that the result *looks* obvious, it is important to realise that it cannot be proved simply by the cancellation

$$\frac{dy}{\cancel{du}} \times \frac{\cancel{du}}{dx} = \frac{dy}{dx},$$

for the symbol du has, on its own, not been given any meaning and so cannot be cancelled.

Example

$y = (x^3 + 1)^{10}$

Let $\quad y = u^{10} \quad$ where $\quad u = x^3 + 1$

Now $\quad \dfrac{dy}{du} = 10u^9 \quad$ and $\quad \dfrac{du}{dx} = 3x^2$

So $\quad \dfrac{dy}{dx} = \dfrac{dy}{du} \times \dfrac{du}{dx} = 10\,(x^3 + 1)^9 \times 3x^2 = 30x^2\,(x^3 + 1)^9.$

Exercise 3.9a

1 Find $\frac{dy}{dx}$ if y is

(*i*) $(3x^2 + 1)^4$; (*ii*) $(9x^2 + 4)^{10}$; (*iii*) $(1 - x^2)^6$;

(*iv*) $(x^2 - x)^5$; (*v*) $(3x^2 - 1)^{-1}$; (*vi*) $\frac{1}{x^2 - 1}$;

(*vii*) $\frac{1}{(x^2 + 2)^3}$; (*viii*) $\sqrt{(1 + x^2)}$; (*ix*) $\sqrt[3]{(x^2 - 1)}$;

(*x*) $\frac{1}{\sqrt{(1 - x)}}$; (*xi*) $(\sqrt{x} + 1)^{10}$; (*xii*) $(ax^2 + b)^n$.

2 Find $f'(t)$ when $f(t)$ is

(*i*) $(t^3 + 1)^4$; (*ii*) $(1 - 3t^2)^5$; (*iii*) $\sqrt{(1 - t)}$; (*iv*) $\frac{1}{\sqrt{(t^2 + 1)}}$.

3 (*i*) If $y = x^2$ (where $x > 0$),

(*a*) express x as a function of y;

(*b*) examine whether $\frac{dx}{dy} = 1 \div \frac{dy}{dx}$;

(*c*) examine whether $\frac{d^2x}{dy^2} = 1 \div \frac{d^2y}{dx^2}$.

(*ii*) Repeat these questions if $y = (x - 1)^3$.

4 A man wishes to cross a square ploughed field from one corner to the opposite corner in the shortest possible time. He can walk at 10 km/h when he keeps to the edge of the field and at 6 km/h when he walks on the ploughed land. Prove that he should walk one-quarter of the way along one side and then aim direct for the opposite corner. (O.C.)

Differentiation of products and quotients

In section 3.3 we investigated the derivatives of multiples and sums of given functions. If u and v were functions of x, we found that

$$\frac{d}{dx}(ku) = k\frac{du}{dx} \quad \text{and} \quad \frac{d}{dx}(u + v) = \frac{du}{dx} + \frac{dv}{dx}$$

We were not, however, able to find an obvious way of differentiating their product uv or their quotient $\frac{u}{v}$, so we shall now investigate these.

Let us first consider the product $y = uv$
where u, v are functions of x.

Let x increase by δx.

Then u, v increase by δu, δv, and so y increases by δy.

$$y + \delta y = (u + \delta u)(v + \delta v)$$
$$= uv + u\delta v + v\delta u + \delta u\delta v$$

But $y = uv$

Subtracting, $\delta y = u\delta v + v\delta u + \delta u\delta v$

$$\Rightarrow \qquad \frac{\delta y}{\delta x} = u\frac{\delta v}{\delta x} + v\frac{\delta u}{\delta x} + \delta u\frac{\delta v}{\delta x}$$

As $\delta x \to 0$, $\delta u \to 0$

and $\dfrac{\delta u}{\delta x}, \dfrac{\delta v}{\delta x}, \dfrac{\delta y}{\delta x} \to \dfrac{du}{dx}, \dfrac{dv}{dx}, \dfrac{dy}{dx}$ respectively

So
$$\frac{dy}{dx} = u\frac{dv}{dx} + v\frac{du}{dx}$$

Example

$$y = x^2(x + 1)^5$$
$$\Rightarrow \frac{dy}{dx} = x^2 5(x + 1)^4 + 2x(x + 1)^5$$
$$= x(x + 1)^4[5x + 2(x + 1)]$$
$$= x(7x + 2)(x + 1)^4$$

We now consider the quotient, $y = \dfrac{u}{v}$

Again, let x increase by δx.
Then u, v increase by δu, δv, and so y increases by δy

$$y + \delta y = \frac{u + \delta u}{v + \delta v}$$

But $y = \dfrac{u}{v}$

Subtracting,
$$\delta y = \frac{u + \delta u}{v + \delta v} - \frac{u}{v}$$
$$= \frac{v(u + \delta u) - u(v + \delta v)}{v(v + \delta v)}$$
$$= \frac{v\delta u - u\delta v}{v(v + \delta v)}$$

$$\Rightarrow \qquad \frac{\delta y}{\delta x} = \frac{v\dfrac{\delta u}{\delta x} - u\dfrac{\delta v}{\delta x}}{v(v + \delta v)}$$

Suppose that $\delta x \to 0$ and $\delta v \to 0$;

and $$\frac{\delta u}{\delta x}, \frac{\delta v}{\delta x}, \frac{\delta y}{\delta x} \to \frac{du}{dx}, \frac{dv}{dx}, \frac{dy}{dx} \text{ respectively}$$

Then $$\frac{dy}{dx} = \frac{v\dfrac{du}{dx} - u\dfrac{dv}{dx}}{v^2}$$

Example

$$y = \frac{x}{1 + x^2}$$

$$\Rightarrow \frac{dy}{dx} = \frac{(1 + x)^2 \times 1 - x \times 2x}{(1 + x^2)^2} = \frac{1 - x^2}{(1 + x^2)^2}$$

These two general results can equally be written:

$$\frac{d}{dx}(uv) = u\frac{dv}{dx} + v\frac{du}{dx}$$

$$\frac{d}{dx}\left(\frac{u}{v}\right) = \frac{v\dfrac{du}{dx} - u\dfrac{dv}{dx}}{v^2}$$

Exercise 3.9b

1 Differentiate:

(*i*) $x(1 + x)^5$; (*ii*) $x^2(3x + 1)^4$;

(*iii*) $(x + 1)^2(x + 2)^3$; (*iv*) $x(2x^3 + 1)^5$;

(*v*) $\dfrac{x - 1}{x + 1}$; (*vi*) $\dfrac{x}{2x^2 + 1}$;

(*vii*) $\dfrac{1 - x^2}{1 + x^2}$; (*viii*) $\dfrac{x}{(3x + 1)^2}$;

(*ix*) $\dfrac{1 + \sqrt{x}}{1 - \sqrt{x}}$; (*x*) $\sqrt{\left(\dfrac{1 - x}{1 + x}\right)}$.

2 Investigate the stationary values of (*i*) $\dfrac{x^2}{x - 1}$; (*ii*) $\dfrac{x^3}{1 - x^2}$.

3 Prove the rule for the differentiation of the quotient $\dfrac{u}{v}$ by expressing it as the product uv^{-1}.

*3.10 Rates of change: implicit functions and parameters

The Chain rule is frequently useful when we are considering rates of change. Take, for instance, the following case of an expanding sphere whose radius r, surface area S, and volume V are all increasing with time t.

Example 1

A pump is inflating a spherical balloon whose radius at a certain instant is 2 m and is increasing at a rate of 1 cm s^{-1}.

(*i*) At what rate is the pump working?

Here $V = \frac{4}{3}\pi r^3$, where r is a function of t.

$$\text{So} \quad \frac{dV}{dt} = \frac{dV}{dr} \times \frac{dr}{dt}$$

$$\frac{dV}{dr} = 4\pi r^2 = 4\pi \times 2^2 = 16\pi \text{ m}^2$$

$$\frac{dr}{dt} = 0.01 \text{ m s}^{-1}$$

$$\text{So} \quad \frac{dV}{dt} = 16\pi \times 0.01 \text{ m}^3 \text{ s}^{-1} \approx 0.50 \text{ m}^3 \text{ s}^{-1}$$

So the pump is delivering air at the rate of 0.50 m^3 s^{-1}, or roughly 1 m^3 every 2 s.

(*ii*) If air continues to be pumped into the balloon at this rate, at what rate will the radius be increasing when it is 5 m?

$$\frac{dV}{dt} = \frac{dV}{dr} \times \frac{dr}{dt}$$

$$\frac{dV}{dt} = 4\pi r^2 \times \frac{dr}{dt}$$

$$\text{But} \quad \frac{dV}{dt} = 0.50 \quad \text{and} \quad r = 5$$

$$\text{So} \quad 0.50 = 4\pi \times 5^2 \frac{dr}{dt}$$

$$\frac{dr}{dt} = \frac{0.50}{100\pi} \approx 0.0016 \text{ m s}^{-1}$$

So the rate of increase of radius has now been reduced to 0.0016 m s^{-1}.

(*iii*) At what rate is the surface area of the balloon increasing when $r = 5$?

$$S = 4\pi r^2$$

$$\Rightarrow \quad \frac{dS}{dt} = \frac{dS}{dr} \times \frac{dr}{dt} = 8\pi r \times \frac{dr}{dt}$$

$$\Rightarrow \quad \frac{dS}{dt} = 8\pi \times 5 \times 0.0016 = 0.20$$

So when the radius of the balloon is 5 m, its membrane is being stretched at the rate of 0.20 $m^2\ s^{-1}$.

Example 2

A particle moves from a point O of a straight line and its velocity v is plotted on a graph as a function of its displacement x.

(*i*) Show that its acceleration $a = v\dfrac{dv}{dx}$;

(*ii*) If its acceleration is constant and it starts with velocity u, show that

$$v^2 = u^2 + 2ax$$

(*i*) $$a = \frac{dv}{dt} = \frac{dv}{dx}\times\frac{dx}{dt} = \frac{dv}{dx} \times v$$

So $$a = v\frac{dv}{dx}.$$

(*ii*) Hence $$\frac{dx}{dv} = \frac{v}{a}.$$

If a is constant, $x = \dfrac{v^2}{2a} + A$

When $x = 0, v = u \Rightarrow 0 = \dfrac{u^2}{2a} + A$

So $$x = \frac{v^2}{2a} - \frac{u^2}{2a}$$

$$\Rightarrow \quad v^2 = u^2 + 2ax.$$

Exercise 3.10a

1 The radius of a circle is 6 m.

(*i*) If it is increasing at 0.1 $m\ s^{-1}$, at what rate is its area increasing?

(*ii*) If its area is increasing at the rate of 20 $m^2 s^{-1}$, what is the rate of increase of its radius?

2 A spherical balloon is losing air at a rate of 10 $m^3\ s^{-1}$. When its radius is 3 m at what rate is it diminishing?

3 A street lamp is at a height 5 m and a man of height 2 m is running away from it at 3 $m\ s^{-1}$. How long is his shadow when he is x m away from its base, and at what rate is this length increasing?

4 The top of a 10 m ladder is at a height 8 m up a vertical wall and is slipping down the wall at a rate of 1 $m\ s^{-1}$. At what rate is the foot of the ladder sliding along the ground?

5 Liquid is dripping through a conical funnel at a rate of 2 $cm^3\ s^{-1}$. When the depth of liquid in the funnel is x cm, its volume is $\frac{1}{3}\pi x^3$. Find the rate at which the level of the liquid is falling when $x = 5$.

6 When the gas in a balloon is kept at constant temperature its volume V and pressure p are connected by Boyle's law, $pV =$ constant.

If its volume 100 m^3 is increasing at a rate 10 $m^3\ s^{-1}$, find the rate at which the pressure is decreasing from its present value of 5 atmospheres.

The Chain rule is also useful when y is known only as an *implicit* function of x (i.e., is not given explicitly by a formula in terms of x).

Example 3

Show that the curve

$$x^3 + y^3 = 2xy$$

passes through the point (1, 1), and find its gradient at this point.

The first part follows from the fact

that $\quad 1^3 + 1^3 = 2\times1\times1$

Also $\quad x^3 + y^3 = 2xy$

$$\Rightarrow \quad \frac{d}{dx}(x^3) + \frac{d}{dx}(y^3) = \frac{d}{dx}(2xy)$$

$$\Rightarrow \quad 3x^2 + 3y^2\frac{dy}{dx} = 2x\frac{dy}{dx} + 2\times1\times y$$

$$\Rightarrow \quad (3y^2 - 2x)\frac{dy}{dx} = 2y - 3x^2$$

$$\Rightarrow \quad \frac{dy}{dx} = \frac{2y - 3x^2}{3y^2 - 2x}$$

So at (1, 1) $\quad \dfrac{dy}{dx} = \dfrac{2-3}{3-2} = -1.$

Example 4

Given that $\frac{d}{dx}(x^n) = nx^{n-1}$ when $n \in Z^+$, prove that this is also true whenever $n \in Q$ (i.e., whenever n is rational).

Case 1: $\quad n \in Z^-$ (i.e., n a negative integer).

Let $n = -m$.

$$\text{Then} \quad \frac{d}{dx}(x^n) = \frac{d}{dx}(x^{-m}) = \frac{d}{dx}(x^{-1})^m$$

$$= m(x^{-1})^{m-1}\times -x^{-2}$$

$$= -mx^{-m-1} = nx^{n-1}$$

Case 2: $n = 0 \quad \frac{d}{dx}(x^0) = \frac{d}{dx}(1) = 0 = 0x^{0-1}.$

Case 3: $n \in Q$.

Then $y = x^n = x^{p/q}$, where $p, q \in Z$.

So
$$y^q = x^p$$
$$\Rightarrow \quad qy^{q-1}\frac{dy}{dx} = px^{p-1}$$
$$\Rightarrow \quad \frac{dy}{dx} = \frac{p}{q}\frac{x^{p-1}}{y^{q-1}}$$
$$= \frac{p}{q}\frac{x^{p-1}}{(x^{p/q})^{q-1}}$$
$$= \frac{p}{q}\frac{x^{p-1}}{x^{p-(p/q)}}$$
$$= \frac{p}{q}x^{(p/q)-1} = nx^{n-1}$$

So $\frac{d}{dx}(x^n) = nx^{n-1}$ whenever $x \in Q$.

Sometimes the equation of a curve is given by expressing the coordinates x and y as functions of a third variable t, called a *parameter*:

Example 5

If a stone is thrown out to sea with velocity 10 m s^{-1} horizontally from the top of a cliff, then its coordinates x, y (referred to axes through its point of projection) can be expressed as functions of time t:

$$x = 20t$$
$$y = -5t^2.$$

It is easy to see, by eliminating t, that the equation of the stone's path is

$$y = -\frac{1}{80}x^2$$

and its directions of motion could readily be calculated from this equation.

Alternatively we can use a version of the Chain rule.

For $\frac{dy}{dx} = \lim_{\delta x \to 0}\frac{\delta y}{\delta x} = \lim_{\delta t \to 0}\frac{\delta y/\delta t}{\delta x/\delta t} = \frac{dy/dt}{dx/dt}$

So
$$\boxed{\frac{dy}{dx} = \frac{dy/dt}{dx/dt}}$$

In our particular case, $\frac{dy}{dx} = \frac{-10t}{20} = -\frac{t}{2}.$

These can be summarised in the figure:

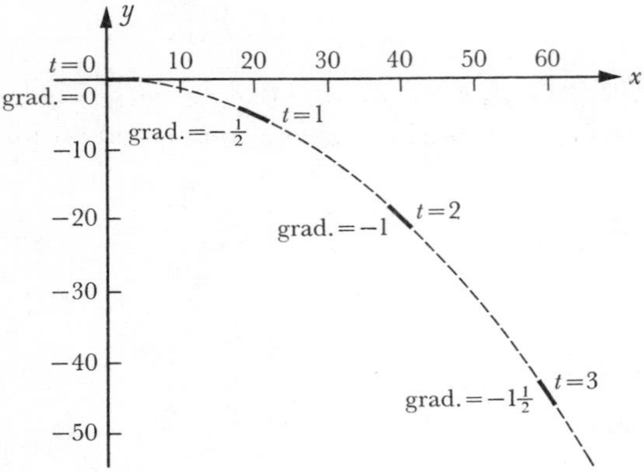

Exercise 3.10b

1 Find expressions for $\mathrm{d}y/\mathrm{d}x$ for the following curves and calculate their gradients at the given points

(*i*) $x^2 - y^2 = 8$, $(3, 1)$;
(*ii*) $x^2 + 4xy + y^2 = 6$, $(1, 1)$;
(*iii*) $x^3 + y^3 = 7$, $(-1, 2)$.

2 Plot the following curves for values of t from -3 to $+3$, and find $\mathrm{d}y/\mathrm{d}x$ in terms of t.

(*i*) $x = 2t^3, \quad y = t^2$;
(*ii*) $x = 4t, \quad y = \dfrac{1}{t}$;
(*iii*) $x = t^3 - t, \quad y = t^2$.

3 At high tide a stone is thrown horizontally with speed $10\ \mathrm{m\ s^{-1}}$ from a cliff of height 80 m. The horizontal and vertical components of the distance it has travelled after t second are given by

$$x = 10t, \quad y = 5t^2$$

Find when, where and what angle it enters the sea.

4 A curve is given by the parametric equations

$$x = t^2 - 3, \quad y = t(t^2 - 3)$$

(*i*) Find its x, y equation, in a form clear of surds and fractions.
(*ii*) Prove that it is symmetrical about the x-axis.
(*iii*) Show that there are no points on the curve for which $x < -3$.
(*iv*) Find $\mathrm{d}y/\mathrm{d}x$ in terms of t, and derive the coordinates of the points on the curve for which $\mathrm{d}y/\mathrm{d}x = 0$.
(*v*) Sketch the form of the curve. (C)

5 Prove that if the velocity of a particle moving in a straight line is always proportional to its displacement from a fixed point, so is its acceleration.

*3.11 Small changes

Differentiation was first introduced by considering the effects of small changes. We can now reverse the process and use differentiation to calculate such effects.

Example 1

Calculate $\sqrt[3]{8.01}$.

$$y = x^{1/3} \quad \Rightarrow \quad \frac{dy}{dx} = \tfrac{1}{3}x^{-2/3}$$

$$\Rightarrow \quad \frac{\delta y}{\delta x} \approx \tfrac{1}{3}x^{-2/3}$$

$$\Rightarrow \quad \delta y \approx \tfrac{1}{3}x^{-2/3}\delta x.$$

If we now let $x = 8$ and $\delta x = 0.01$,

then $\qquad y = \sqrt[3]{8} = 2$

and $\qquad \delta y \approx \frac{1}{3} \times 8^{-2/3} \times 0.01$

$$\approx \tfrac{1}{3} \times \tfrac{1}{4} \times 0.01$$

$$\approx 0.000\,83$$

So $\qquad y + \delta y \approx 2.000\,83$

Hence $\quad \sqrt[3]{8.01} \approx 2.000\,83$

Example 2

The time T taken by a planet to revolve round the Sun and its mean distance r from the Sun are such that $T = kr^{3/2}$.

If the Earth's distance from the Sun were to be increased by 1%, how much longer would a year become?

$$T = kr^{3/2}$$

$$\Rightarrow \quad \frac{dT}{dr} = \frac{3}{2}kr^{1/2}$$

$$\Rightarrow \quad \delta T \approx \frac{3}{2}kr^{1/2}\,\delta r$$

Hence $\quad \dfrac{\delta T}{T} \approx \dfrac{3}{2}\dfrac{\delta r}{r}$

But $\quad \dfrac{\delta r}{r} = \dfrac{1}{100}$, so $\quad \dfrac{\delta T}{T} \approx \dfrac{3}{200} = 1.5\%$

So a year would be increased by $1.5\% \approx 5.5$ days.

Exercise 3.11a

1 Find, without use of logarithms or long multiplication, approximate values for:
(*i*) $(20.1)^3$; (*ii*) $\sqrt[3]{27.2}$; (*iii*) $\sqrt[2]{15.9}$.

2 The lanes of a circular running track are 1 m wide. What is the difference in length of adjacent lanes?

3 When a hemispherical bowl of radius r contains liquid to a depth x, its volume V is $\frac{1}{3}\pi(3rx^2 - x^3)$.

Find a relationship between slight changes δx and δV in depth and volume respectively.

If a liquid is originally of depth 1 m in a bowl of radius 2 m, find the decrease in depth when 0.1 m^3 has evaporated.

4 An error of 2% is made in measuring the radius of a sphere. What are the resulting errors in the calculation of its surface area and volume?

5 A cube of metal is heated so that its temperature is raised by 1° C and each side expands by p%. What are the percentage expansions of its surface area and its volume?

6 When a wire is slightly stretched, its length l and diameter x vary in such a way that its volume remains constant. Find the percentage decrease in diameter in terms of the percentage increase in length.

Numerical solution of equations: the Newton–Raphson method

In the last chapter we saw how to find the zeros of a quadratic function by means of an algebraic method. For most functions, however, this is impossible and we can only find their zeros approximately by a numerical method.

Suppose that the given function is $f(x)$ and we wish to solve the equation

$$f(x) = 0$$

Let us also suppose (as we can perhaps discover from a graph) that the equation has a root in the neighbourhood of a particular value, $x = a$.

If the exact root is $x = a + h$, then $f(a + h) = 0$

But $$\frac{f(a+h) - f(a)}{h} \approx f'(a)$$

So $f(a + h) - f(a) \approx hf'(a)$

$\Rightarrow \quad -f(a) \approx hf'(a)$

$$\Rightarrow \quad h \approx -\frac{f(a)}{f'(a)}$$

We therefore expect

$$\boxed{x = a - \frac{f(a)}{f'(a)}}$$

to be a more accurate value of the root and can then calculate the root by means of successive adjustments.

Example

$f(x) = x^3 - 4x - 1$

We can see from a sketch graph that $x^3 - 4x - 1 = 0$ has a root near $x = 2$.

Let us call the successive adjustments h_1, h_2, etc.

Now $f(x) = x^3 - 4x - 1$ and $f(2) = -1$

$\Rightarrow$ $f'(x) = 3x^2 - 4$ and $f'(2) = 8$

If the exact root is $2 + h_1$,

then $h_1 \approx -\dfrac{f(2)}{f'(2)} = +\dfrac{1}{8} = 0.125$

So the next approximation is 2.125

If the exact root is $2.125 + h_2$,

then $h_2 \approx -\dfrac{f(2.125)}{f'(2.125)} = -\dfrac{0.092}{9.545} \approx -0.009$

Hence we should expect a closer approximation to be $x = 2.116$

The basis of the method can be conveniently illustrated in a graph:

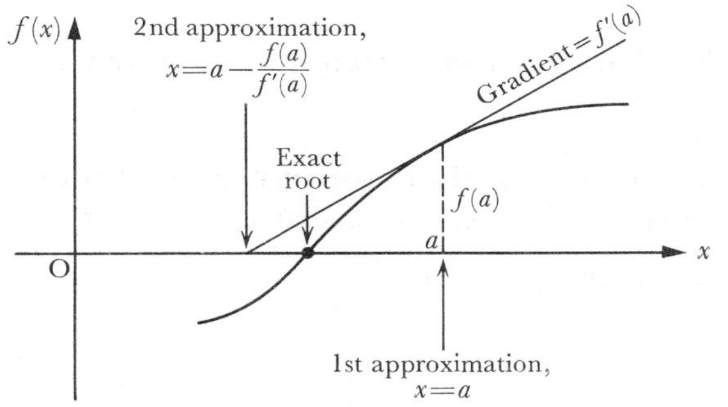

Exercise 3.11b

1 Apply the Newton–Raphson method to the equation x^3
starting with $x = 4$ as the first approximation. Prove

approximation is $x = 4.2$ and that the equation has a root between 4 and 4.2. (O.C.)

2 The equation $x^3 - 3x - 1 = 0$ has a root between 1.8 and 2.0. Find the value of this root to 3 decimal places. (O.C.)

3 Show that the equation

$$x^4 - 3x^2 - 3x + 1 = 0$$

has real roots between 0 and 1 and between 2 and 3. Find the larger of these roots correct to three significant figures and check that three figure accuracy has been achieved. (O.C.)

4 Use any method to locate the three roots of the equation

$$x^3 - 5x + 3 = 0$$

and evaluate the numerically smallest root to two significant figures. (O.C.)

5 If k is sufficiently large, show that one root of the equation

$$x^{10} - kx + k = 0$$

is approximately $(k - 9)/(k - 10)$. (L.)

3.12 Further kinematics

Area under velocity–time curve

When we considered velocity–time curves, we were usually given the velocity as an algebraic function of time, and were able to find displacements simply by reversing the process of differentiation. Frequently, however, we are less fortunate and only know the velocity as a set of readings for different values of t. In such cases we must find a new procedure for the calculation of distances.

As a simple example, let us consider the case of a car which is in the hands of a learner driver and accelerates by a series of jolts:

for 2 s it is moving at 3 m s^{-1},
then for 4 s it is moving at 5 m s^{-1},
and for 3 s it is moving at 6 m s^{-1}.

The total distance travelled $= 3\times2 + 5\times4 + 6\times3 = 44$ m, and this can be represented by the area beneath the velocity–time graph

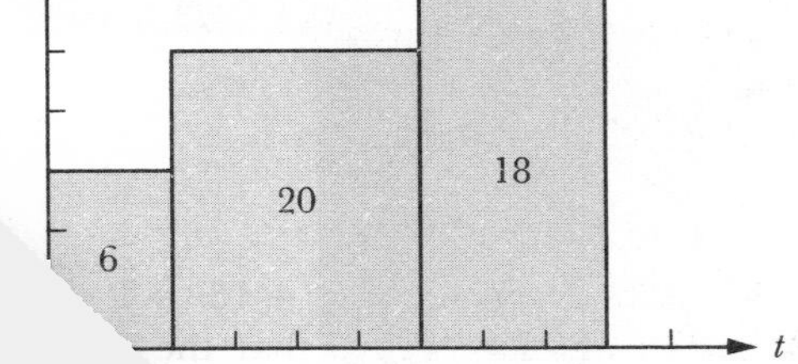

Clearly the distance travelled by a body which moves rather less jerkily can be found in just the same way, by dividing the area beneath its velocity–time graph into narrow strips and estimating their total area.

For instance, the velocity–time graph of an accelerating sports car is:

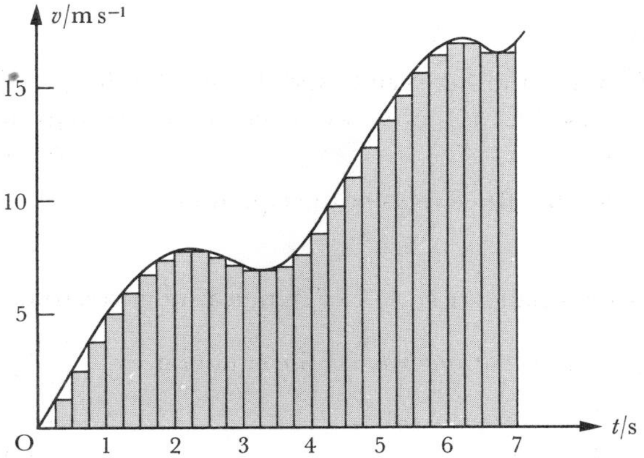

and the distance travelled in the first 7 seconds can be estimated by calculating the area of the rectangles shown. This will be a slight underestimate but, if greater accuracy is required, more rectangles can be drawn. Or, alternatively, a corresponding overestimate can be made by taking rectangles above, rather than below, the curves.

In any event, it is apparent that *the distance travelled in a certain interval of time is represented by the area beneath the corresponding section of the velocity–time curve.*

Moreover, just as distances (or changes of displacement) can be calculated from areas beneath velocity–time curves, so changes of velocity can be calculated from areas beneath acceleration-time curves.

Motion with constant acceleration

It is now instructive to use both methods for the particular case when the acceleration of a body moving in a straight line is constant.

A particle moves from the origin with velocity u, and has *constant* acceleration a. If its displacement and velocity after time t are s and v respectively, we shall find the relationships between s, u, v, a, and t.

Method 1

$$\frac{dv}{dt} = a, \text{ which is a constant}$$

$$\Rightarrow \quad v = at + c, \quad \text{where } c \text{ is a constant.}$$

When $t = 0$, $v = u$,

so $u = a0 + c \Rightarrow c = u$

$\Rightarrow v = u + at$

$\Rightarrow \dfrac{ds}{dt} = u + at$

$\Rightarrow s = ut + \frac{1}{2}at^2 + d$, where d is a constant.

When $t = 0$, $s = 0$,

so $0 = 0 + 0 + d \Rightarrow d = 0$

$\Rightarrow s = ut + \frac{1}{2}at^2$

Eliminating a: $s = \frac{1}{2}(u + v)t$

Eliminating t: $v^2 = u^2 + 2as$

Eliminating u: $s = vt - \frac{1}{2}at^2$

Method 2

Since acceleration is constant, the velocity–time curve has constant gradient and so is a straight line

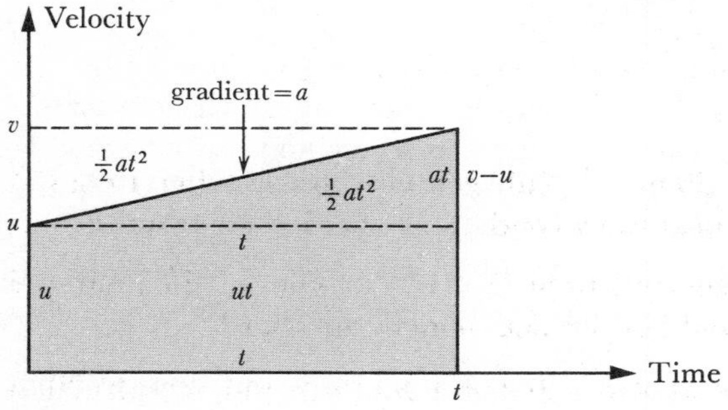

From the figure,

gradient of line $= a = \dfrac{v - u}{t} \Rightarrow v = u + at$

area under line $= s = \frac{1}{2}(u + v)t = ut + \frac{1}{2}at^2 = vt - \frac{1}{2}at^2$

Eliminating t from any pair of equations:

$v^2 = u^2 + 2as$

So both methods lead to the five relationships:

$$
\begin{aligned}
v &= u + at \\
s &= \tfrac{1}{2}(u + v)t \\
s &= ut + \tfrac{1}{2}at^2 \\
s &= vt - \tfrac{1}{2}at^2 \\
v^2 &= u^2 + 2as
\end{aligned}
$$

It must, however, be emphasised that *these relationships can only be used when it has already been established that the acceleration is constant.* This is not usually so, except in certain special (but important) cases, such as the free fall of a stone under gravity, and even then only under certain limitations.

Example

A stone dropped from a high cliff is falling with speed 20 m s^{-1} and acceleration 10 m s^{-2}. How long does it take to drop another 160 m and what is then its speed?

In this case, $u = 20, \quad a = 10, \quad$ and $\quad s = 160$

So $\quad s = ut + \frac{1}{2}at^2$

$\Rightarrow \quad 160 = 20t + 5t^2$

$\Rightarrow \quad t^2 + 4t - 32 = 0$

$\Rightarrow \quad (t + 8)(t - 4) = 0 \quad \Rightarrow \quad t = 4$ or -8

So the stone takes another 4 s to reach the required level (which it might also have passed 8 s previously had it been catapulted upwards).

Also $\quad v = u + at$

$\Rightarrow \quad v = 20 + 10 \times 4 = 60$

So its speed is then 60 m s^{-1}.

Exercise 3.12

1 A stone is falling at 30 m s^{-1} with constant acceleration 10 m s^{-2}. How far does it go in the next two seconds, and what is then its speed?

2 A train accelerates uniformly from 12 m s^{-1} to 36 m s^{-1} in 2 minutes. What is its acceleration and how far does it go in this time?

3 A cyclist travelling at 20 m s^{-1} brakes uniformly and stops in 50 m. What is its rate of deceleration and how long does this take?

4 A car travelling at 20 m s^{-1} accelerates uniformly so that in the next 2 s it covers 42 m. What is its acceleration and its final speed?

5 The driver of a train, travelling at 80 km h^{-1} on a straight level track, sees a signal against him at a distance of 1 km and, putting on the brakes, comes to rest at the signal. He stops for one minute and then resumes the journey, attaining the original speed in a distance of 4 km. Assuming that retardation and acceleration are uniform, find how much time has been lost owing to the stoppage. (O.C.)

6 A car approaches a corner at a speed of 100 km h^{-1} and has to reduce this speed to 50 km h^{-1} in a distance of 200 m in order to take the corner. Find the requisite deceleration in m s^{-2}.

After the corner the car regains its former speed in 20 seconds. Find the distance it travels in so doing.

[Assume that both acceleration and deceleration are uniform.]

7 A train, slowing down with uniform retardation, passes three telegraph posts spaced at intervals of 100 m. It takes 8 seconds between the first and second posts, and 12 seconds between the second and third. Find the retardation of the train, and the distance beyond the third post of the point where the train comes to rest.

8 Given that the maximum acceleration of a lift is 4 m s^{-2} and the maximum deceleration is 5 m s^{-2}, find the minimum time for a journey of 30 m

(*i*) if the maximum speed is 5 m s^{-1};
(*ii*) if there is no restriction on speed.

Miscellaneous problems 3

1 Discuss the existence of the derivative at $x = 0$ of the function

$$f: x \mapsto x\,|x|,$$

showing that you have considered both positive and negative values as x tends to 0.

State with reasons which of the following are true and which are false:

(*i*) f is an even function;
(*ii*) f is continuous at $x = 0$;
(*iii*) f is differentiable at $x = 0$.

Sketch a graph of the function f and find the derived function f'.
Is f' differentiable at $x = 0$? (S.M.P.)

2 Show that the curve $y = x^4 - 6a^2x^2$ has two points of inflection H and K. Prove that the tangents at H and K meet on the y-axis at the point $(0, 3a^4)$. Find the turning points of the curve, and sketch it. (O.C.)

3 A cylinder of radius x is inscribed in a cone of height h and base radius a, the two figures having a common axis of symmetry. Find an expression for the height of the cylinder in terms of x, a, and h.

Prove that the cylinder of greatest volume that can be so inscribed has a volume $\frac{4}{9}$ of the volume of the cone. (O.C.)

4 A cylindrical tin of radius r and height h is constructed from a fixed area of sheeting so as to have maximum capacity.

Find the relationship between h and r

(*i*) if the tin has a base but no lid;
(*ii*) if it has both base and lid;
(*iii*) if it has a slip-on lid with a flange of depth a.

5 Show that if a is an approximation to $2^{1/3}$, then

$$\frac{2}{3}\left(a + \frac{1}{a^2}\right) \text{ should be a better approximation.}$$

Hence determine $2^{1/3}$ to four places of decimals. (C.S.)

6 Plot a graph of the function

$$y = \frac{x(x - 4)}{\sqrt{(x^2 + 3)}}$$

for values of x between -1 and 5 and show that the graph has only one turning point. Find the value of x corresponding to this turning point, correct to two places of decimals. (O.C.)

7 Sketch the function

$$f(x) = |x|^{-1}$$

and find the radius of the circle which touches both its branches and also the x-axis at the origin. (C.S.)

8 Without actually plotting to scale, find out all you can about the curves

$$y^2 = \frac{x}{1 + x^2} \quad \text{and} \quad y^2 = \frac{x}{1 - x^2}$$

and make sketches of them. (O.S.)

9 A slice of bread is in the shape of a rectangle surmounted by a semi-circle. Find the ratio of its total height to width if a given area is to have minimum crust.

4

Areas: integration

One of the first uses of geometry was for surveying and the calculation of areas of land. This was clearly simplest with squares and rectangles, but the Greeks were also familiar with means of finding areas of triangles. When, however, an area was wholly or partly bounded by a curve, the problem became more difficult, and they were successful only in a limited number of cases.

The simplest of these was, of course, the circle. The fact that

$$\frac{\text{area}}{(\text{radius})^2}$$

remains constant was well known, and the evaluation of this constant proceeded with increasing accuracy. It is still instructive to carry out such an evaluation of π by drawing a circle of radius 10 cm on graph paper and counting squares. What percentage accuracy would you expect, and what do you achieve? Can it be improved by averaging a number of similar results?

In the first half of the seventeenth century, mathematicians became increasingly concerned with areas bounded by other kinds of curve. Kepler's Second Law of Planetary Motion (1609) stated that the speed of a

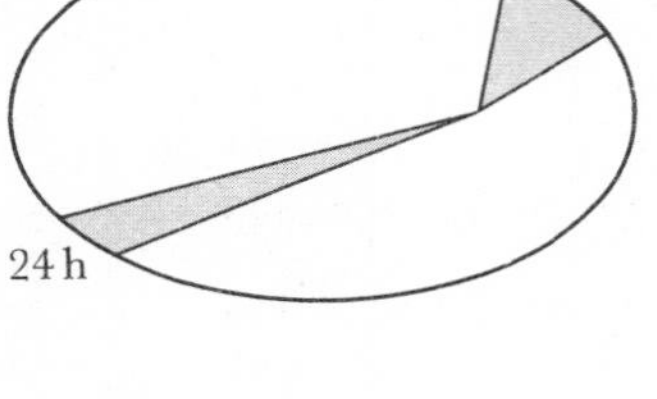

planet in its orbit varies in such a way that a line joining the planet to the Sun sweeps out equal *areas* in equal periods of time. This law and, more simply, the calculation of areas lying beneath parabolas and similar curves were soon to lead Newton to another use for his newly discovered Method of Fluxions, or *calculus* as it was later to be called.

4.1 The area beneath a curve: the integral

Our first need is for a notation with which to describe an area bounded by curves, and we begin by considering that beneath the graph of a function.

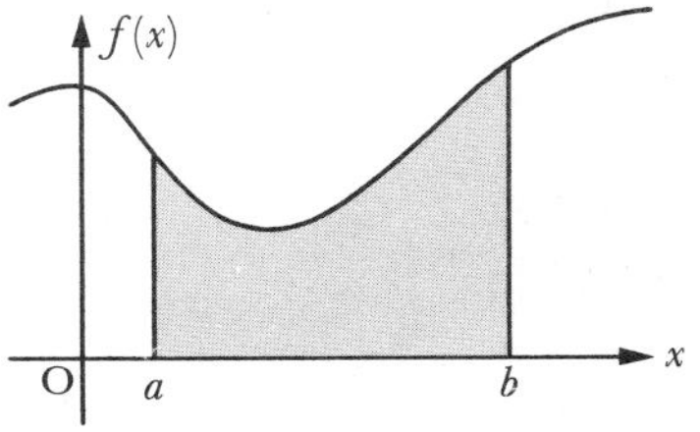

If $f(x)$ is a continuous function which is positive over the domain $a \leqslant x \leqslant b$, we call the area beneath its curve the *integral* of $f(x)$ from $x = a$ to $x = b$, and write it

$$\int_a^b f(x)\,\mathrm{d}x$$

The symbol $\int$ arose as an elongated form of the letter S (for Sum), simply because splitting the area into approximately rectangular strips, each of small width δx and area $f(x)\,\delta x$, leads us to:

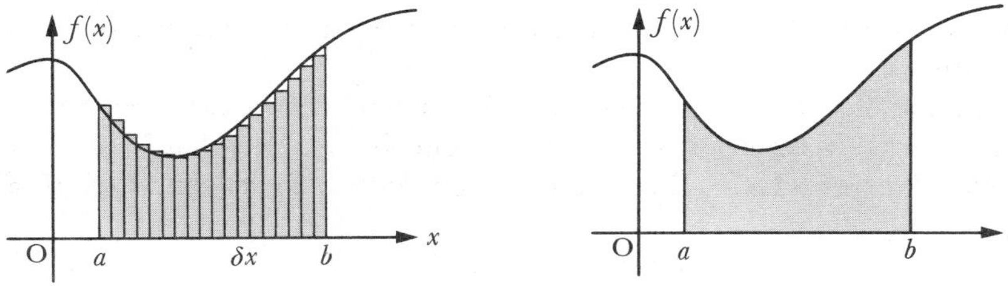

$$\text{Area} \approx \underset{x=a}{\overset{x=b}{\text{Sum}}}\, f(x)\,\delta x \qquad \text{and so to the notation:} \qquad \text{Area} = \int_a^b f(x)\,\mathrm{d}x$$

Example

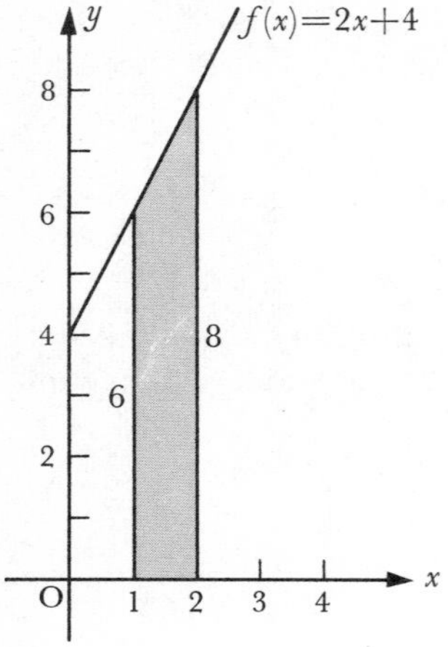

$$\int_1^2 (2x + 4)\, dx$$

indicates the shaded area, which is clearly 7 square units.

So $\displaystyle\int_1^2 (2x + 4)\, dx = 7$

It is apparent that the letter x is simply taken from the name of the axis and has no influence on the result; for this reason it is sometimes called a *dummy*, and it is equally valid to write

$$\int_1^2 (2t + 4)\, dt = 7 \quad \text{or} \quad \int_1^2 (2\theta + 4)\, d\theta = 7$$

Exercise 4.1

Sketch and evaluate the following integrals:

1 $\displaystyle\int_0^5 7\, dx$ **2** $\displaystyle\int_0^{10} 5x\, dx$ **3** $\displaystyle\int_2^4 (3x + 2)\, dx$

4 $\displaystyle\int_3^6 \frac{7 - t}{2}\, dt$ **5** $\displaystyle\int_{-1}^2 (4 - u)\, du$ **6** $\displaystyle\int_{-2}^2 |v|\, dv$

4.2 The two staircases

Each of the above examples and exercises concerned the area beneath a straight line. They were simply to make the symbol $\int$ rather more familiar, but we must now meet the challenge of a genuine curve:

Example

Find $\int_0^1 x^3 \, dx$

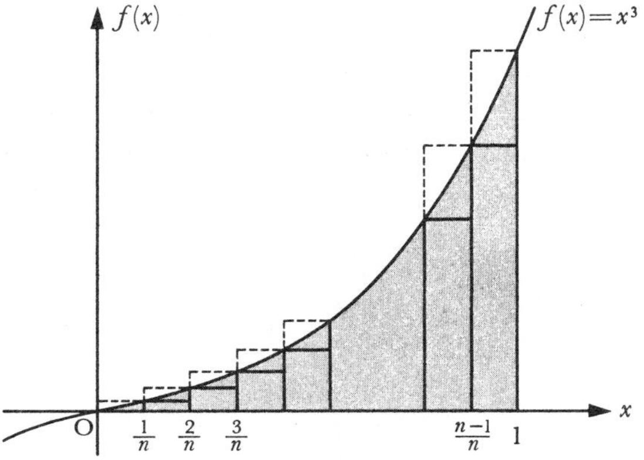

First we set up the two staircases shown, one above the given curve and the other below it, and each consisting of n steps of width $1/n$. For convenience, we shall call the required area A. It is clearly greater than the area beneath the lower staircase and less than the area beneath the upper staircase.

But the heights above the x-axis of the several steps are:

$$\left(\frac{1}{n}\right)^3, \left(\frac{2}{n}\right)^3, \left(\frac{3}{n}\right)^3, \ldots \left(\frac{n-1}{n}\right)^3, 1.$$

So $\frac{1}{n}\left(\frac{1}{n}\right)^3 + \frac{1}{n}\left(\frac{2}{n}\right)^3 + \cdots + \frac{1}{n}\left(\frac{n-1}{n}\right)^3$

$$< A < \frac{1}{n}\left(\frac{1}{n}\right)^3 + \frac{1}{n}\left(\frac{2}{n}\right)^3 + \cdots + \frac{1}{n}\left(\frac{n-1}{n}\right)^3 + \frac{1}{n}1^3$$

$$\Rightarrow \quad \frac{1}{n^4}\{1^3 + 2^3 + 3^3 + \cdots + (n-1)^3\}$$

$$< A < \frac{1}{n^4}\{1^3 + 2^3 + \cdots + (n-1)^3 + n^3\}.$$

Now it can be shown (see section 6.4) that the sum of the cubes of the first n positive integers is $\frac{1}{4}n^2(n+1)^2$.

(This is easily verified for particular cases, such as $1^3 + 2^3$, $1^3 + 2^3 + 3^3$, $1^3 + 2^3 + 3^3 + 4^3$, etc.)

So the sum of the first $n - 1$ positive integers will be $\frac{1}{4}(n - 1)^2 n^2$

$$\text{Hence} \quad \frac{1}{n^4}\frac{1}{4}n^2 (n - 1)^2 < A < \frac{1}{n^4}\frac{1}{4}n^2 (n + 1)^2$$

$$\Rightarrow \quad \frac{(n - 1)^2}{4n^2} < A < \frac{(n + 1)^2}{4n^2}$$

$$\Rightarrow \quad \frac{1}{4}\left(1 - \frac{1}{n}\right)^2 < A < \frac{1}{4}\left(1 + \frac{1}{n}\right)^2$$

We now make the two staircases draw closer and closer to the original curve simply by reducing the width of each step and by making many more of them, i.e., by increasing the value of n.

Now as n grows larger and larger, both $\frac{1}{4}(1 - 1/n)^2$ and $\frac{1}{4}(1 + 1/n)^2$ approach $\frac{1}{4}$.

But A always lies between them,

$$\text{so} \quad A = \tfrac{1}{4}$$

$$\text{Hence} \quad \int_0^1 x^3 \, dx = \tfrac{1}{4}$$

Exercise 4.2

The sum of the squares of the first n positive integers is $\frac{1}{6} n (n + 1) (2n + 1)$.

Check this for $n = 10$ and $n = 20$, and then use it to calculate $\int_0^1 x^2 \, dx$.

4.3 The generation of area

The reader may have been irritated by this last section. First, he may suspect our good fortune in having a ready-made formula for $1^3 + 2^3 + 3^3 + \cdots + n^3$ and, though glad to accept it, he might well wonder whether the choice of example was entirely accidental. An attempt to repeat the process for the curve $f(x) = \sqrt{x}$, and to find a corresponding formula for $\sqrt{1} + \sqrt{2} + \sqrt{3} + \cdots + \sqrt{n}$, would probably deepen his suspicion.

Secondly (a more creative irritation), on seeing the simplicity of the final result, he might wonder whether some key can be found to unlock the problem more swiftly.

In the search for such a key, he might recall velocity–time curves in section 3.12, and remember that distances were obtained *both* by finding areas beneath curves *and* by reversing the process of differentiation. Might it not be possible that these are the same process, and that in order to find areas beneath curves we simply have to reverse the process of differentiation? At very least, we can begin by looking at area under a curve in a more dynamic way, not as something presented once and for all as a single snapshot, but rather as a quantity which is gradually growing as our eyes sweep along the curve.

Let us start by looking again at the problem of 4.1:

Example 1

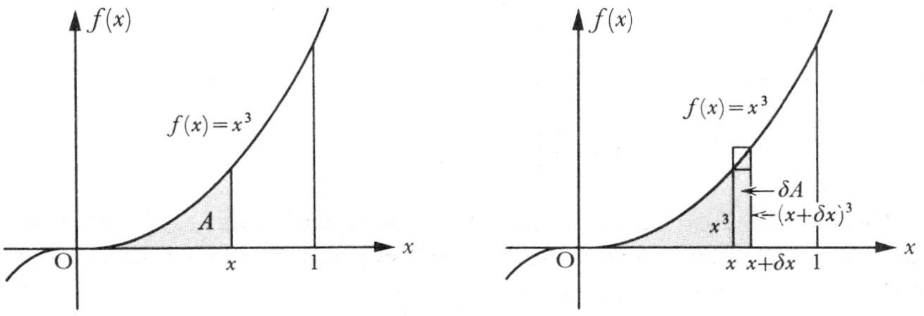

(i) (ii)

Let A be the shaded area shown in Fig. (*i*) and let us see how it increases as its boundary gradually sweeps to the right.

If, as in Fig. (*ii*), x increases by δx, A will increase by δA and this additional area lies between two rectangles of height x^3 and $(x + \delta x)^3$.

So $\quad x^3\delta x < \delta A < (x + \delta x)^3\,\delta x$

$$\Rightarrow \qquad x^3 < \frac{\delta A}{\delta x} < (x + \delta x)^3$$

Now as $\delta x \to 0$, $(x + \delta x)^3 \to x^3$.

$$\Rightarrow \qquad \frac{\mathrm{d}A}{\mathrm{d}x} = x^3$$

$\Rightarrow \qquad A = \frac{1}{4}x^4 + c$, for some value of c.

But when $x = 0$, $A = 0$

$\Rightarrow \quad 0 = 0 + c \quad \Rightarrow \quad c = 0$

So $\quad A = \frac{1}{4}x^4$

We have, therefore, by this method discovered the area beneath the curve *up to any point*.

In particular, taking $x = 1$, we confirm that the required area between $x = 0$ and $x = 1$ is $\frac{1}{4} \times 1^4 = \frac{1}{4}$.

Example 2

Let us now consider the problem mentioned in 4.1 of finding the area beneath the curve $f(x) = \sqrt{x}$, and for the sake of variety we shall take this between the limits $x = 1$ and $x = 2$.

Let A be the shaded area of Fig. (*i*).

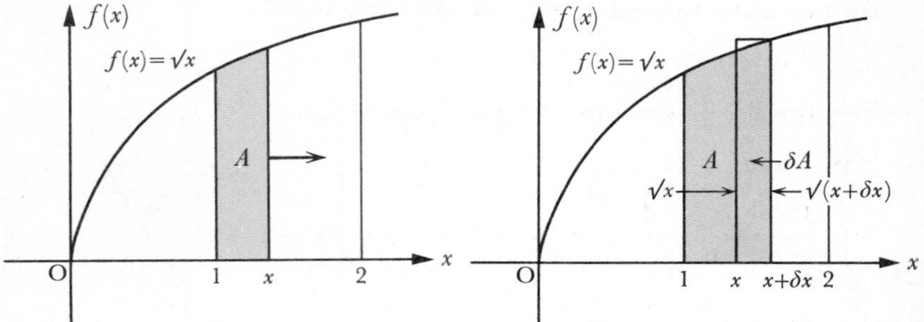

(i) (ii)

As before, we see that

$$\delta A \text{ lies between } \sqrt{x}\delta x \text{ and } \sqrt{(x+\delta x)}\delta x$$

$$\Rightarrow \quad \frac{\delta A}{\delta x} \text{ lies between } \sqrt{x} \text{ and } \sqrt{(x+\delta x)}$$

Again letting $\delta x \to 0$, we see that

$$\frac{\mathrm{d}A}{\mathrm{d}x} = \sqrt{x} = x^{1/2}$$

$$\Rightarrow \quad A = \tfrac{2}{3}x^{3/2} + c, \text{ for some value of } c$$

But the generation of A begins when $x = 1$

So $\quad 0 = \frac{2}{3}1^{3/2} + c \quad \Rightarrow \quad c = -\frac{2}{3}$

$\Rightarrow \quad A = \frac{2}{3}x^{3/2} - \frac{2}{3}$

In particular, taking $x = 2$ we see that the required area is

$$A = \tfrac{2}{3}2^{3/2} - \tfrac{2}{3} = \tfrac{2}{3}(2\sqrt{2} - 1) \approx 1.22$$

So, compared with the method of the last section, our new approach is not only swifter, but also more powerful.

Further, if the area A is denoted by the function $A(x)$, we note that in each case

$A'(x) = f(x)$, and so are led to:

4·4 The fundamental theorem of calculus

If $f(x)$ is continuous, and $A(x)$ is the area developed beneath the curve $f(x)$ from a starting line $x = a$, then $A'(x) = f(x)$.

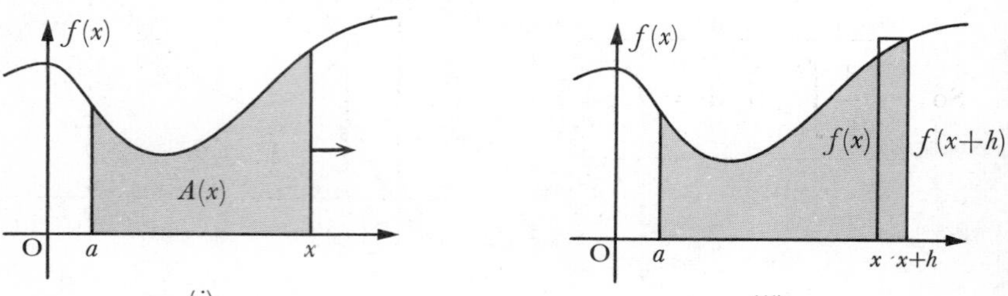

(i) (ii)

When x is increased to $x + h$, the new area added is

$$A(x + h) - A(x)$$

Now this lies between the two rectangles of areas

$$hf(x) \quad \text{and} \quad hf(x + h)$$

$$\Rightarrow \quad \frac{A(x + h) - A(x)}{h} \text{ lies between } f(x) \text{ and } f(x + h).$$

When $h \to 0$, $\dfrac{A(x + h) - A(x)}{h} \to A'(x)$,

and as $f(x + h)$ squeezes closer and closer to $f(x)$, it is clear that

$$\boxed{A'(x) = f(x)}$$

In the language of integrals, the first temptation is to write this result as

$$\frac{\mathrm{d}}{\mathrm{d}x}\int_a^x f(x)\,\mathrm{d}x = f(x)$$

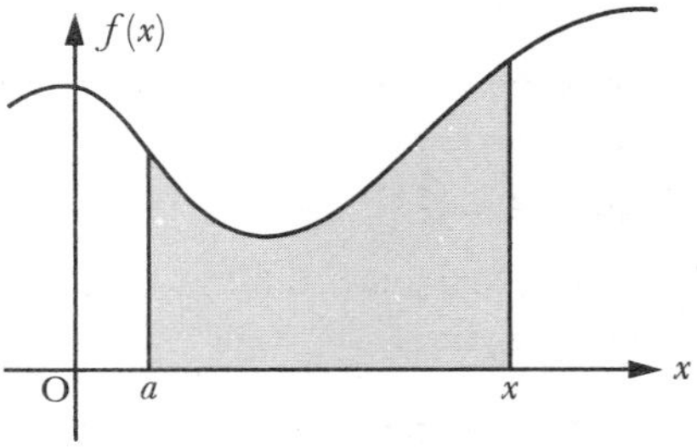

But it will be seen that in this equation, as in the corresponding graph, x is being used in two senses: firstly, as a variable (or name of the x-axis), and secondly, as a particular value of this variable (or particular x-coordinate) at the right-hand boundary of the area. This confusion is very easily avoided by renaming the general variable (or coordinate) as t:

So $$\boxed{\frac{\mathrm{d}}{\mathrm{d}x}\int_a^x f(t)\,\mathrm{d}t = f(x)}$$

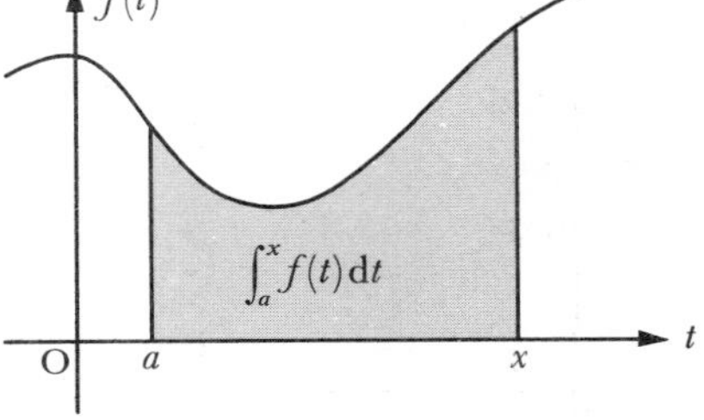

which is the usual form of the *fundamental theorem of calculus*.

The significance of the result is immediately clear: that, as was foreshadowed in the last section, the process of finding the area beneath the curve of a given function $f(x)$ is that of discovering another function which, *when differentiated*, yields the given function; i.e., finding a *primitive*, as we called it in the last chapter.

Such a function is usually written without any limits, as

$$\int f(x)\,\mathrm{d}x$$

So $\displaystyle\int x^3\,\mathrm{d}x = \tfrac{1}{4}x^4, \quad \tfrac{1}{4}x^4 + 3, \quad \tfrac{1}{4}x^4 - 2\tfrac{1}{2}$, etc.

More generally,

$$\int x^3\,\mathrm{d}x = \tfrac{1}{4}x^4 + c, \quad \text{where } c \text{ is a constant.}$$

This is usually known as the *indefinite integral* and the constant c, which can take any value, is called an *arbitrary constant*.

Because of the importance of indefinite integrals, we must first gain some familiarity with them.

Example 1

$$\int x\sqrt{x}\,\mathrm{d}x = \int x^{3/2}\,\mathrm{d}x = \tfrac{2}{5}x^{5/2} + c$$

$$\left[\text{Check: } \frac{\mathrm{d}}{\mathrm{d}x}(\tfrac{2}{5}x^{5/2} + c) = \tfrac{2}{5} \times \tfrac{5}{2}x^{3/2} + 0 = x^{3/2}\right]$$

Example 2

$$\int \frac{3 + t^2}{t^2}\,\mathrm{d}t = \int \left(\frac{3}{t^2} + 1\right)\mathrm{d}t = \int (3t^{-2} + 1)\,\mathrm{d}t$$

$$= -3t^{-1} + t + c$$

$$= -\frac{3}{t} + t + c$$

$$\left[\text{Check: } \frac{\mathrm{d}}{\mathrm{d}t}\left(-\frac{3}{t} + t + c\right) = \frac{3}{t^2} + 1\right]$$

Exercise 4·4

Find the following integrals:

1 (*i*) $\displaystyle\int x^2\,\mathrm{d}x$; (*ii*) $\displaystyle\int (3x^2 - 4x + 2)\,\mathrm{d}x$;

(*iii*) $\displaystyle\int (x^2 - 2x^3)\,\mathrm{d}x$; (*iv*) $\displaystyle\int (x^3 - 2x^2)\,\mathrm{d}x$.

2 (*i*) $\int (t^2 - 2t + 3)\,dt$; (*ii*) $\int t(t + 5)\,dt$;

(*iii*) $\int (t + 1)(t + 2)\,dt$; (*iv*) $\int (2t + 1)^2\,dt$.

3 (*i*) $\int u(u^4 + 1)\,du$; (*ii*) $\int (u^2 + 1)^2\,du$;

(*iii*) $\int 2u^2(u^3 - 1)\,du$; (*iv*) $\int (u^4 + 2)^2\,du$.

4 (*i*) $\int x^2\sqrt{x}\,dx$; (*ii*) $\int \sqrt[3]{x}\,dx$;

(*iii*) $\int (x + 1)\sqrt{x}\,dx$; (*iv*) $\int (\sqrt{x} + 3)^2\,dx$.

5 (*i*) $\int \frac{3}{x^2}\,dx$; (*ii*) $\int \frac{x + 3}{x^4}\,dx$;

(*iii*) $\int \frac{x + 1}{\sqrt{x}}\,dx$; (*iv*) $\int \frac{\sqrt{x} + 5}{x^2}\,dx$.

6 (*i*) $\int (ax + b)\,dx$; (*ii*) $\int (ax^2 + bx + c)\,dx$;

(*iii*) $\int (a\sqrt{x} + b)\,dx$; (*iv*) $\int \frac{ax + b}{\sqrt{x}}\,dx$.

7 $\int x^n\,dx$, where n is a constant.

Is your result true, without exception, for all values of n?

8 $\int \sqrt[n]{u}\,du$.

4.5 The calculation of definite integrals

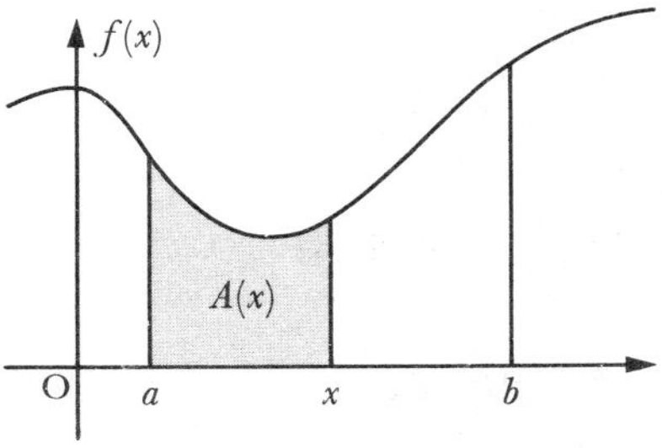

Let us now return to the problem of finding $\int_a^b f(x)\,\mathrm{d}x$, i.e., the area under the curve $f(x)$ between $x = a$ and $x = b$.

Let $A(x)$ be the area shown.

Then $A'(x) = f(x)$

Now suppose that the indefinite integral of $f(x)$ is $F(x) + c$.

Then, for some value of c, $A(x) = F(x) + c$.

Putting $x = b$, required area $= A(b) = F(b) + c$

Putting $x = a$, $0 = A(a) = F(a) + c$

Subtracting, we see that

required area $= F(b) - F(a)$,

which is usually written:

$$\int_a^b f(x)\,\mathrm{d}x = [F(x)]_a^b$$

So, for instance, we can now write:

$$\int_1^2 x^4\,\mathrm{d}x = \left[\frac{x^5}{5}\right]_1^2 = \tfrac{32}{5} - \tfrac{1}{5} = 6\tfrac{1}{5}$$

Again we see that the value of $\int_a^b f(x)\,\mathrm{d}x$ depends only on the function f which gives rise to F, and on the limits of integration a and b. The integral does not depend on x, and could equally well be written

$$\int_a^b f(t)\,\mathrm{d}t \quad \text{or} \quad \int_a^b f(\theta)\,\mathrm{d}\theta$$

Further, we notice that integrals (as we would expect from their definition as areas beneath curves) are additive:

$$\int_a^b f(x)\,\mathrm{d}x + \int_b^c f(x)\,\mathrm{d}x = \{F(b) - F(a)\} + \{F(c) - F(b)\}$$
$$= F(c) - F(a)$$

So $$\int_a^b f(x)\,\mathrm{d}x + \int_b^c f(x)\,\mathrm{d}x = \int_a^c f(x)\,\mathrm{d}x$$

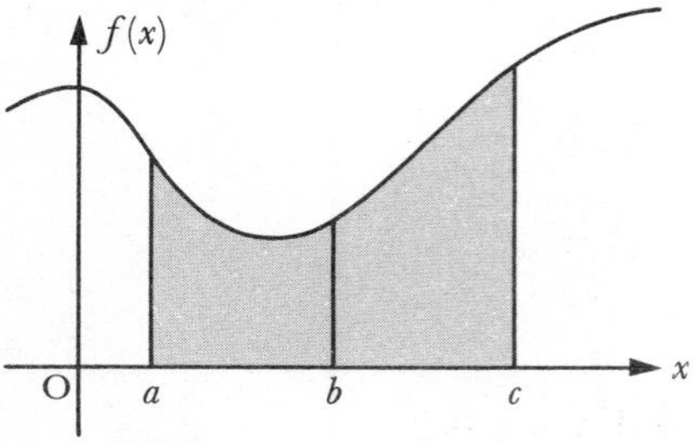

Lastly, $\int_b^a f(x)\,dx = F(a) - F(b)$.

So $\int_b^a f(x)\,dx = -\int_a^b f(x)\,dx$.

The examples of 4.3 can now be abbreviated:

Example 1

To find the area beneath $f(x) = x^3$ between $x = 0$ and $x = 1$:

$$\text{Required area} = \int_0^1 x^3\,dx$$
$$= \left[\tfrac{1}{4}x^4\right]_0^1$$
$$= \tfrac{1}{4}1^4 - \tfrac{1}{4}0^4 = \tfrac{1}{4}.$$

Example 2

To find the area beneath $f(x) = \sqrt{x}$ between $x = 1$ and $x = 2$:

$$\text{Required area} = \int_1^2 \sqrt{x}\,dx$$
$$= \int_1^2 x^{1/2}\,dx$$
$$= \left[\tfrac{2}{3}x^{3/2}\right]_1^2$$
$$= \tfrac{2}{3}2^{3/2} - \tfrac{2}{3}1^{3/2}$$
$$= \tfrac{2}{3}(2\sqrt{2} - 1).$$

Example 3

To find the area enclosed between $f(t) = 3 + 2t - t^2$ and the t-axis:

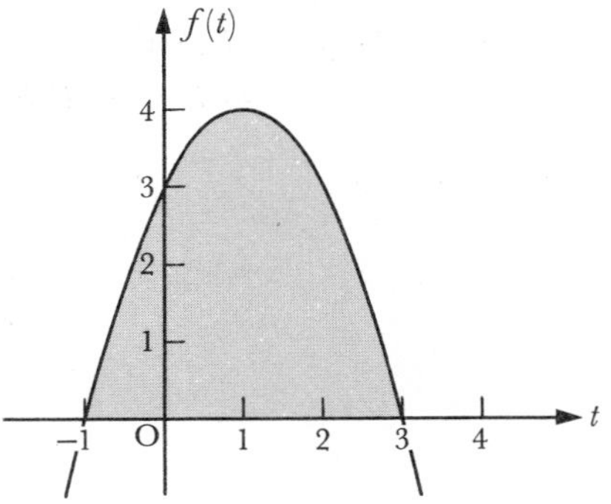

$3 + 2t - t^2 = 0 \quad \Rightarrow \quad t = -1$ or 3,

and so the curve cuts the t-axis at $t = -1$ and $t = 3$.

$$\text{Required area} = \int_{-1}^{3} (3 + 2t - t^2)\, dt$$
$$= \left[3t + t^2 - \tfrac{1}{3}t^3\right]_{-1}^{3}$$
$$= \{3\times3 + 3^2 - \tfrac{1}{3}\times3^3\} - \{3\times-1 + (-1)^2 - \tfrac{1}{3}\times(-1)^3\}$$
$$= 9 - (-1\tfrac{2}{3}) = 10\tfrac{2}{3}.$$

Exercise 4.5

1 Evaluate:

(*i*) $\int_{1}^{3} x^2\, dx$; (*ii*) $\int_{2}^{4} (3t^2 - 2t)\, dt$;

(*iii*) $\int_{-2}^{-1} u^4\, du$; (*iv*) $\int_{1}^{2} (x^3 - x)\, dx$.

2 Evaluate:

(*i*) $\int_{1}^{4} \sqrt{x}\, dx$; (*ii*) $\int_{1}^{8} \sqrt[3]{t}\, dt$;

(*iii*) $\int_{1}^{2} \frac{du}{u^2}$; (*iv*) $\int_{4}^{9} \frac{dx}{x\sqrt{x}}$.

3 Calculate the area

(*i*) beneath $f(x) = x^3$ and between $x = 1$ and $x = 2$;

(*ii*) beneath $f(x) = \frac{1}{x^2}$ and between $x = 2$ and $x = 3$;

(*iii*) beneath $f(x) = (x + 1)(x + 2)$ and between $x = 0$ and $x = 1$;

(*iv*) beneath $f(x) = x^2 - 2x$ and between $x = -1$ and $x = 0$.

4 Sketch graphs of the following functions and find the areas between them:

(*i*) $y = x^2$ and $y = 4x$;

(*ii*) $y = 4 - x^2$ and $y = 3$;

(*iii*) $y = x^2 - 1$ and $y = 1 - x^2$;

(*iv*) $y = \frac{1}{x^2}$, $y = x^2$ and $y = \frac{x^2}{16}$.

*4.6 Negative areas

We have, so far, always considered areas *beneath* curves and have restricted ourselves to curves, or sections of curves, which lie above the x-axis. But it is clearly possible to calculate definite integrals when a curve lies below the x-axis and we must now examine the meaning of such integrals.

Example

$$\int_0^1 (-x^3)\, dx = \left[-\frac{1}{4}x^4\right]_0^1 = -\frac{1}{4}1^4 - 0 = -\frac{1}{4}.$$

and it therefore appears that the integral still represents the shaded area:

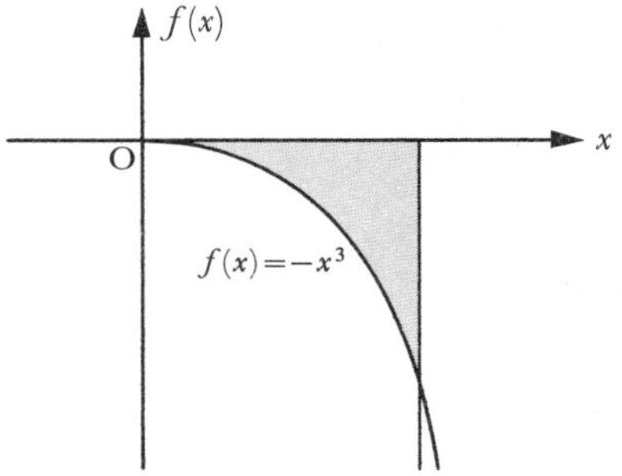

provided that such an area *below* the x-axis is reckoned as *negative*. This is also apparent from the relationship

$$A'(x) = f(x)$$

When $f(x) > 0$, $A(x)$ is clearly an increasing function and the area being generated is positive.

When $f(x) < 0$, $A(x)$ is a decreasing function and so we must regard the area being generated as negative:

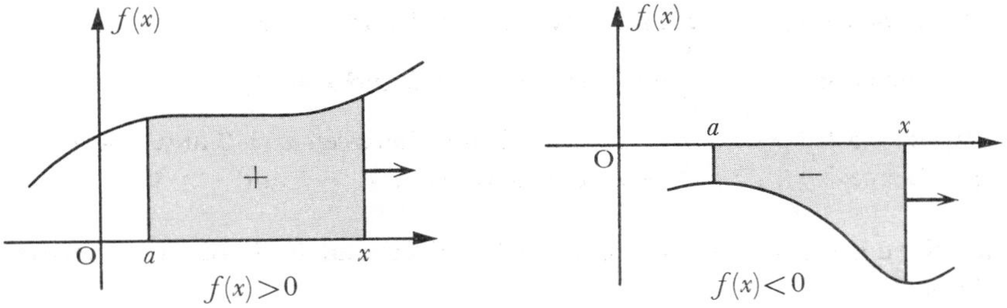

It is therefore clear that a definite integral $\int_a^b f(x)\, dx$ represents the *algebraic* sum of areas between a curve and the x-axis, those above the axis being counted positive and those below the axis being counted negative.

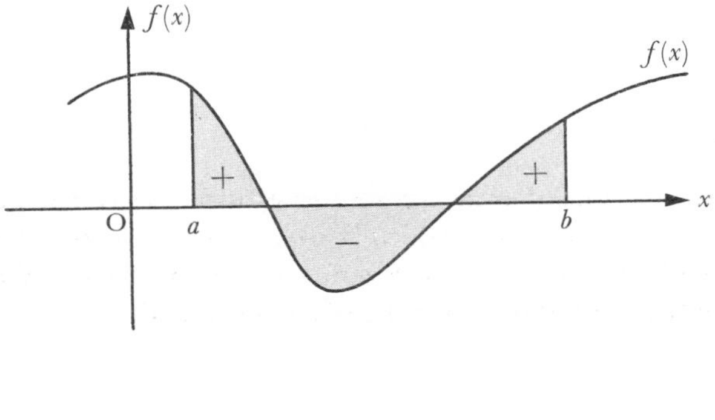

Example

What is the area between $f(x) = x^3$, the x axis and the lines $x = -1$ and $x = 2$?

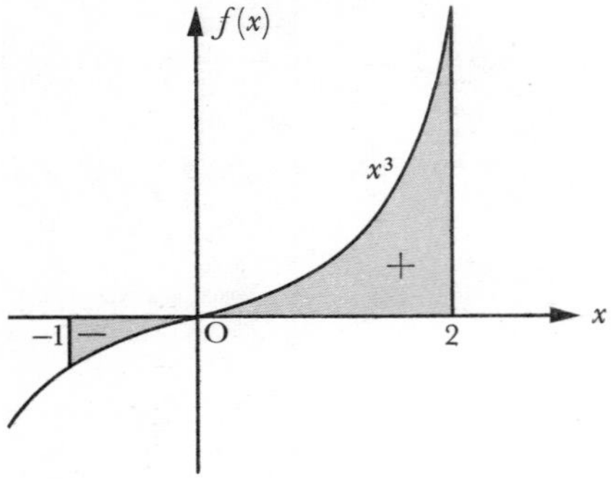

$$\int_{-1}^{2} x^3 \, dx = [\tfrac{1}{4} x^4]_{-1}^{2}$$
$$= \tfrac{1}{4}2^4 - \tfrac{1}{4}(-1)^4$$
$$= 4 - \tfrac{1}{4} = 3\tfrac{3}{4}$$

But this is the *algebraic* sum of the shaded areas.

If their *numerical* sum is required, the two parts must be calculated separately:

$$\int_{-1}^{0} x^3 \, dx = [\tfrac{1}{4} x^4]_{-1}^{0} = 0 - \tfrac{1}{4} = -\tfrac{1}{4}$$
$$\int_{0}^{2} x^3 \, dx = [\tfrac{1}{4} x^4]_{0}^{2} = 4$$

So the total *numerical* area is $4\tfrac{1}{4}$.

Exercise 4.6

1 Calculate: (*i*) $\int_0^4 (x^2 - 4x)\,dx$. (*ii*) $\int_2^6 (x^2 - 4x)\,dx$, and illustrate your answers on a sketch of $y = x^2 - 4x$.

2 (*i*) Calculate $\int_{-2}^{2} x \; (x^2 - 4)\,dx$ and illustrate your answer.
(*ii*) What is the numerical value of the area formed between $y = x(x^2 - 4)$ and the x-axis?

3 If a stone is catapulted upwards with velocity 30 m s^{-1}, its velocity v upwards after time t is given by

$v = 30 - 10t$

Calculate: (*i*) $\int_0^3 v\,dt$, (*ii*) $\int_3^6 v\,dt$, (*iii*) $\int_0^6 v\,dt$, (*iv*) $\int_0^8 v\,dt$, and show the corresponding areas on the graph of $v = 30 - 10t$.

4 Sketch the graphs of:

(*i*) $y = (x^2 - 1)(x^2 - 4)$, (*ii*) $y = x(x^2 - 1)(x^2 - 4)$,

and in each case calculate the total numerical area enclosed by the curve beneath the x-axis.

*4.7 Numerical integration

Faced with the problem of finding the area beneath a curve, we frequently
either (*i*) do not know an algebraic expression $f(x)$ for the curve,
or (*ii*) know $f(x)$ but are unable to integrate it,
or (*iii*) find it simpler, and sufficient for our purpose, to obtain an approximate numerical value.

We could, of course, simply plot the curve on graph-paper and count the number of squares beneath the required section; or we could (as in section 4.2) regard the curve as squeezed between two staircases. But more frequently we approximate to the curve by a succession of straight lines or a succession of parabolas, and from these approaches we derive two widely-used results, the Trapezium rule and Simpson's rule:

The Trapezium rule

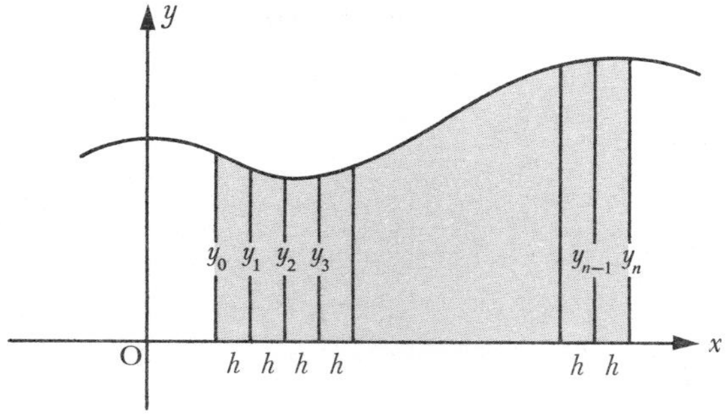

Divide the area into a number of parallel strips, each of width h, and let the ordinates of the curve at successive points be

$y_0, y_1, y_2, y_3 \dots y_n$

Area under curve $\approx$ Sum of the areas of trapezia

$$\approx h\frac{y_0 + y_1}{2} + h\frac{y_1 + y_2}{2} + h\frac{y_2 + y_3}{2} + \cdots + \cdots + h\frac{y_{n-1} + y_n}{2}$$

$$\approx h\left\{\tfrac{1}{2}y_0 + y_1 + y_2 + y_3 + \cdots + y_{n-1} + \tfrac{1}{2}y_n\right\}$$

$$\approx h\left\{\tfrac{1}{2}(y_0 + y_n) + (y_1 + y_2 + \cdots + y_{n-1})\right\}$$

So $\boxed{\text{Area} \approx \text{Strip width} \times \begin{Bmatrix}\text{Average of extreme ordinates} \\ + \text{ Sum of remaining ordinates}\end{Bmatrix}}$

Simpson's rule

Just as any *two* successive points can always be joined by a straight line $y = a + bx$, so any *three* successive points can always be joined by a parabola $y = a + bx + cx^2$. But before we can assemble a collection of areas beneath successive parabolas we first need to know the area beneath a single specimen.

Suppose that a curve has ordinates y_0, y_1, y_2 at points whose successive horizontal separations are h, and we take axes Ox, Oy as follows:

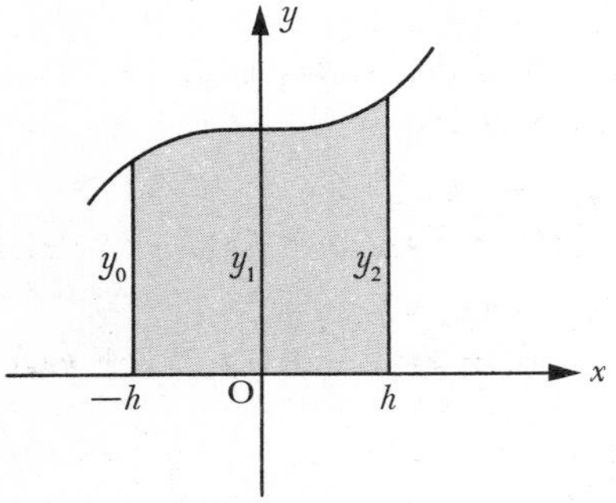

Let the parabolic section joining $(-h, y_0)$, $(0, y_1)$, (h, y_2) be

$$y = a + bx + cx^2$$

$$\begin{aligned}\text{Area under parabola} &= \int_{-h}^{h} (a + bx + cx^2)\,\mathrm{d}x \\ &= \left[ax + \tfrac{1}{2}bx^2 + \tfrac{1}{3}cx^3\right]_{-h}^{h} \\ &= 2ah + \tfrac{2}{3}ch^3 \\ &= \tfrac{1}{3}h(6a + 2ch^2)\end{aligned}$$

But since the parabola goes through $(-h, y_0)$, $(0, y_1)$, (h, y_2),

$$\left.\begin{aligned} a - bh + ch^2 &= y_0 \\ a \qquad\qquad &= y_1 \\ a + bh + ch^2 &= y_2 \end{aligned}\right\}$$

from which we see that

$$6a + 2ch^2 = y_0 + 4y_1 + y_2$$

So Area under parabola $= \frac{1}{3}h(y_0 + 4y_1 + y_2)$.

We can now divide our total area into an *even* number of strips ($2n$) and let successive heights of the curve be

$y_0, y_1, y_2, y_3, y_4 \ldots y_{2n}$.

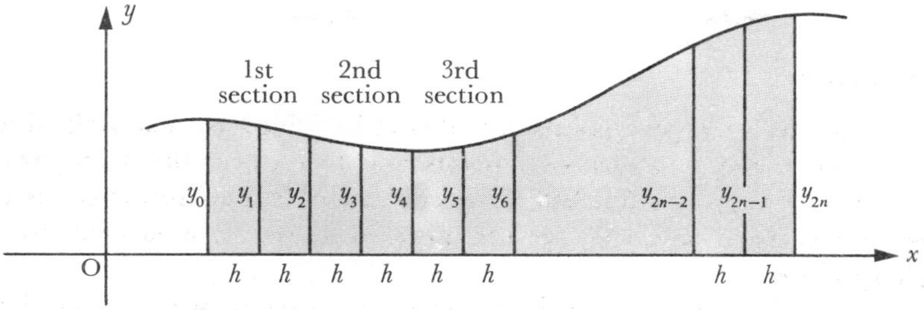

Approximating to the curve by a succession of parabolic sections, we see that

$$\text{Area under curve} \approx \tfrac{1}{3}h\,(y_0 + 4y_1 + y_2) + \tfrac{1}{3}h\,(y_2 + 4y_3 + y_4) + \cdots$$
$$\cdots + \tfrac{1}{3}h\,(y_{2n-2} + 4y_{2n-1} + y_{2n})$$
$$\approx \tfrac{1}{3}h\,\{(y_0 + 4y_1 + y_2) + (y_2 + 4y_3 + y_4) + \cdots$$
$$\cdots + (y_{2n-2} + 4y_{2n-1} + y_{2n})\}$$
$$\approx \tfrac{1}{3}h\,\{y_0 + 4y_1 + 2y_2 + 4y_3 + 2y_4 + \cdots$$
$$\cdots + 2y_{2n-2} + 4y_{2n-1} + y_{2n}\}$$

$$\approx \tfrac{1}{3}h \begin{Bmatrix} (y_0 + y_{2n}) \\ + 4(y_1 + y_3 + y_5 + \cdots + y_{2n-1}) \\ + 2(y_2 + y_4 + \cdots + y_{2n-2}) \end{Bmatrix}$$

This is Simpson's rule, that if a curve be divided into an *even* number of strips of equal width,

$$\text{Area beneath curve} \approx \tfrac{1}{3} \times \text{Strip width} \times \begin{Bmatrix} \text{Sum of end ordinates} \\ +4\times\text{Sum of odd ordinates} \\ +2\times\text{Sum of remaining} \\ \text{even ordinates} \end{Bmatrix}$$

Example

Use both the Trapezium rule and Simpson's rule to calculate

$$\int_0^6 \sqrt{(1 + x^2)}\,\mathrm{d}x$$

The values of $\sqrt{(1 + x^2)}$ at unit intervals can be tabulated, and the calculation set out as follows:

x	$1 + x^2$	$\sqrt{(1 + x^2)}$	Trapezium rule		Simpson's rule		
0	1	1.000	1.000		1.000		
1	2	1.414		1.414		1.414	
2	5	2.236		2.236			2.236
3	10	3.162		3.162		3.162	
4	17	4.123		4.123			4.123
5	26	5.099		5.099		5.099	
6	37	6.083	6.083		6.083		
			2)7.083	16.034	7.083	9.675	6.359
			3.5415		38.700	4	2
			16.034		12.718	38.700	12.718
			19.5755		3)58.501		
					19.500		

So, to 4 significant figures,

$$\int_0^6 \sqrt{(1 + x^2)}\,dx = \begin{cases} 19.58 \text{ by the Trapezium rule} \\ 19.50 \text{ by Simpson's rule} \end{cases}$$

[We shall later be able to show that its actual value to 4 significant figures is 19.50.]

Exercise 4·7

1 The depth of a river of width 80 m is measured at 10-m intervals of its cross-section as follows:

Distance (m)	0	10	20	30	40	50	60	70	80
Depth (m)	0	1.31	3.17	6.24	3.79	1.16	5.25	3.76	2.48

Estimate the area of this cross-section by (*i*) the Trapezium rule; (*ii*) Simpson's rule.

2 The speed of an accelerating car is recorded at second intervals as follows:

t/s	0	1	2	3	4	5	6	7	8	9	10
v/m s^{-1}	0	5.3	8.1	10.3	11.9	13.0	14.1	14.8	15.3	15.6	15.8

Estimate the distance it travels in 10 s using (*i*) the Trapezium rule; (*ii*) Simpson's rule.

3 The acceleration, at half-second intervals, of a stone dropped from the surface of a deep lake is given by

t/s	0	0.5	1.0	1.5	2.0	2.5	3.0
a/m s^{-2}	3.37	1.54	0.93	0.68	0.53	0.41	0.30

Estimate its speed after 3 seconds using (*i*) the Trapezium rule; (*ii*) Simpson's rule.

4 Estimate $\int_1^2 \frac{dx}{x}$ by means of (*i*) the Trapezium rule; (*ii*) Simpson's rule; using only two intervals.

Given that its correct value, to 4 decimal places, is 0.6931, calculate the percentage errors of each estimate.

5 Repeat the last question, using ten intervals.

6 A quadrant of a circle of radius 10 cm divided into ten parallel strips. Use (*i*) the Trapezium rule, and (*ii*) Simpson's rule, to calculate π.

7 It can be shown that

$$\int_0^1 \frac{dx}{1 + x^2} = \frac{\pi}{4}$$

Use Simpson's rule (10 strips) to calculate π.

8 Tabulate to 3 decimal places the values of the function $f(x) = \sqrt{(1 + x^2)}$ for values of x from 0 to 0.8 at intervals of 0.1. Use these values to estimate

$\int_0^{0.8} f(x)\, dx$ by Simpson's method,

(*i*) using all the ordinates, and
(*ii*) using only ordinates at intervals of 0.2.
Draw any conclusions you can about the accuracy of the results. (M.E.I.)

*4.8 Other summations

We now return to our observation in 4.1, that an integral is simply the limit of the sum of a large number of small quantities.

For example,

As $\delta x \to 0$

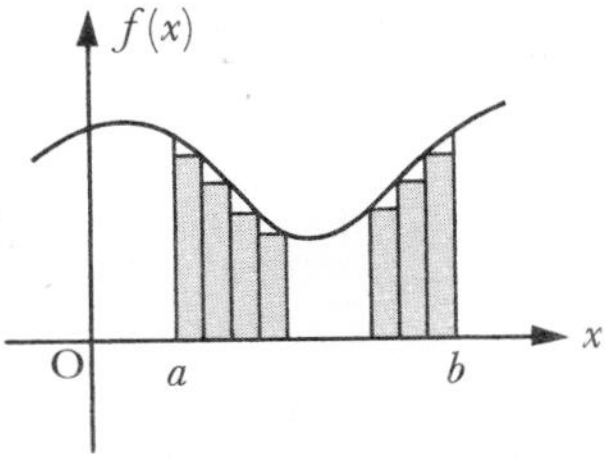

tends to

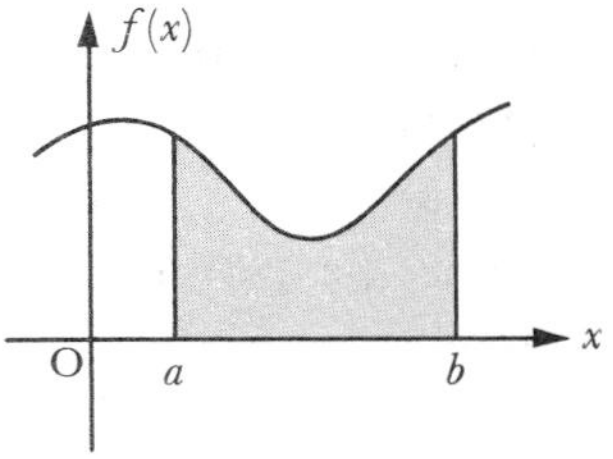

and
$$\underset{x=a}{\overset{x=b}{\text{Sum}}}\, f(x)\delta x \to \int_a^b f(x)\, dx.$$

This is an idea which has wide applications, and we shall usually denote such a sum by $\sum$ (Sigma, the Greek capital S).

So $\boxed{\sum f(x)\,\delta x \to \int_a^b f(x)\,dx.}$

Example 1

If $f(x) > 0$ when $a \leqslant x \leqslant b$, find the volume formed when the area between $y = f(x)$, $y = 0$, $x = a$ and $x = b$ is rotated about Ox.

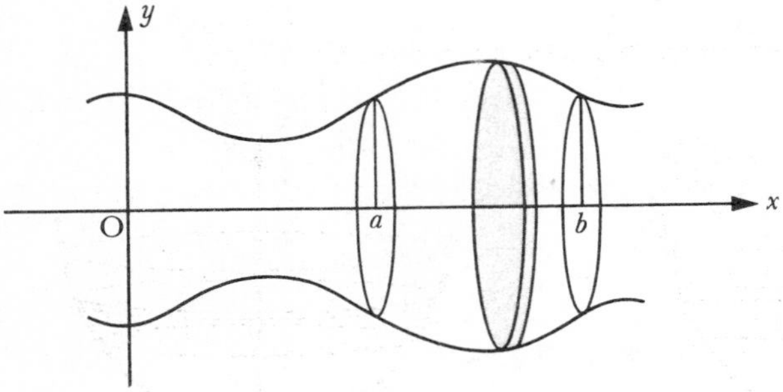

The volume of a typical disc of thickness δx is $\pi y^2 \delta x$.

So the sum of all these volumes is $\sum_{x=a}^{x=b} \pi y^2 \delta x$

$$\Rightarrow \boxed{\text{Required volume} = \int_a^b \pi y^2\,dx}$$

If, for example, the area bounded by $y = x^2$, $y = 0$, $x = 1$, and $x = 2$ is rotated about Ox, we see that

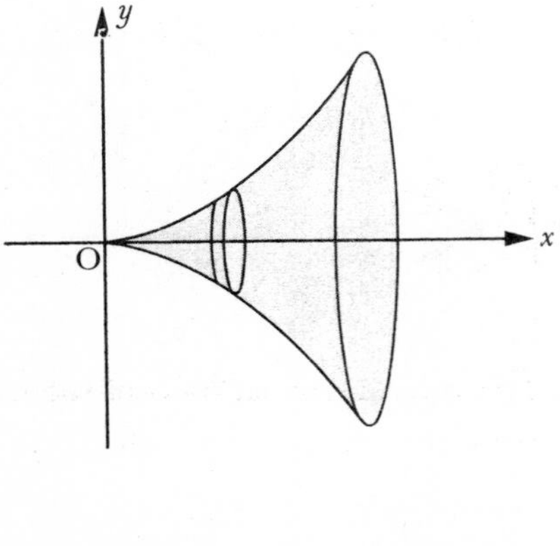

$$\text{Volume} = \int_1^2 \pi y^2 \, dx$$

$$= \int_1^2 \pi x^4 \, dx$$

$$= \left[\frac{1}{5}\pi x^5\right]_1^2 = \frac{32\pi}{5} - \frac{\pi}{5} = \frac{31\pi}{5}$$

Example 2

Find: (*i*) the area enclosed by

$y = x^2$, $y = 1$, $y = 2$ and $x = 0$;

(*ii*) the volume formed when this area is rotated about Oy.

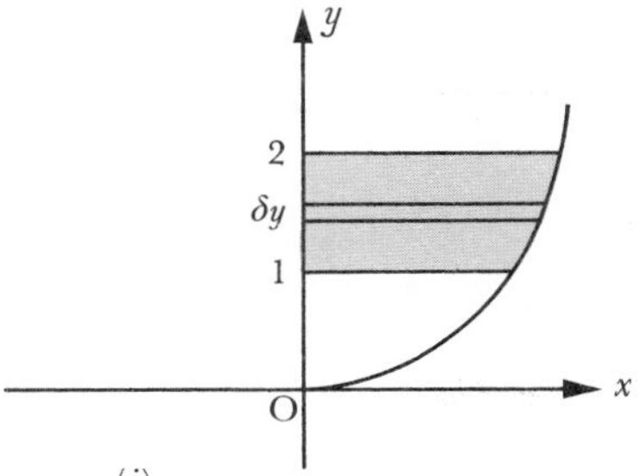

(*i*)

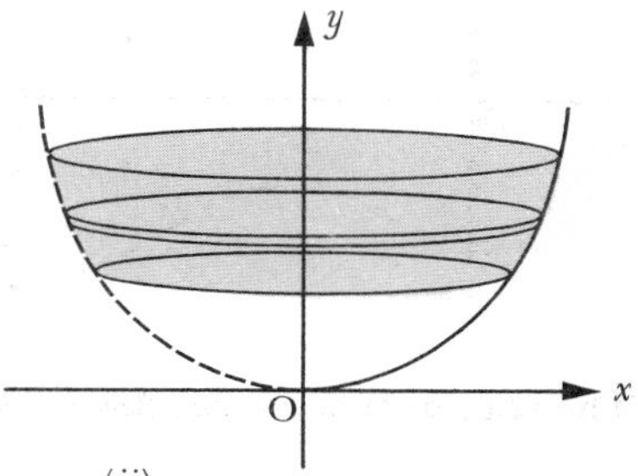

(*ii*)

(*i*) Area of a typical rectangular element $= x\delta y$

Sum of such areas $= \sum x\delta y$

$$\text{Required area} = \int_1^2 x \, dy = \int_1^2 \sqrt{y} \, dy$$

$$= \left[\frac{2}{3} y^{3/2}\right]_1^2 = \frac{2}{3}(2\sqrt{2} - 1)$$

(*ii*) Volume of a typical circular disc $= \pi x^2 \delta y$

Sum of such volumes $= \sum \pi x^2 \delta y$

$$\text{Required volume} = \int_1^2 \pi x^2 \, dy$$

$$= \int_1^2 \pi y \, dy = \pi\left[\frac{1}{2}y^2\right]_1^2 = \frac{3\pi}{2}$$

Example 3

Find the volume of water required to fill a hemispherical bowl of radius 5 cm to a depth of 3 cm.

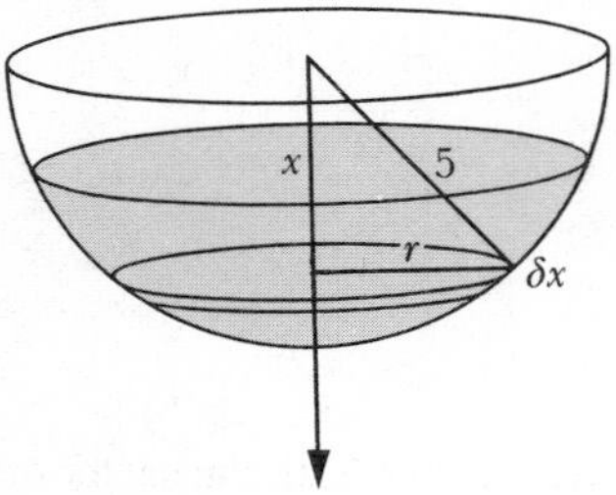

At depth x below centre, radius of cross-section is r,

where $r^2 = 25 - x^2$

So volume of a typical disc of thickness δx is

$\pi r^2 \delta x = \pi (25 - x^2) \delta x$

Sum of such volumes $= \sum \pi (25 - x^2) \delta x$

So total volume $= \int_2^5 \pi (25 - x^2) \, dx$

$$= \pi \left[25x - \tfrac{1}{3}x^3\right]_2^5$$
$$= \pi (25 \times 5 - \tfrac{1}{3} \times 5^3) - \pi (25 \times 2 - \tfrac{1}{3} \times 2^3)$$
$$= \pi (\tfrac{2}{3} \times 125 - \tfrac{142}{3})$$
$$= 36\pi$$
$$= 113 \text{ cm}^3$$

Example 4

A cone has base-area A and height h. Show that its volume is $\frac{1}{3}Ah$.

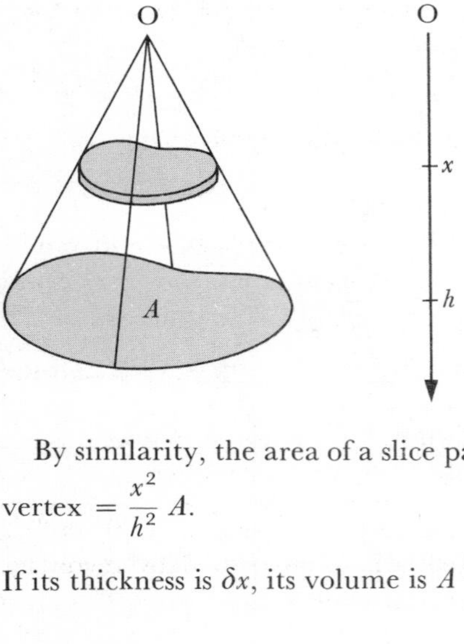

By similarity, the area of a slice parallel to the base at depth x below the vertex $= \dfrac{x^2}{h^2} A$.

If its thickness is δx, its volume is $A \dfrac{x^2}{h^2} \delta x$

So total volume of all such slices is $\sum \frac{Ax^2}{h^2}\,\delta x$

$$\Rightarrow \quad \text{Volume of cone is } \int_0^h \frac{Ax^2}{h^2}\,\mathrm{d}x = \left[\frac{Ax^3}{3h^2}\right]_0^h = \frac{1}{3}Ah$$

Exercise 4.8

1 Calculate the volumes formed by rotating about Ox the areas bounded by:

(*i*) $y = x,\ y = 0;\ x = 1,\ x = 2;$
(*ii*) $y = x^2 + 1,\ y = 0;\ x = 0,\ x = 1;$
(*iii*) $y = \frac{1}{x},\ y = 0;\ x = 2,\ x = 3;$
(*iv*) $y = \pm\sqrt{x},\ x = 4;$
(*v*) $y = x,\ y = x^2;$
(*vi*) $y = x - x^2,\ y = 0.$

2 Calculate the areas bounded by the following curves:

(*i*) $y = \frac{1}{2}x,\ y = 1,\ y = 2;\ x = 0;$
(*ii*) $y = x^2,\ y = 1,\ y = 4;\ x = 0;$
(*iii*) $y = x^3,\ y = 8;\ x = 0;$
(*iv*) $y = \frac{1}{x^4},\ y = 1,\ y = 16;$
(*v*) $y = \sqrt{x},\ y = 2;\ x = 0.$

3 Calculate the volumes formed when the areas in no. **2** are rotated about Oy.

4 Calculate the volume of a cone of base-radius r and height h by revolving the line $y = \frac{r}{h}x$ about Ox.

5 Calculate the volume of a sphere of radius r by revolving $x^2 + y^2 = r^2$ about Ox.

6 Draw a rough sketch of the curve $y^2 = 16x$ between $x = 0$ and $x = 4$. Calculate the area contained between the curve and the line $x = 4$, and the volume obtained by completely rotating that area about the axis of x (O.C.)

7 Find the points of intersection of the curve $y = x(4 - x)$ and the line $y = 2x$.

Find the volume generated when the area enclosed between the curve and the line makes one complete revolution about the x-axis. (O.C.)

8 The curves $y^2 = 2x$, $x^3 = 4y$ intersect at the points (0, 0) and (2, 2). Find the area they enclose and the volume generated by revolving this area through four right angles about the axis of x. (O.C.)

9 O is the origin and P is the point $(a, 2a)$ on the parabola $y^2 = 4ax$. Prove that the area bounded by the chord OP and the arc of the parabola between O and P is $\frac{1}{3}a^2$.

Find the volume generated if this area is revolved about the x-axis through four right angles (O.C.)

10 Find the area enclosed between the curves $y^2 = 4x$ and $x^2 = 4y$.

Find the volume generated when this area is rotated through four right angles about the x-axis. (O.C.)

11 A container 9 m high is such that when the depth of liquid in it is x m, the area of the surface of the liquid is $(10x - x^2)$ m^2. What is its capacity when full? (O.C.)

12 A hole of radius 3 cm is bored symmetrically through a sphere of radius 5 cm. Find the volume of the part which remains.

*4.9 Mean values

Suppose that a particle is moving so that its velocity v (m s^{-1}) after time t (s) is given by $v = 6t^2$

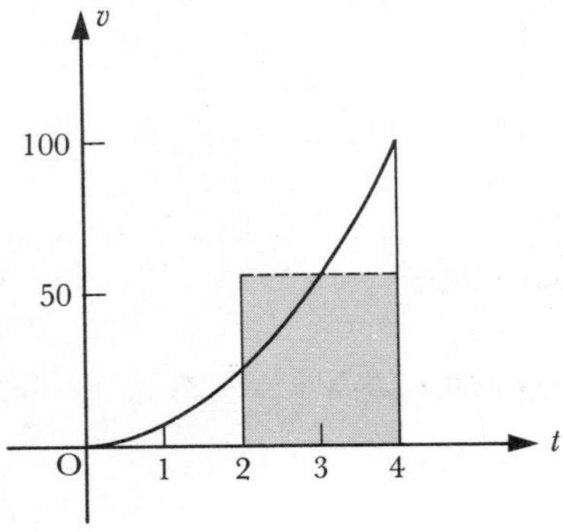

The distance travelled between $t = 2$ and $t = 4$ is

$$\int_2^4 6t^2 \, dt = [2t^3]_2^4 = 128 - 16 = 112 \text{ m.}$$

So the average (or *mean*) velocity in this interval is

$$\frac{112}{2} \text{ m s}^{-1} = 56 \text{ m s}^{-1}$$

(Is this the same as the average of the velocities when $t = 2$ and $t = 4$?)

More generally, the *mean value of a function $f(x)$ between $x = a$ and $x = b$* can be defined as the height of the line beneath which there is the same area:

$$\frac{1}{b-a}\int_a^b f(x)\,dx$$

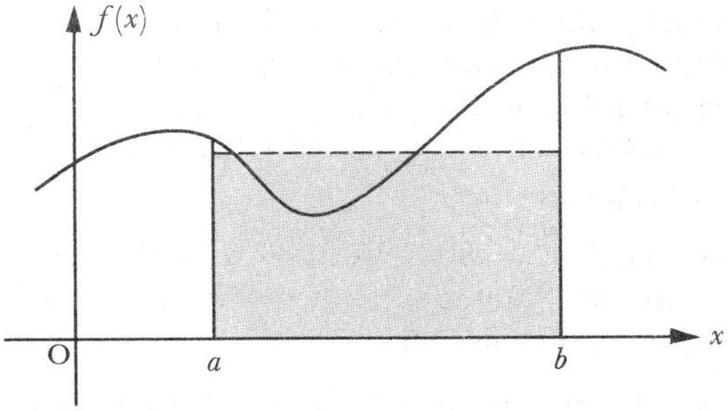

Example 1

The population of a circular parish of radius 1 km is uniformly distributed, and the church is at its centre. Find the mean distance of a parishioner from the church.

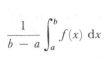

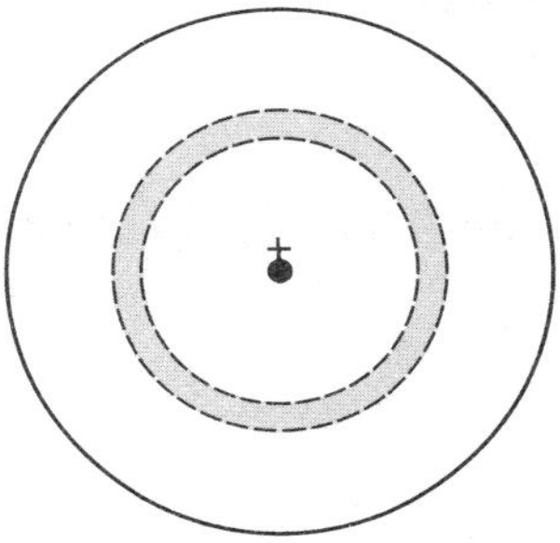

Let the total population be N.

Then density of population $= \dfrac{N}{\pi}$ parishioners/km^2

So number of parishioners at a distance between x and $x + \delta x$ is

$$\frac{N}{\pi}\{\pi(x+\delta x)^2 - \pi x^2\} \approx 2Nx\delta x$$

The sum of their distances from church

$$\approx 2Nx\delta x \times x = 2Nx^2\delta x$$

$$\Rightarrow \quad \text{Mean distance} \approx \frac{\sum 2Nx^2\delta x}{N}$$

Taking limits, their mean distance from church

$$= \frac{\int_0^1 2Nx^2\,dx}{N}$$

$$= \frac{[\frac{2}{3}Nx^3]_0^1}{N} = \frac{2}{3}$$

So the mean distance is $\frac{2}{3}$ km.

Exercise 4.9

1 What is the mean value of

(*i*) $f(x) = 3x - x^2$ between $x = 0$ and 3;
(*ii*) $f(x) = x^3$ between $x = 1$ and 5;
(*iii*) $f(x) = \sqrt{x}$ between $x = 1$ and 2.

2 A sphere of radius 1 m has density which increases linearly from 0 at its centre to 1 000 kg m^{-3} at its surface.
Find: (*i*) the mass of the sphere;
(*ii*) its mean density.

3 A hive of bees swarms round its queen in a uniform sphere of radius 10 cm. What is their mean distance from her?

*4.10 Integration by substitution

So far, when we have been faced with the problem of integrating a given function, we have had to guess another function which when differentiated yields the original. It may be difficult, or even impossible, to find an answer to the problem; but when we think we have found one it is always possible to check its truth by differentiation.

Though integration remains a speculative venture which often calls for considerable powers of imagination, there are some general methods which are of assistance. In particular, because it is the reverse of differentiation, we might expect any general result about differentiation to yield a corresponding result about integration. In this section we shall look again at the chain rule and, unlike our usual practice, shall go directly to the general case which we shall then seek to illustrate by particular examples.

Suppose that the indefinite integral of $f(x)$ is $F(x)$, so that

$$\int f(x)\,dx = F(x) \quad \text{and} \quad \frac{dF}{dx} = f(x)$$

Let us now introduce a new variable t, and suppose that $x = \phi(t)$. Then $f(x)$ can be written $f\phi(t)$, and $F(x)$ can be written $F\phi(t)$.

$$\text{So} \quad \frac{d}{dt}F\{\phi(t)\} = \frac{d}{dt}F(x) = \frac{dF(x)}{dx} \times \frac{dx}{dt} \quad \text{(the Chain rule)}$$

$$= f(x)\phi'(t)$$

$$= f\phi(t)\phi'(t)$$

$$\text{So} \quad \int f\phi(t)\phi'(t)\,dt = F\phi(t) = F(x)$$

$$\Rightarrow \quad \boxed{\int f(x)\,dx = \int f\phi(t)\;\phi'(t)\,dt}$$

So in the given integral

$$\int f(x)\,\mathrm{d}x,$$

we can replace x by $\phi(t)$ providing we also replace $\mathrm{d}x$ by $\phi'(t)\,\mathrm{d}t$.

This is most easily understood from a number of examples.

Example 1

Find $\int x\sqrt{(x+2)}\,\mathrm{d}x$

As the integral is not immediately obvious we try a substitution, and as the most disagreeable part of the integral is the square root, we put

$$\sqrt{(x+2)} = t \qquad \Rightarrow \quad x = t^2 - 2,$$

so that $\mathrm{d}x$ is replaceable by $2t\,\mathrm{d}t$

$$\text{So} \quad \int x\sqrt{(x+2)}\,\mathrm{d}x = \int (t^2-2)t\,2t\,\mathrm{d}t$$

$$= \int (2t^4 - 4t^2)\,\mathrm{d}t = \tfrac{2}{5}t^5 - \tfrac{4}{3}t^3 + A$$

Hence, returning to the original variable x,

$$\int x\sqrt{(x+2)}\,\mathrm{d}x = \tfrac{2}{5}(x+2)^{5/2} - \tfrac{4}{3}(x+2)^{3/2} + A$$

$$= \frac{2}{15}(x+2)^{3/2}\,[3(x+2) - 10] + A$$

$$= \frac{2}{15}(3x-4)\,(x+2)^{3/2} + A$$

It is hardly surprising that we were unable to guess this answer, but we can at least check that it is correct:

$$\frac{\mathrm{d}}{\mathrm{d}x}\left\{\frac{2}{15}(3x-4)\,(x+2)^{3/2}\right\} = \frac{2}{15}\left\{(3x-4)\,\frac{3}{2}\,(x+2)^{1/2} + 3\,(x+2)^{3/2}\right\}$$

$$= \frac{2}{15}(x+2)^{1/2}\left\{\frac{9x}{2} - 6 + 3\,(x+2)\right\}$$

$$= \frac{2}{15}(x+2)^{1/2}\,\frac{15x}{2} = x\,(x+2)^{1/2}$$

$$\text{So} \qquad \int x\sqrt{(x+2)}\,\mathrm{d}x = \frac{2}{15}(3x-4)\,(x+2)^{3/2} + A$$

Example 2

Calculate $$I = \int_1^2 \frac{x\,dx}{\sqrt[3]{(x-1)}}$$

Here we try $t = \sqrt[3]{(x-1)}$

$$\Rightarrow \quad x = t^3 + 1 \quad \Rightarrow \quad dx = 3t^2\,dt$$

So $$\int \frac{x\,dx}{\sqrt[3]{(x-1)}} = \int \frac{(t^3+1)3t^2\,dt}{t}$$

$$= \int (3t^4 + 3t)\,dt$$

Now the limits of integration were $x = 1$ and $x = 2$, and so become

$t = \sqrt[3]{(1-1)} = 0$ and $t = \sqrt[3]{(2-1)} = 1$

Hence, $$I = \int_0^1 (3t^4 + 3t)\,dt$$

$$= \left[\frac{3}{5}t^5 + \frac{3}{2}t^2\right]_0^1 = \frac{3}{5} + \frac{3}{2} = \frac{21}{10} = 2.1$$

Sometimes it is easier to express the connection between x and t with t as a function of x:

Example 3

$$\int x\sqrt{(1-x^2)}\,dx$$

Here we put $t = 1 - x^2$

$$\Rightarrow \quad dt = -2x\,dx$$

Now $$\int x\sqrt{(1-x^2)}\,dx = \int \sqrt{(1-x^2)}\,x\,dx$$

$$= \int \sqrt{t}\,\frac{-dt}{2}$$

$$= \int -\tfrac{1}{2}t^{1/2}\,dt = -\tfrac{1}{3}t^{3/2} + A$$

So $$\int x\sqrt{(1-x^2)}\,dx = -\tfrac{1}{3}(1-x^2)^{3/2} + A$$

Example 4

Calculate $$I = \int_0^1 \frac{x^2\,dx}{(x^3+1)^2}$$

If $u = x^3 + 1$, then $\mathrm{d}u = 3x^2\,\mathrm{d}x$. Also $x = 0 \Rightarrow u = 1$ and $x = 1 \Rightarrow u = 2$.

So
$$I = \int_1^2 \frac{\frac{1}{3}\,\mathrm{d}u}{u^2} = \int_1^2 \frac{1}{3u^2}\,\mathrm{d}u$$
$$= \left[-\frac{1}{3u}\right]_1^2 = \left(-\frac{1}{6}\right) - \left(-\frac{1}{3}\right) = \frac{1}{6}$$

Hence $\displaystyle\int_0^1 \frac{x^2}{(x^3+1)^2}\,\mathrm{d}x = \frac{1}{6}$

Exercise 4.10

1 Find:

(*i*) $\displaystyle\int (x+1)^5\,\mathrm{d}x$ (put $x = t - 1$); (*ii*) $\displaystyle\int \frac{\mathrm{d}x}{(1-2x)^2}$ $\left(\text{put } x = \frac{1-t}{2}\right)$;

(*iii*) $\displaystyle\int \sqrt{(2-x)}\,\mathrm{d}x$; (*iv*) $\displaystyle\int (2x+1)^{3/2}\,\mathrm{d}x$;

(*v*) $\displaystyle\int \frac{x\,\mathrm{d}x}{\sqrt[3]{(1+x)}}$

2 Find:

(*i*) $\displaystyle\int x(x^2+1)^4\,\mathrm{d}x$ (put $t = x^2 + 1$);

(*ii*) $\displaystyle\int x^2(2x^3+3)^4\,\mathrm{d}x$ (put $t = 2x^3 + 3$); (*iii*) $\displaystyle\int x\sqrt{(1-x^2)}\,\mathrm{d}x$;

(*iv*) $\displaystyle\int \frac{1+2x}{\sqrt{(x+x^2)}}\,\mathrm{d}x$; (*v*) $\displaystyle\int \frac{x}{\sqrt[3]{(x^2-1)}}\,\mathrm{d}x$

3 Evaluate:

(*i*) $\displaystyle\int_0^1 (1-x)^4\,\mathrm{d}x$; (*ii*) $\displaystyle\int_2^3 \frac{\mathrm{d}x}{(2x+1)^2}$;

(*iii*) $\displaystyle\int_0^3 x\sqrt{(x^2+1)}\,\mathrm{d}x$; (*iv*) $\displaystyle\int_1^2 \frac{x^2\,\mathrm{d}x}{\sqrt{(x^3-1)}}$;

(*v*) $\displaystyle\int_1^4 \sqrt{\frac{1+\sqrt{x}}{x}}\,\mathrm{d}x$ (put $x = (u-1)^2$)

Miscellaneous problems 4

1 It is known that the area beneath the curve $y = 1/x$ from $x = 1$ to $x = 10^6$ is approximately 13.8. Use the staircase method to calculate the sum of the reciprocals of the integers from 1 to 1 000 000 as accurately as you can.

2 Show that $\int_{-a}^{a} f(x)\,dx$ is

$$2\int_{0}^{a} f(x)\,dx \quad \text{when } f(x) \text{ is an even function,}$$

and $\quad 0 \qquad$ when $f(x)$ is an odd function.

Hence, calculate:

(*i*) $\displaystyle\int_{-1}^{1} \frac{x^3\,dx}{(1+x^2)^2}$; (*ii*) $\displaystyle\int_{-2}^{2} \frac{1+x^3+x^4}{1+x^4}\,dx$

3 Draw a rough sketch of the graph of $y^2 = x(1-x)^2$, and find the area enclosed by the loop.

Find also the volume of the solid formed by rotating this area about the x-axis. (O.C.)

4 Find the coordinates of the points of intersection of the two parabolas $y^2 = 4ax$ and $x^2 = 4ay$ $(a > 0)$. Show that the area between these parabolas is $16a^2/3$.

This area is rotated through four right angles about the x-axis. Show that the volume generated is $96\pi a^3/5$.

5 Sketch the parabolas

$$y^2 = 4x \quad \text{and} \quad y^2 = 5x - 4$$

and find their points of intersection.

A fruit-bowl is made by rotating completely about the x-axis the area enclosed by the curves. Find the volume of the material required to make the bowl. (S.M.P.)

6 A napkin-ring is formed by boring a cylindrical hole symmetrically through a sphere of radius a. If the height of the ring is h when resting on one of its circular ends, find its volume.

7 An ellipse of length $2a$ and breadth $2b$ has the equation

$$\frac{x^2}{a^2} + \frac{y^2}{b^2} = 1$$

Find the volume of the solid formed when it is rotated about the x-axis.

8 An oil-tanker has a water-line length of 200 m and at a point x m from its bow its underwater cross-sectional area is

$$\frac{3}{1\,000}x^2(200 - x)\ \mathrm{m}^2$$

Find its displacement in tonne.

9 The figure shown is the cross-section of a tyre lying on the ground.

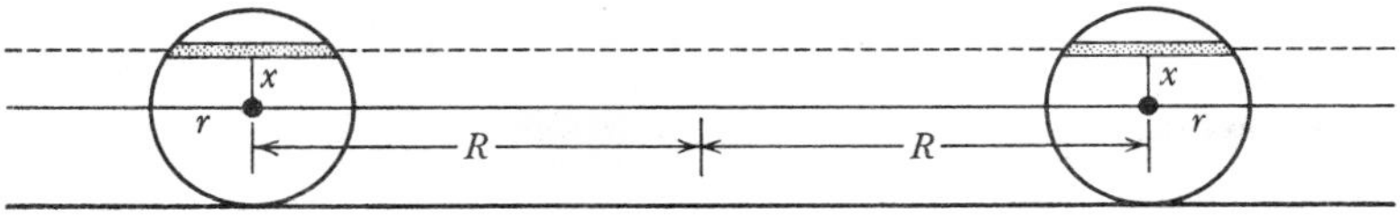

Each circular cross-section has radius r and a centre which is at distance R from the centre of the tyre. The dotted plane is at a height x above the horizontal plane of symmetry. Show that it intersects the tyre in a ring whose outer and inner radii are

$R + \sqrt{(r^2 - x^2)}$ and $R - \sqrt{(r^2 - x^2)}$,

and which has area

$2\pi R\sqrt{(r^2 - x^2)}$

Hence show that the volume of the tyre is $2\pi^2 Rr^2$.

5

Trigonometric functions

5.1 Sine, cosine, and tangent

On a sheet of graph-paper, draw rectangular axes Ox, Oy together with a circle which has centre O and a convenient radius (say 5 or 10 cm), which will be used as the unit of length. Label the point of the circle which lies on the positive x-axis, i.e., the point (1, 0), as 0°. Then move round the circle in an anti-clockwise direction and label the 35 other points at 10°

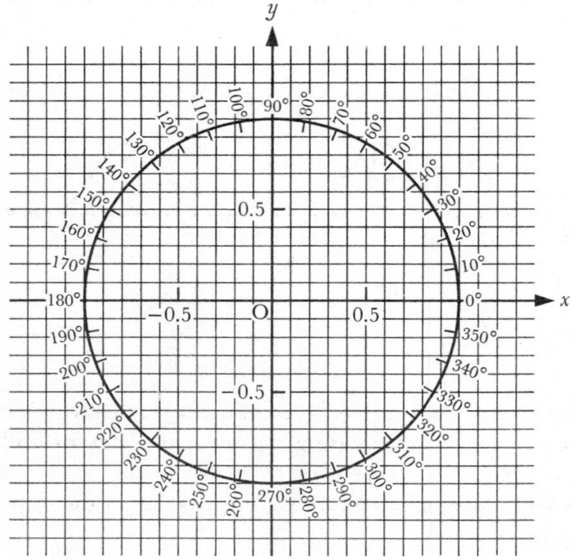

intervals as

10°, 20°, 30°...350° (360° clearly coinciding with 0°)

Next, copy and complete the following table at 10° intervals, marking in the y and x coordinates of successive points:

θ	0°	10°	20°	. . . .	350°	360°
y	0	+0.17	+0.34	. . . .	−0.17	0
x	+1.00	+0.98	+0.94	. . . .	+0.98	+1.00

It will quickly be seen that from $\theta = 0°$ to $\theta = 90°$ the values of y and x which are being registered are precisely the values of $\sin\theta$ and $\cos\theta$ respectively.
We therefore *define* the sine and cosine of *any angle* as the y and x coordinates of the corresponding point.

So $\sin 20° \approx +0.34$ and $\sin 350° \approx -0.17$

$\cos 20° \approx +0.94$ and $\cos 350° \approx +0.98$

Even more generally, if θ is an angle of any size (perhaps greater than 360°, or negative) we define $\sin\theta$ and $\cos\theta$ as the y and x coordinates of the point reached by moving anti-clockwise through angle θ from the point (1, 0). It is clear that a negative rotation will be effectively clockwise, and also that any two angles which differ by a multiple of 360° must have the same sine and cosine. We therefore say that $\sin\theta$ and $\cos\theta$ are *periodic functions*, with period 360°.
The *periodicity* of $\sin\theta$ and $\cos\theta$ is conveniently displayed by plotting their graphs:

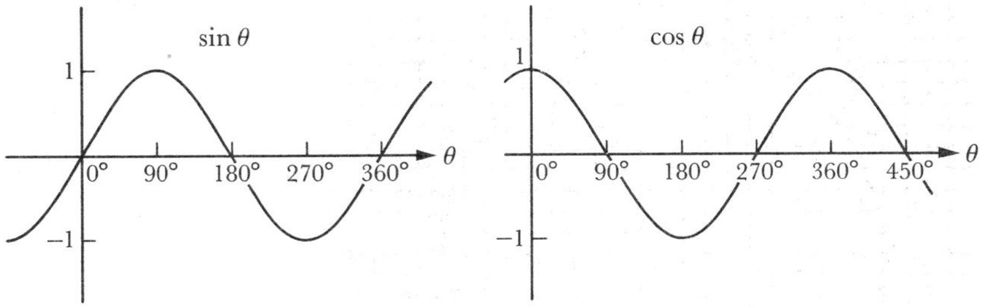

It is this feature of $\sin\theta$ and $\cos\theta$ which is fundamental to their wide importance in the theory of waves, alternating currents, oscillations and vibrations of all kinds.

We can obtain a number of useful results simply by looking at the symmetries of the following figure:

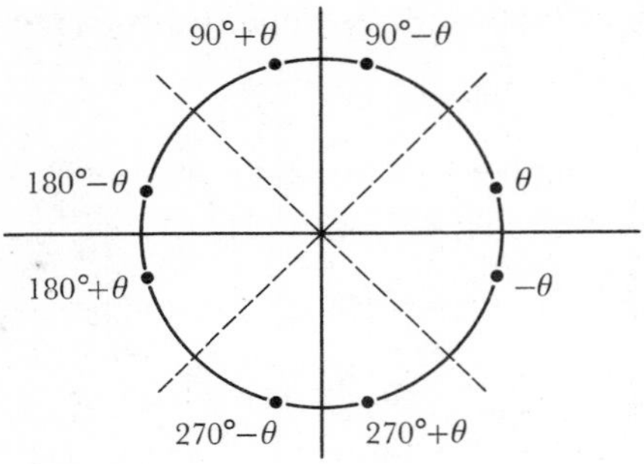

In particular:

$$\sin(-\theta) = -\sin\theta \qquad \cos(-\theta) = +\cos\theta$$

i.e., $\sin\theta$ is an odd function and $\cos\theta$ is an even function

$$\sin(90° - \theta) = +\cos\theta \qquad \cos(90° - \theta) = +\sin\theta$$

i.e., the sine of an angle θ is the *co*sine of its *co*mplementary angle $90° - \theta$

$$\sin(90° + \theta) = +\cos\theta \qquad \cos(90° + \theta) = -\sin\theta$$

Each of these has an interpretation for the graphs of the functions; e.g., $\sin(90° + \theta) = \cos\theta$ shows that the graph of $\sin\theta$ is simply the graph of $\cos\theta$ displaced 90° to the right:

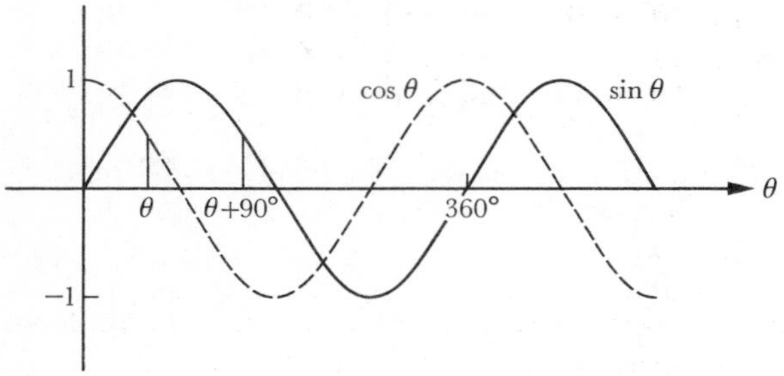

Tangent

The other most frequently used trigonometric function is $\tan\theta$, defined by

$$\boxed{\tan\theta = \frac{\sin\theta}{\cos\theta}}$$

(and so undefined when $\cos\theta = 0$).

In the original diagram, $\tan\theta$ is represented by $\dfrac{\sin\theta}{\cos\theta} = \dfrac{y}{x}$, i.e., by the gradient of the radius.

It, too, can be plotted from the above table and (as would be expected of a gradient) is periodic with period 180°.

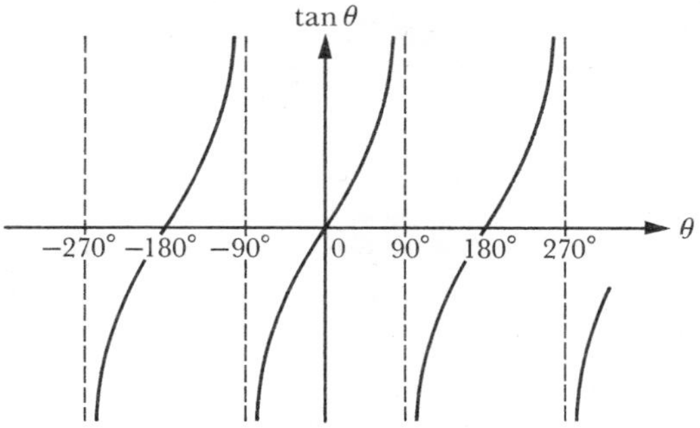

Exercise 5.1a

1 Use the figure to find the approximate values of $\sin\theta$ and $\cos\theta$, where θ is

(*i*) 35°; (*ii*) 167°; (*iii*) 214°; (*iv*) 304°; (*v*) 400°; (*vi*) −32°.

2 Use the above answers to find the corresponding values of $\tan\theta$.

3 Express in terms of $\sin\theta$, $\cos\theta$, and $\tan\theta$:
(*i*) $\sin(180° - \theta)$, $\cos(180° - \theta)$, $\tan(180° - \theta)$;
(*ii*) $\sin(180° + \theta)$, $\cos(180° + \theta)$, $\tan(180° + \theta)$;
(*iii*) $\sin(270° + \theta)$, $\cos(270° + \theta)$, $\tan(270° + \theta)$.

'Set-square' angles: 45°, 30°, 60°

We can easily calculate the trigonometric functions of these angles from the dimensions of two set-squares, each of height 1 unit.

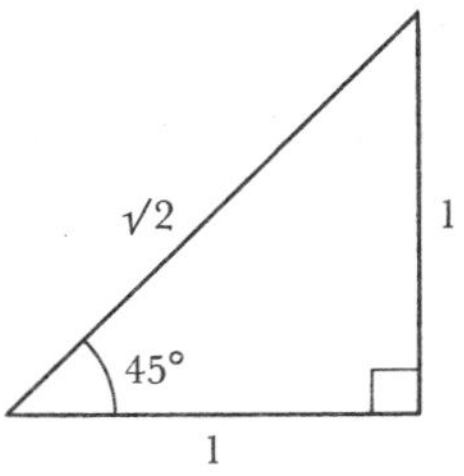

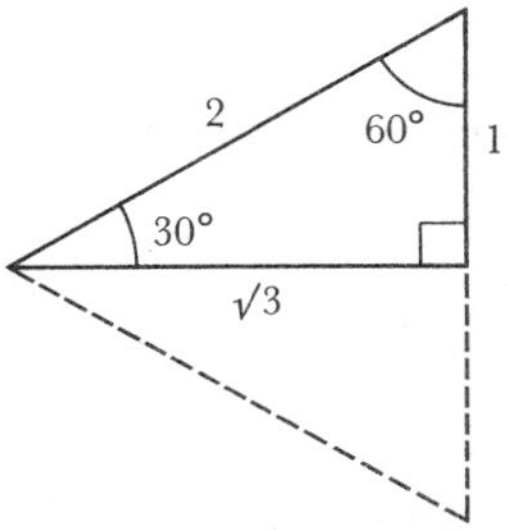

If the second set square is reflected in its base, it forms an equilateral triangle, so that the hypotenuse of the set square must be 2 and its base $\sqrt{3}$.

So $\sin 45^\circ = \frac{1}{\sqrt{2}}$ $\sin 30^\circ = \frac{1}{2}$ $\sin 60^\circ = \frac{\sqrt{3}}{2}$

$\cos 45^\circ = \frac{1}{\sqrt{2}}$ $\cos 30^\circ = \frac{\sqrt{3}}{2}$ $\cos 60^\circ = \frac{1}{2}$

$\tan 45^\circ = 1$ $\tan 30^\circ = \frac{1}{\sqrt{3}}$ $\tan 60^\circ = \sqrt{3}$

Exercise 5.1b

1 What are the exact values of $\sin\theta$, $\cos\theta$, and $\tan\theta$ (in terms of square roots where necessary) if θ is
(*i*) 120°; (*ii*) 225°; (*iii*) 330°.

5.2 Simple equations

We are now able to find the value of the sine, cosine and tangent of any angle. But very frequently we need to ask the inverse question: for what value (or values) of θ is $\sin\theta = 0.27$?
or $\cos\theta = -0.73$?
or $\tan\theta = -2$?

These can usually be solved most simply by use of tables combined with a careful study of either the original circle or the graph of the particular function.

Example 1

$\sin\theta = 0.27$

From our tables, we find that $\theta = 15^\circ\ 40'$ is a solution.

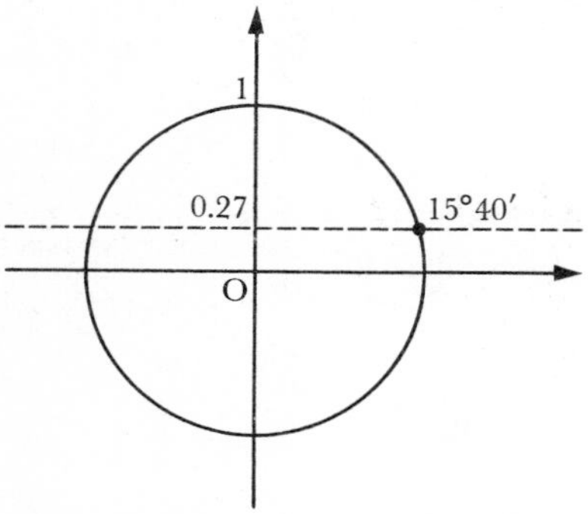

But it is also clear from the figure that another possible solution is $\theta = 164^\circ\ 20'$, and that we can obtain further solutions from these two by adding or subtracting multiples of 360°.

So $\theta = \cdots -344^\circ\ 20',\ 15^\circ\ 40',\ 375^\circ\ 40' \ldots$

and $\theta = \cdots -195^\circ\ 40',\ 164^\circ\ 20',\ 524^\circ\ 20' \ldots$

are all solutions of the given equation.

This can also be seen from the graph of sin θ:

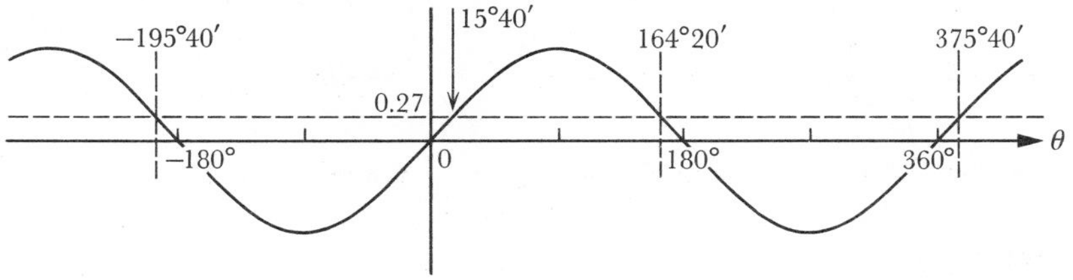

Example 2

$\cos\theta = -0.73$

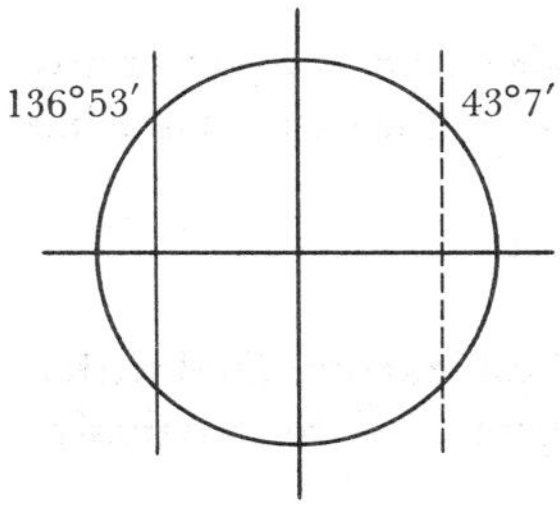

From tables, we find that $\cos\theta = +0.73$ has a solution $\theta = 43°\ 7'$. Hence, from the figure, we see that $\cos\theta = -0.73$ has solutions $\theta = \pm 136°\ 53'$.

Finally, by adding multiples of 360°, we obtain the full set of solutions

$$\theta = \cdots -223°\ 7',\ 136°\ 53',\ 136°\ 53',\ 223°\ 7',\ \cdots$$

Example 3

$\tan\theta = -2$

In this case we look for the line whose gradient is -2:

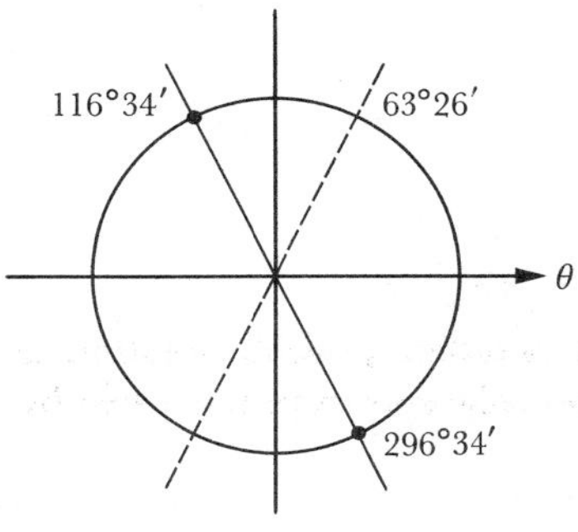

Now $\tan 63°\ 26' = +2$, and the first two solutions of $\tan\theta = -2$ are $\theta = 116°\ 34'$ and $296°\ 34'$, which clearly differ by 180°. So the full list of solutions is

$$\theta = \cdots -243^\circ\ 26',\ -63^\circ\ 26',\ 116^\circ\ 34',\ 296^\circ\ 34',\ 476^\circ\ 34',\ 656^\circ\ 34' \ldots$$

which can also be seen from the graph of $\tan\theta$:

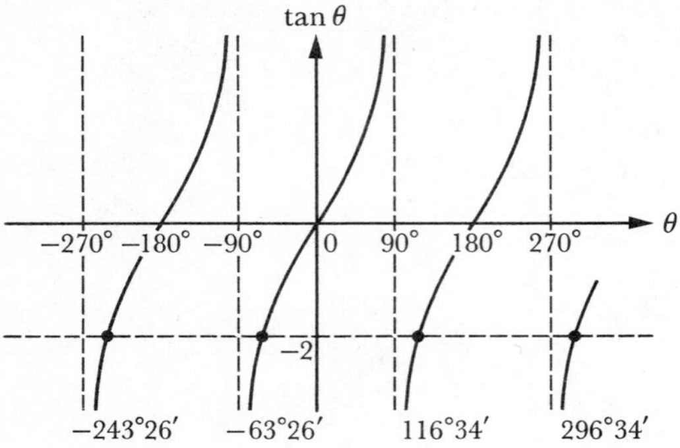

More generally, we can express the solutions to simple equations as follows:

(*i*) $\sin x = \sin\alpha$

$\Rightarrow \quad x = \alpha + 360n^\circ$
or $\quad (180^\circ - \alpha) + 360n^\circ$,
where n is an integer (or $n \in Z$).

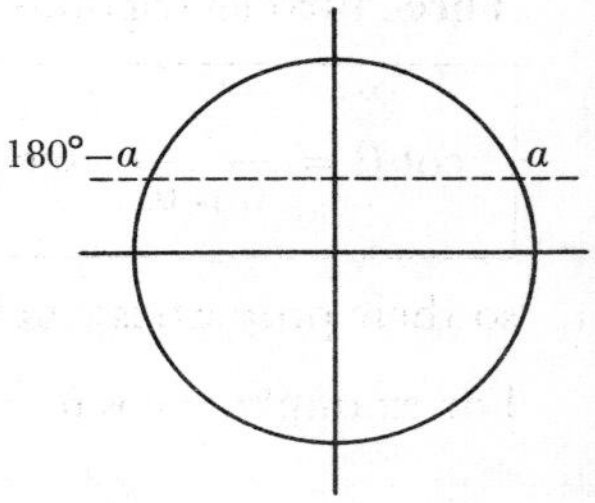

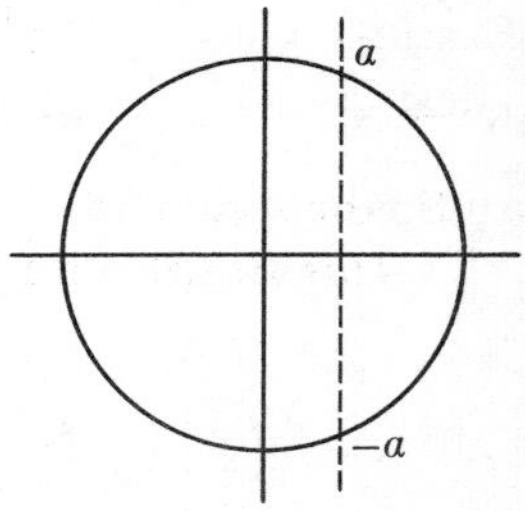

(*ii*) $\cos x = \cos\alpha$

$\Rightarrow \quad x = \pm\alpha + 360n^\circ \ (n \in Z)$.

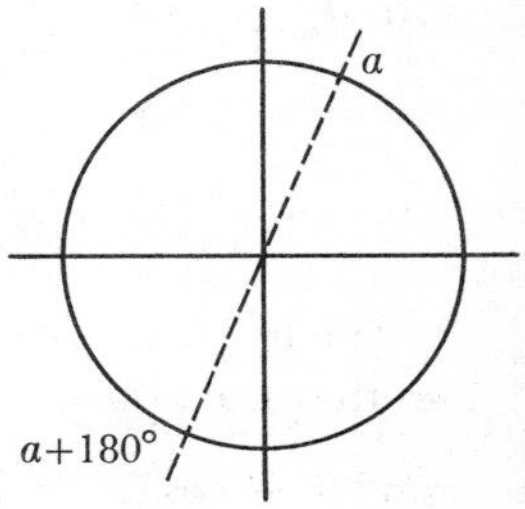

(*iii*) $\tan x = \tan\alpha$

$\Rightarrow \quad x = \alpha + 180n^\circ \ (n \in Z)$.

Exercise 5.2

Find the values of θ between 0° and 360° for which

1 (*i*) $\sin\theta = 0.37$; (*ii*) $\sin\theta = 1$; (*iii*) $\sin\theta = -0.74$.

2 (*i*) $\cos\theta = 0.45$; (*ii*) $\cos\theta = -0.61$; (*iii*) $\cos\theta = -1$.

3 (*i*) $\tan\theta = -1$; (*ii*) $\tan\theta = 0$; (*iii*) $\tan\theta = 2.72$.

4 (*i*) $\sin^2\theta + \sin\theta = 0$;
(*ii*) $2\cos^2\theta - 3\cos\theta + 1 = 0$;
(*iii*) $3\tan^2\theta = 1$.

5 (*i*) $\sin\frac{3\theta}{2} = -0.4266$;
(*ii*) $\cos\frac{1}{2}\theta = -0.5230$;
(*iii*) $\tan 2\theta = +0.4550$.

5.3 Cotangent, secant, cosecant, and the use of Pythagoras' theorem

Three further trigonometric functions are defined by

$$\cot\theta = \frac{1}{\tan\theta}, \quad \sec\theta = \frac{1}{\cos\theta}, \quad \operatorname{cosec}\theta = \frac{1}{\sin\theta}$$

so their properties can be deduced from those of the original functions.

For example, $\cos\theta = \sin(90^\circ - \theta)$

$$\Rightarrow \quad \frac{1}{\cos\theta} = \frac{1}{\sin(90^\circ - \theta)}$$

$$\Rightarrow \quad \sec\theta = \operatorname{cosec}(90^\circ - \theta)$$

So the secant of an angle is the *co*secant of its *co*mplementary angle.

It is clear from Pythagoras' theorem that, for any angle θ,

$$(\sin\theta)^2 + (\cos\theta)^2 = 1$$

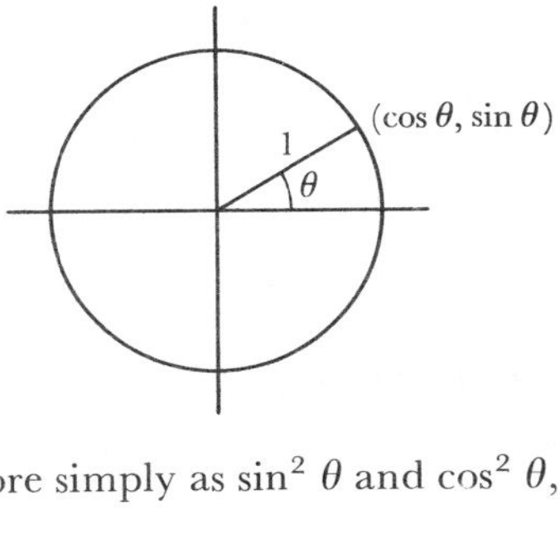

We usually write $(\sin\theta)^2$ and $(\cos\theta)^2$ more simply as $\sin^2\theta$ and $\cos^2\theta$, so that the above result is written

$$\sin^2\theta + \cos^2\theta = 1$$

Dividing by $\cos^2\theta$ and $\sin^2\theta$, we obtain:

$$\frac{\sin^2\theta}{\cos^2\theta}+\frac{\cos^2\theta}{\cos^2\theta}=\frac{1}{\cos^2\theta} \quad\text{and}\quad \frac{\sin^2\theta}{\sin^2\theta}+\frac{\cos^2\theta}{\sin^2\theta}=\frac{1}{\sin^2\theta}$$

$$\Rightarrow \quad \tan^2\theta+1=\sec^2\theta \qquad\text{and}\qquad 1+\cot^2\theta=\operatorname{cosec}^2\theta$$

Summarising:

$$\boxed{\begin{aligned}\sin^2\theta+\cos^2\theta&=1\\ \tan^2\theta+1&=\sec^2\theta\\ \cot^2\theta+1&=\operatorname{cosec}^2\theta\end{aligned}}$$

Example 1

Simplify $\dfrac{1}{\sec A-\tan A}$.

$$\frac{1}{\sec A-\tan A}=\frac{\sec^2 A-\tan^2 A}{\sec A-\tan A}$$

$$=\frac{(\sec A+\tan A)(\sec A-\tan A)}{\sec A-\tan A}=\sec A+\tan A$$

Example 2

Solve the equation

$$3\sin^2 x=3\cos^2 x-\cos x+1$$

for values of x between 0° and 360°.

$$3\sin^2 x=3\cos^2 x-\cos x+1$$

$$\Rightarrow \quad 3-3\cos^2 x=3\cos^2 x-\cos x+1$$

$$\Rightarrow \quad 6\cos^2 x-\cos x-2=0$$

$$\Rightarrow \quad (3\cos x-2)(2\cos x+1)=0$$

$$\Rightarrow \quad \cos x=\tfrac{2}{3} \quad\text{or}\quad -\tfrac{1}{2}$$

$$\Rightarrow \quad x=48^\circ\,11',\quad 311^\circ\,49',\quad\text{or}\quad 120^\circ,\quad 240^\circ$$

So $x=48^\circ\,11',\ 120^\circ,\ 240^\circ$, or $311^\circ\,49'$.

Example 3

Eliminate θ from the equations

$$x=2+3\cos\theta;\quad y=1+3\sin\theta.$$

As $3\cos\theta=x-2$ and $3\sin\theta=y-1$

$$(x-2)^2+(y-1)^2=9\cos^2\theta+9\sin^2\theta$$

$$\Rightarrow \quad (x-2)^2+(y-1)^2=9$$

So the original equations define a point (x, y) which always lies on the circle

$$(x-2)^2+(y-1)^2=9,$$

with centre (2, 1) and radius 3.

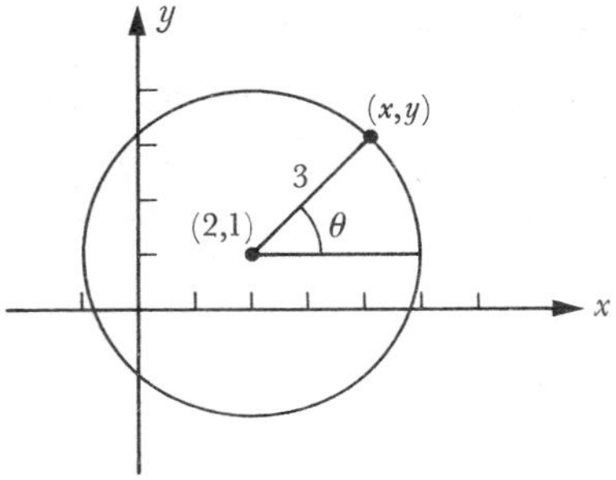

Exercise 5·3

1 Sketch the graph of
(*i*) $\cot x$, (*ii*) $\sec x$, (*iii*) $\operatorname{cosec} x$, for values of x from $-180°$ to $+540°$.

2 Simplify:
(*i*) $\cos\theta\tan\theta$; (*ii*) $\sin\theta\sec\theta$;
(*iii*) $\operatorname{cosec}\theta\tan\theta$; (*iv*) $\cot\theta\sec\theta$.

3 Simplify:

(*i*) $\dfrac{1-\sin^2\theta}{1-\cos^2\theta}$; (*ii*) $\dfrac{\cos^2\theta}{1+\sin\theta}+\dfrac{\cos^2\theta}{1-\sin\theta}$;

(*iii*) $\dfrac{\tan^2\theta}{\sec\theta+1}$; (*iv*) $\dfrac{1}{\cot\theta+\tan\theta}$;

(*v*) $(\operatorname{cosec}\theta+1)(\operatorname{cosec}\theta-1)$; (*vi*) $\dfrac{\operatorname{cosec}\theta}{\operatorname{cosec}\theta-\sin\theta}$;

(*vii*) $\dfrac{\sec^2\theta+2\tan\theta}{(\cos\theta+\sin\theta)^2}$; (*viii*) $\dfrac{1+2\sin\theta\cos\theta}{\sin\theta+\cos\theta}$.

4 Solve the following equations for values of θ between 0° and 360°:
(*i*) $\sec\theta=2$; (*ii*) $\cot\theta=-1$;
(*iii*) $2\sin^2\theta+\cos\theta=1$; (*iv*) $\sec^2\theta=1+\tan\theta$;
(*v*) $\tan\theta+4\cot\theta=4\sec\theta$; (*vi*) $3\tan^2\theta-5\sec\theta+1=0$.

5 (*i*) If $\sec x-\tan x=\frac{1}{3}$, find the values of $\sec x$ and $\tan x$.
(*ii*) Given that

$$x=\tan\theta-\sin\theta,\quad y=\tan\theta+\sin\theta,$$

prove that $(x^2-y^2)^2=16xy$.

*5.4 Sine and cosine rules

If we know the sizes a, b, c of the three sides of a triangle ABC, then we can construct it, and the angles A, B, C are automatically determined. We therefore expect that there will be some relationships between a, b, c and A, B, C. The most important of these, which enable us to calculate unknown angles (or sides) are:

The sine rule:

$$\frac{a}{\sin A} = \frac{b}{\sin B} = \frac{c}{\sin C}$$

and the cosine rules:

$$a^2 = b^2 + c^2 - 2bc \cos A$$
$$b^2 = c^2 + a^2 - 2ca \cos B$$
$$c^2 = a^2 + b^2 - 2ab \cos C$$

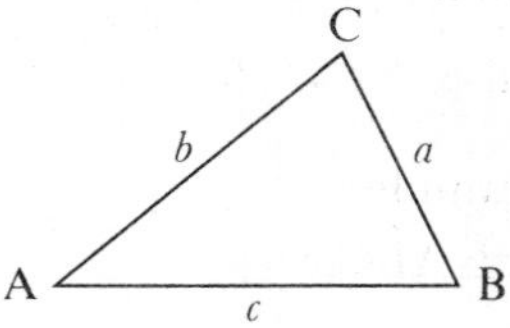

These results are conveniently proved by moving the triangle on to a framework of axes so that A lies at (0, 0) and B at (c, 0), with C above the x-axis. (If the original triangle is lettered clockwise rather than anti-clockwise this can be achieved by first turning the triangle upside-down.)

The triangle is then in the following position, and in every possible case the coordinates of C are

($b \cos A$, $b \sin A$), even when A is obtuse.

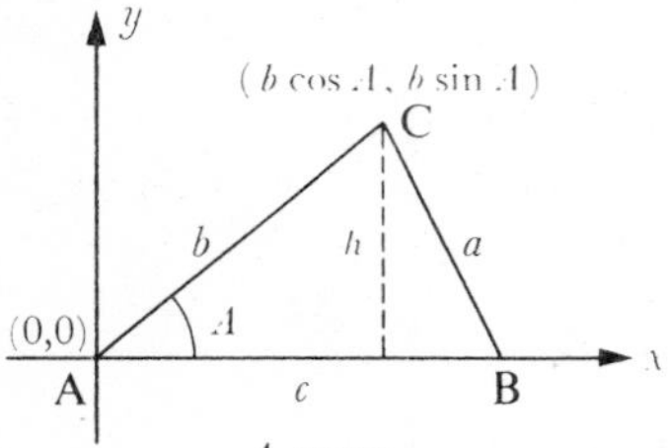

A acute

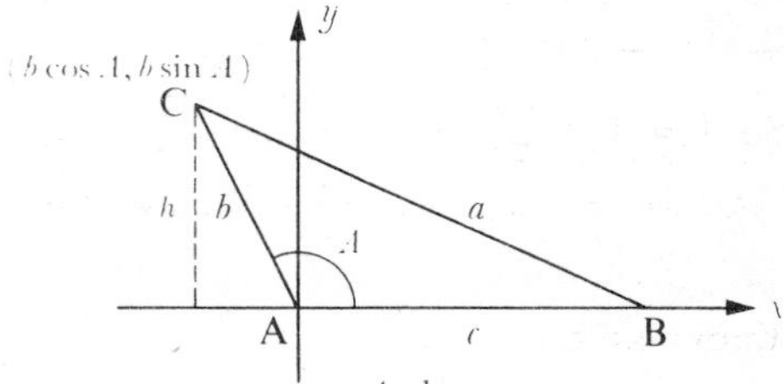

A obtuse

Sine rule

If the height of C above AB is h, then in each case $h = b \sin A$

Similarly $h = a \sin B$

So $\quad a \sin B = b \sin A$

$$\Rightarrow \quad \frac{a}{\sin A} = \frac{b}{\sin B}$$

Similarly $\quad \dfrac{a}{\sin A} = \dfrac{c}{\sin C}$

So $\quad \dfrac{a}{\sin A} = \dfrac{b}{\sin B} = \dfrac{c}{\sin C}$

Cosine rules

It is also clear in the above triangle that

$$a^2 = BC^2 = (b \cos A - c)^2 + (b \sin A - 0)^2$$

$$\Rightarrow \quad a^2 = b^2 \cos^2 A - 2bc \cos A + c^2 + b^2 \sin^2 A$$

$$\Rightarrow \quad a^2 = b^2 + c^2 - 2bc \cos A$$

Similarly, $b^2 = c^2 + a^2 - 2ca \cos B$

and $c^2 = a^2 + b^2 - 2ab \cos C$

Example

In $\triangle$ABC, $a = 4$, $b = 3$, and $c = 2$. Find the angles A, B, C.

Using the cosine rule, $4^2 = 3^2 + 2^2 - 2{\times}3{\times}2 \cos A$

$$\Rightarrow \quad \cos A = -\frac{3}{12} = -0.25 \quad \Rightarrow \quad A = 104^\circ\ 29'$$

By the Sine rule, $\dfrac{4}{\sin 104^\circ\ 29'} = \dfrac{3}{\sin B} = \dfrac{2}{\sin C}$

$\Rightarrow \quad \sin B = \frac{3}{4} \sin 104^\circ\ 29' = \frac{3}{4} \times 0.968\ 2 = 0.726\ 15$

$\Rightarrow \quad B = 46^\circ\ 34'$ (since B cannot be obtuse)

and $\sin C = \frac{1}{2} \sin 104^\circ\ 29' = 0.484\ 1$

$\Rightarrow \quad C = 28^\circ\ 57'$.

So $A = 104^\circ\ 29'$, $B = 46^\circ\ 34'$, $C = 28^\circ\ 57'$

(and we check that $A + B + C = 180^\circ$).

Exercise 5.4

Calculate the unknown sides and angles of $\triangle$ABC, where

1 (*i*) $a = 5$, $A = 80^\circ$, $B = 30^\circ$;
(*ii*) $b = 4.17$, $A = 106^\circ\ 5'$, $B = 36^\circ\ 41'$.

2 (*i*) $a = 4$, $b = 5$, $c = 6$;
(*ii*) $a = 3.49$, $b = 4.62$, $c = 6.93$.

3 (*i*) $b = 3$, $c = 4$, $A = 50^\circ$;
(*ii*) $a = 4.71$, $b = 3.62$, $C = 103^\circ\ 41'$.

4 $a = 5$, $b = 3$, $B = 30^\circ$.
(In this question, begin by drawing $\triangle$ABC to scale; and state if there is more than one possible set of answers.)

5 A destroyer and a cruiser leave harbour at 0900 hours, the destroyer at 24 knots on course 037° and the cruiser at 15 knots on course 139°. Find

the bearing, and the distance in nautical miles, of the destroyer from the cruiser at 1400 hours. (1 knot is a speed of 1 nautical mile per hour.) (O.C.)

6 (*i*) $\triangle$ABC has $BC = 26$ cm, $CA = 14$ cm, $AB = 30$ cm. Calculate the angle BAC.
(*ii*) $\triangle$ABC has $AB = 30$ cm, $AC = 14$ cm, $BAC = 60°$. **BX** and **CY** are perpendiculars drawn to CA and AB respectively. Calculate the length of XY. (O.C.)

7 AM is a median of $\triangle$ABC (i.e., M is the mid-point of BC). Use the cosine rule to prove Apollonius' theorem, that:

$$AB^2 + AC^2 = 2(AM^2 + BM^2).$$

8 A triangle has sides 5 cm, 8 cm, 9 cm. Calculate:

(*i*) the size of its smallest angle;
(*ii*) the length of its shortest median (using Apollonius' theorem). (O.C.)

5.5 Functions of compound angles

Suppose we are told that

$\sin 10° = 0.173\,6$ and $\cos 10° = 0.984\,8$
$\sin 1° = 0.017\,5$ and $\cos 1° = 0.999\,4$

Can we use these values in order to calculate the sines and cosines of 11° and 9°?

The reader will probably find that the answers are not obvious. But similar questions will recur very frequently and it will soon become extremely important to know how

$\sin(\theta \pm \phi)$ and $\cos(\theta \pm \phi)$

are connected with

$\sin\theta$, $\cos\theta$, $\sin\phi$, and $\cos\phi$.

We shall begin by investigating the rotation of a point P(x, y) about O through an angle θ.

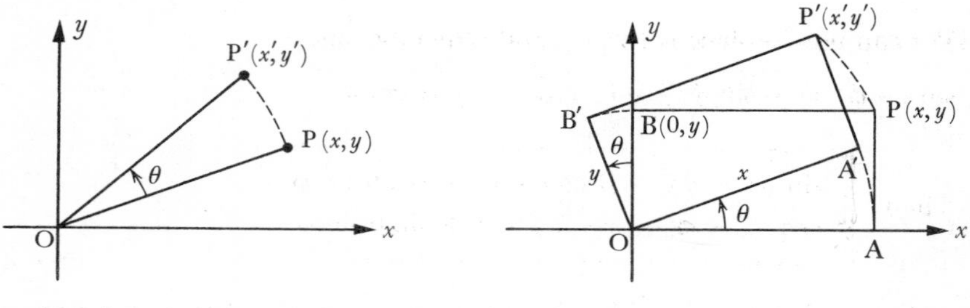

This can be achieved by rotating the rectangle OAPB through an angle θ, so that it takes the position OA′P′B′.

If (as in section 0.6) we denote P by $\begin{pmatrix} x \\ y \end{pmatrix}$,

then A $\begin{pmatrix} x \\ 0 \end{pmatrix}$ moves to A′ $\begin{pmatrix} x\cos\theta \\ x\sin\theta \end{pmatrix}$

and B $\begin{pmatrix} 0 \\ y \end{pmatrix}$ moves to B′ $\begin{pmatrix} -y\sin\theta \\ y\cos\theta \end{pmatrix}$

But P′ is given by

$$x' = x_{A'} + x_{B'} = x\cos\theta - y\sin\theta$$

$$y' = y_{A'} + y_{B'} = x\sin\theta + y\cos\theta$$

Hence $\begin{pmatrix} x' \\ y' \end{pmatrix} = \begin{pmatrix} x\cos\theta & -y\sin\theta \\ x\sin\theta & +y\cos\theta \end{pmatrix} = \begin{pmatrix} \cos\theta & -\sin\theta \\ \sin\theta & \cos\theta \end{pmatrix}\begin{pmatrix} x \\ y \end{pmatrix}$

and the rotation θ is a linear transformation with matrix

$$\boldsymbol{T}(\theta) = \begin{pmatrix} \cos\theta & -\sin\theta \\ \sin\theta & \cos\theta \end{pmatrix}$$

But rotation $\theta + \phi$ is the same as rotation ϕ followed by rotation θ, and so has matrix

$$\boldsymbol{T}(\theta + \phi) = \boldsymbol{T}(\theta)\boldsymbol{T}(\phi)$$

Hence $\begin{pmatrix} \cos(\theta+\phi) & -\sin(\theta+\phi) \\ \sin(\theta+\phi) & \cos(\theta+\phi) \end{pmatrix}$

$$= \begin{pmatrix} \cos\theta & -\sin\theta \\ \sin\theta & \cos\theta \end{pmatrix}\begin{pmatrix} \cos\phi & -\sin\phi \\ \sin\phi & \cos\phi \end{pmatrix}$$

$$= \begin{pmatrix} \cos\theta\cos\phi - \sin\theta\sin\phi & -\sin\theta\cos\phi - \cos\theta\sin\phi \\ \sin\theta\cos\phi + \cos\theta\sin\phi & \cos\theta\cos\phi - \sin\theta\sin\phi \end{pmatrix}$$

So

$$\boxed{\begin{aligned} \sin(\theta+\phi) &= \sin\theta\cos\phi + \cos\theta\sin\phi \\ \cos(\theta+\phi) &= \cos\theta\cos\phi - \sin\theta\sin\phi \end{aligned}}$$

We can now replace ϕ by $-\phi$ and remember that

$\sin(-\phi) = -\sin\phi$ and $\cos(-\phi) = \cos\phi$

Then

$$\boxed{\begin{aligned} \sin(\theta-\phi) &= \sin\theta\cos\phi - \cos\theta\sin\phi \\ \cos(\theta-\phi) &= \cos\theta\cos\phi + \sin\theta\sin\phi \end{aligned}}$$

These, therefore, are the formulae which were sought at the start of this section, and the reader can now use them to calculate the sines and cosines of 11° and 9°, checking his answers from tables.

They are all extremely important results which should be thoroughly understood and committed to memory.

Lastly,

$$\tan(\theta + \phi) = \frac{\sin(\theta + \phi)}{\cos(\theta + \phi)} = \frac{\sin\theta\cos\phi + \cos\theta\sin\phi}{\cos\theta\cos\phi - \sin\theta\sin\phi}$$

$$= \frac{\dfrac{\sin\theta\cos\phi + \cos\theta\sin\phi}{\cos\theta\cos\phi}}{\dfrac{\cos\theta\cos\phi - \sin\theta\sin\phi}{\cos\theta\cos\phi}}$$

So
$$\tan(\theta + \phi) = \frac{\tan\theta + \tan\phi}{1 - \tan\theta\tan\phi}$$

Similarly
$$\tan(\theta - \phi) = \frac{\tan\theta - \tan\phi}{1 + \tan\theta\tan\phi}$$

Example

Find expressions for the values of sin 75°, cos 75°, and tan 75°.

$$\sin 75^\circ = \sin(45^\circ + 30^\circ)$$

$$= \sin 45^\circ \cos 30^\circ + \cos 45^\circ \sin 30^\circ$$

$$= \frac{1}{\sqrt{2}}\frac{\sqrt{3}}{2} + \frac{1}{\sqrt{2}}\frac{1}{2} = \frac{\sqrt{3}+1}{2\sqrt{2}}$$

$$\cos 75^\circ = \cos(45^\circ + 30^\circ)$$

$$= \cos 45^\circ \cos 30^\circ - \sin 45^\circ \sin 30^\circ$$

$$= \frac{1}{\sqrt{2}}\frac{\sqrt{3}}{2} - \frac{1}{\sqrt{2}}\frac{1}{2} = \frac{\sqrt{3}-1}{2\sqrt{2}}$$

$$\tan 75^\circ = \tan(45^\circ + 30^\circ)$$

$$= \frac{\tan 45^\circ + \tan 30^\circ}{1 - \tan 45^\circ \tan 30^\circ} = \frac{1 + \dfrac{1}{\sqrt{3}}}{1 - 1\dfrac{1}{\sqrt{3}}}$$

$$= \frac{1 + \dfrac{1}{\sqrt{3}}}{1 - \dfrac{1}{\sqrt{3}}} = \frac{\sqrt{3}+1}{\sqrt{3}-1} = \frac{(\sqrt{3}+1)^2}{(\sqrt{3}-1)(\sqrt{3}+1)}$$

$$= \frac{4 + 2\sqrt{3}}{2} = 2 + \sqrt{3}$$

Exercise 5.5

1 Use the above formulae to calculate, without using tables:
(*i*) $\sin(60° + 45°)$, $\cos(60° + 45°)$, $\tan(60° + 45°)$;
(*ii*) $\sin(60° - 45°)$, $\cos(60° - 45°)$, $\tan(60° - 45°)$.

2 Use the compound angle formulae to simplify:
(*i*) $\sin(90° + x)$, $\cos(90° + x)$, $\tan(90° + x)$;
(*ii*) $\sin(180° - x)$, $\cos(180° - x)$, $\tan(180° - x)$.

3 Express in terms of $\sin x$, $\cos x$, $\tan x$:
(*i*) $\sin(45° + x)$, $\cos(45° + x)$, $\tan(45° + x)$;
(*ii*) $\sin(60° - x)$, $\cos(60° - x)$, $\tan(60° - x)$.

4 Simplify:
(*i*) $\sin 45° \cos 15° + \cos 45° \sin 15°$;
(*ii*) $\cos 35° \cos 15° - \sin 35° \sin 15°$;
(*iii*) $\sin 2\alpha \cos\beta - \cos 2\alpha \sin\beta$;
(*iv*) $\cos 2\theta \cos 3\phi + \sin 2\theta \sin 3\phi$;
(*v*) $\dfrac{\sqrt{3} + \tan\theta}{1 - \sqrt{3}\tan\theta}$;
(*vi*) $\dfrac{\tan\theta - 1}{1 + \tan\theta}$.

5 Simplify:
(*i*) $\sin(A + B) + \sin(A - B)$;
(*ii*) $\cos(A + B) - \cos(A - B)$;
(*iii*) $\dfrac{\cos(A + B) + \cos(A - B)}{\sin(A + B) - \sin(A - B)}$;
(*iv*) $\dfrac{\sin(A - B)}{\cos A \cos B}$;
(*v*) $\dfrac{\cos(A + B)}{\cos A \sin B}$;
(*vi*) $\tan(A + 45°)\tan(A - 45°)$.

6 Express $\tan(A + B + C)$ in terms of $\tan A$, $\tan B$, $\tan C$. Hence:
(*i*) If A, B, C are acute and $\tan A = \frac{1}{2}$, $\tan B = \frac{1}{5}$, $\tan C = \frac{1}{8}$, calculate $A + B + C$.
(*ii*) If A, B, C are the angles of a triangle, show that the sum of their tangents is equal to their product.

5.6 Multiple angles

If we put $\phi = \theta$ in the formulae of the last section, we obtain

$$\sin 2\theta = 2 \sin \theta \cos \theta$$
$$\cos 2\theta = \cos^2 \theta - \sin^2 \theta$$
$$= 2 \cos^2 \theta - 1 = 1 - 2 \sin^2 \theta$$
$$\tan 2\theta = \frac{2 \tan \theta}{1 - \tan^2 \theta}$$

This last formulae can be used to express $\tan \theta$ in terms of $\tan \frac{1}{2}\theta$. If we let $t = \tan \frac{1}{2}\theta$, then

$$\tan \theta = \frac{2t}{1 - t^2}$$

and it is also easy to express $\sin \theta$ and $\cos \theta$ in terms of t.

For $\sin \theta = 2 \sin \frac{1}{2}\theta \cos \frac{1}{2}\theta = \dfrac{2 \tan \frac{1}{2}\theta}{\sec^2 \frac{1}{2}\theta} = \dfrac{2t}{1 + t^2}$

Similarly $\cos \theta = \cos^2 \frac{1}{2}\theta - \sin^2 \frac{1}{2}\theta$

$$= \frac{1 - \tan^2 \frac{1}{2}\theta}{\sec^2 \frac{1}{2}\theta} = \frac{1 - t^2}{1 + t^2}$$

So $\sin \theta = \dfrac{2t}{1 + t^2}, \quad \cos \theta = \dfrac{1 - t^2}{1 + t^2}, \quad \tan \theta = \dfrac{2t}{1 - t^2}$

which are easily memorable from the triangle

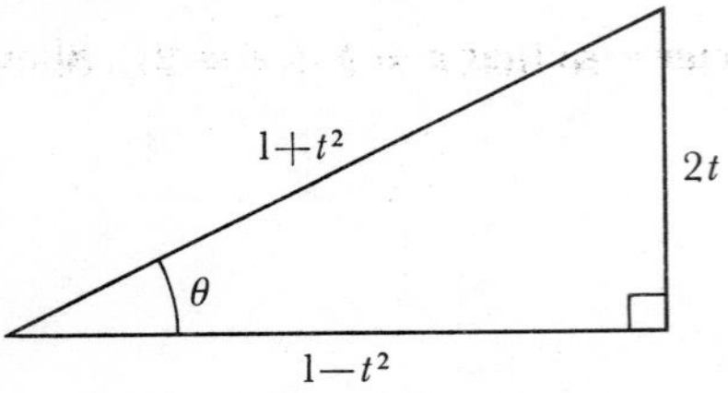

in which $(1 - t^2)^2 + (2t)^2 = (1 + t^2)^2$.

Example 1

Simplify $\dfrac{1 - \cos A}{1 + \cos A}$

$$1 - \cos A = 1 - (1 - 2 \sin^2 \tfrac{1}{2}A) = 2 \sin^2 \tfrac{1}{2}A$$

and $1 + \cos A = 1 + (2 \cos^2 \frac{1}{2}A - 1) = 2 \cos^2 \frac{1}{2}A$

So $$\frac{1 - \cos A}{1 + \cos A} = \frac{2 \sin^2 \frac{1}{2}A}{2 \cos^2 \frac{1}{2}A} = \tan^2 \tfrac{1}{2}A$$

Example 2

Express tan $3A$ in terms of tan A

$$\tan 3A = \tan (2A + A) = \frac{\tan 2A + \tan A}{1 - \tan 2A \tan A}$$

$$= \frac{\dfrac{2 \tan A}{1 - \tan^2 A} + \tan A}{1 - \dfrac{2 \tan A}{1 - \tan^2 A} \tan A}$$

$$= \frac{3 \tan A - \tan^3 A}{1 - 3 \tan^2 A}$$

Example 3

Find the angles between 0° and 360° which satisfy $\cos 2\theta = 7 \cos \theta + 3$

Now $\cos 2\theta = 7 \cos \theta + 3$

$\Rightarrow \quad 2 \cos^2 \theta - 1 = 7 \cos \theta + 3$

$\Rightarrow \quad 2 \cos^2 \theta - 7 \cos \theta - 4 = 0$

$\Rightarrow \quad (2 \cos \theta + 1) (\cos \theta - 4) = 0$

$\Rightarrow \quad \cos \theta = -\frac{1}{2}$ or $+4$ (which is impossible)

$\Rightarrow \quad \theta = 120°$ or $240°$

Example 4

$\triangle ABC$ has sides a, b, c and semi-perimeter s (so that $a + b + c = 2s$). Show that its area is

$$\triangle = \sqrt{\{s(s - a)(s - b)(s - c)\}}$$

We know that

$$\begin{aligned}
\triangle &= \tfrac{1}{2} bc \sin A \\
&= \tfrac{1}{2} bc \sqrt{(1 - \cos^2 A)} \\
&= \frac{1}{2} bc \sqrt{\left\{1 - \left(\frac{b^2 + c^2 - a^2}{2bc}\right)^2\right\}} \\
&= \tfrac{1}{4} \sqrt{\{4b^2c^2 - (b^2 + c^2 - a^2)^2\}} \\
&= \tfrac{1}{4} \sqrt{[\{2bc + (b^2 + c^2 - a^2)\} \{2bc - (b^2 + c^2 - a^2)\}]} \\
&= \tfrac{1}{4} \sqrt{[\{(b + c)^2 - a^2\} \{a^2 - (b - c)^2\}]}
\end{aligned}$$

$$= \tfrac{1}{4}\sqrt{\{(a+b+c)(b+c-a)(a-b+c)(a+b-c)\}}$$
$$= \tfrac{1}{4}\sqrt{\{2s(2s-2a)(2s-2b)(2s-2c)\}}$$
$$\Rightarrow \quad \Delta = \sqrt{\{s(s-a)(s-b)(s-c)\}}$$

Exercise 5.6

1 Simplify:

(*i*) $2 \sin 15^\circ \cos 15^\circ$; (*ii*) $2 \cos^2 15^\circ - 1$;

(*iii*) $1 - 2 \sin^2 15^\circ$; (*iv*) $\cos^2 15^\circ - \sin^2 15^\circ$;

(*v*) $\cos^2 15^\circ + \sin^2 15^\circ$; (*vi*) $\dfrac{2 \tan 15^\circ}{1 - \tan^2 15^\circ}$.

2 If $\tan \theta = \frac{3}{4}$, evaluate (without using tables):
(*i*) $\sin 2\theta$; (*ii*) $\cos 2\theta$; (*iii*) $\tan 2\theta$.

3 Use double-angle formulae to find expressions for the values of:
(*i*) $\sin 22\frac{1}{2}^\circ$; (*ii*) $\cos 22\frac{1}{2}^\circ$; (*iii*) $\tan 22\frac{1}{2}^\circ$.

4 Express:
(*i*) $\sin 3A$ in terms of $\sin A$;
(*ii*) $\cos 3A$ in terms of $\cos A$.

5 Simplify:

(*i*) $\dfrac{\sin A}{1 + \cos A}$; (*ii*) $\operatorname{cosec} 2x - \cot 2x$;

(*iii*) $\cos^4 \theta - \sin^4 \theta$; (*iv*) $\dfrac{\sin 2A + \sin 4A}{1 + \cos 2A + \cos 4A}$;

(*v*) $\cot \theta - 2 \cot 2\theta$; (*vi*) $\tan 2A (\cot A - \tan A)$.

6 (*i*) Express $\cos 4A$ in terms of $\cos A$;
(*ii*) Express $\tan 4A$ in terms of $\tan A$.

7 Find an expression for the value of $\tan 7\frac{1}{2}^\circ$.

8 Find the values of x between 0° and 360° such that
(*i*) $4 \cos^3 x + 2 \cos x = 5 \sin 2x$; (*ii*) $\cos 3x - 3 \cos x = \cos 2x + 1$;
(*iii*) $\cot 2x = 2 + \cot x$; (*iv*) $\sin 3x = \sin^2 x$.

9 By expressing each side in terms of $t = \tan \dfrac{x}{2}$, prove that

$$\sec x + \tan x = \tan\left(45^\circ + \frac{x}{2}\right).$$

*5.7 The combination of waves

The sound-curve of a tuning-fork which has been lightly tapped is a simple sine-wave (see Fig. 1). But it frequently happens that a wave does not have such a simple curve. If, for example, the tuning-fork were struck with a piece of metal, the sound-wave of the resulting clang would be as shown in

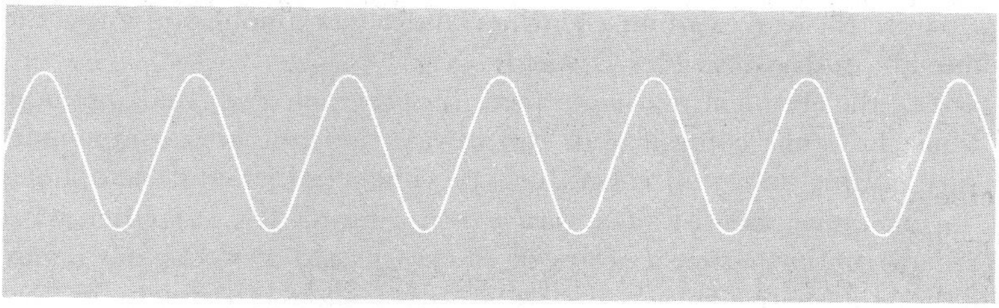

Fig. 1

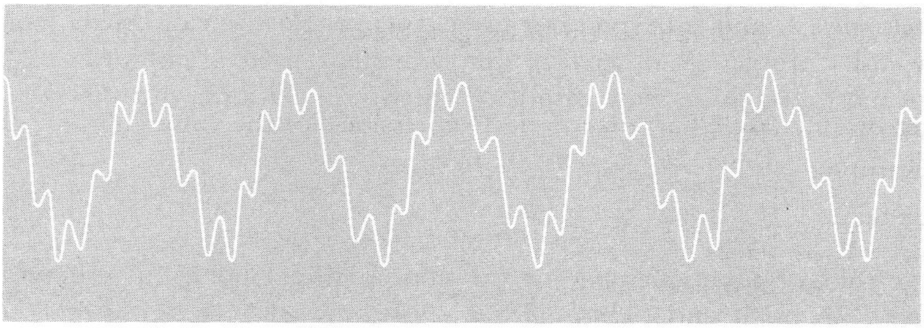

Fig. 2

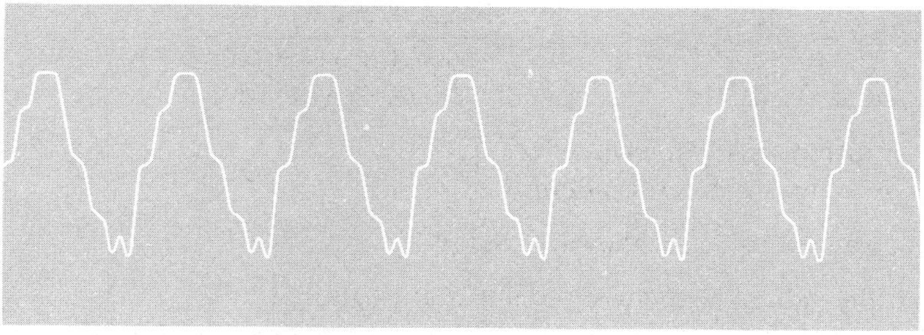

Fig. 3

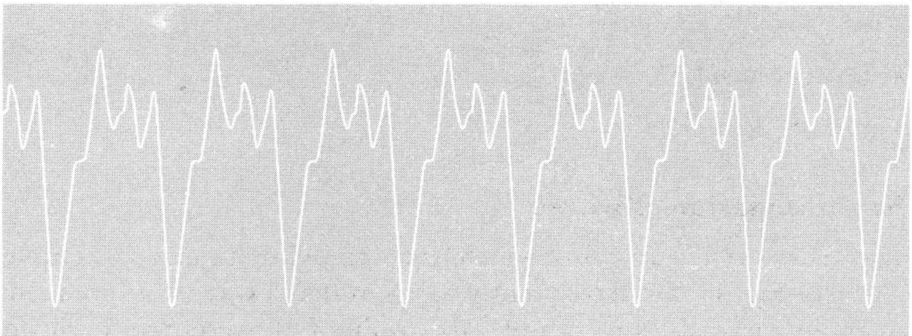

Fig. 4

Fig. 2, whilst the waves from a clarinet and a saxophone sounding the same note are as shown in Figs. 3 and 4.

In 1822, the French mathematician Fourier showed that periodic oscillations, however complicated, can always be split into components which are simple sine and cosine curves. The clarinet and saxophone notes, for instance, consist of a *fundamental* (middle C), together with smaller contributions from a series of *harmonics*, and it is the particular combination of these components which gives each instrument its distinctive tone. In a similar way, engineers studying the vibrations of a gas-turbine blade or the frame of an aircraft are able to separate their complicated behaviour into series of simpler components. This is known as *harmonic analysis* and is beyond our present scope. But we can briefly look at the easier problem, and see how simple oscillations are combined.

The most straightforward case is when two oscillations with the same period are superimposed.
For example,

$$f(x) = 3\sin x + 4\cos x$$

Here the two components have the same period (360°), but different *amplitudes* (3 and 4) and a *phase difference* (90°) between successive peaks:

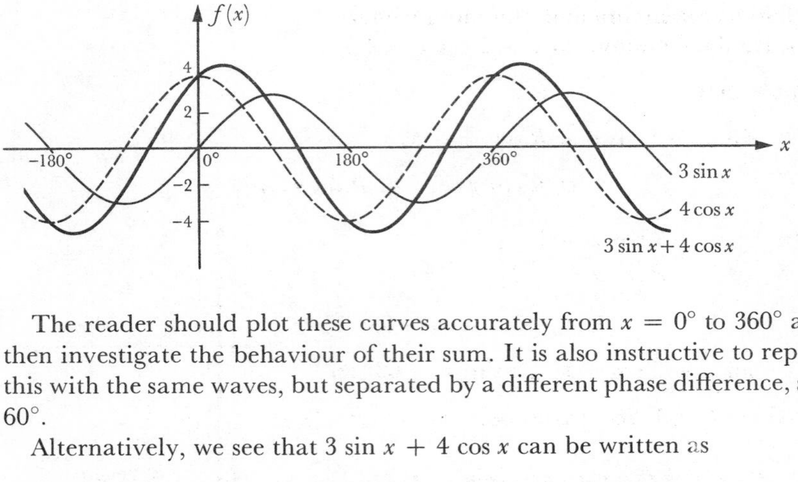

The reader should plot these curves accurately from $x = 0°$ to 360° and then investigate the behaviour of their sum. It is also instructive to repeat this with the same waves, but separated by a different phase difference, say 60°.

Alternatively, we see that $3\sin x + 4\cos x$ can be written as

$$R\sin(x + \alpha) \equiv R\cos\alpha\sin x + R\sin\alpha\cos x,$$

provided that $R\cos\alpha = 3$

and $R\sin\alpha = 4$

$$\Rightarrow \quad R = \sqrt{(3^2 + 4^2)} = 5$$

and $\tan\alpha = \dfrac{4}{3} \quad \Rightarrow \quad \alpha \approx 53°$

So $\quad 3\sin x + 4\cos x \equiv 5\sin(x + 53°)$

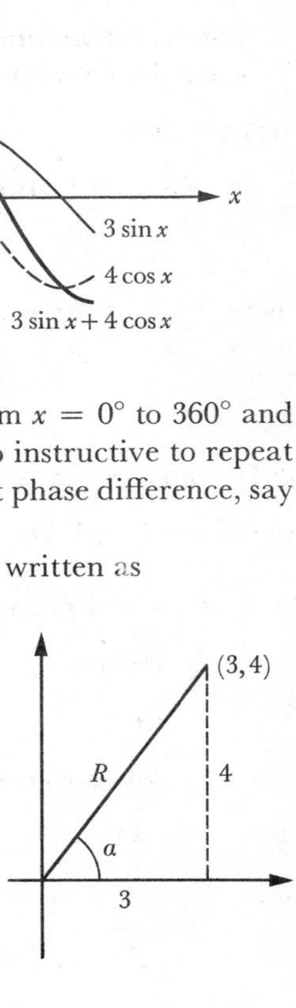

More generally, $a \sin x + b \cos x$ can be written as

$R \sin (x + \alpha) \equiv R \cos \alpha \sin x + R \sin \alpha \cos x$,

provided that $R \cos \alpha = a$

and $R \sin \alpha = b$

$\Rightarrow \quad R = \sqrt{a^2 + b^2}$

$\tan \alpha = \dfrac{b}{a}$

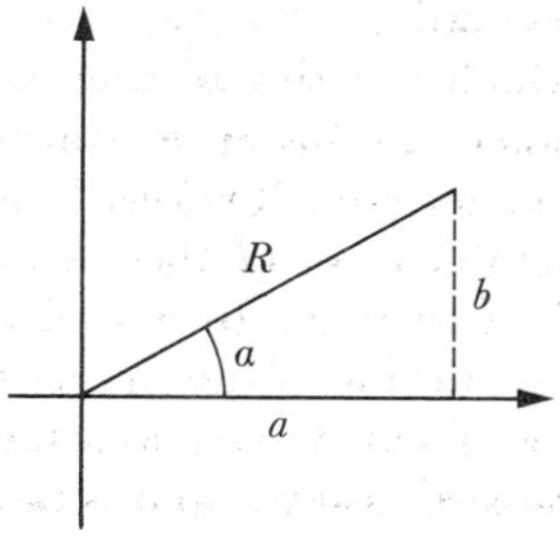

So in every case the sum of two such oscillations with the same period is another oscillation with this period: or, in musical terms, two pure notes of the same frequency always combine perfectly even though their oscillations have different amplitudes and are out of phase.

Example

Express $\sin x - 2 \cos x$ in the form $R \sin (x + \alpha)$, and so

(*i*) find its maximum and minimum values;
(*ii*) solve the equation $\sin x - 2 \cos x = 1.5$

Suppose that

$$\sin x - 2 \cos x \equiv R \sin (x + \alpha)$$

$$\equiv R \cos \alpha \sin x + R \sin \alpha \cos x$$

Then $\left.\begin{array}{l} R \cos \alpha = 1 \\ R \sin \alpha = -2 \end{array}\right\}$

$\Leftrightarrow \quad R = \sqrt{5}, \quad \alpha \approx -63° \, 26'$

So $\quad \sin x - 2 \cos x \equiv \sqrt{5} \sin (x - 63° \, 26')$

Now $\sin (x - 63° \, 26')$ varies between $+1$ and -1.

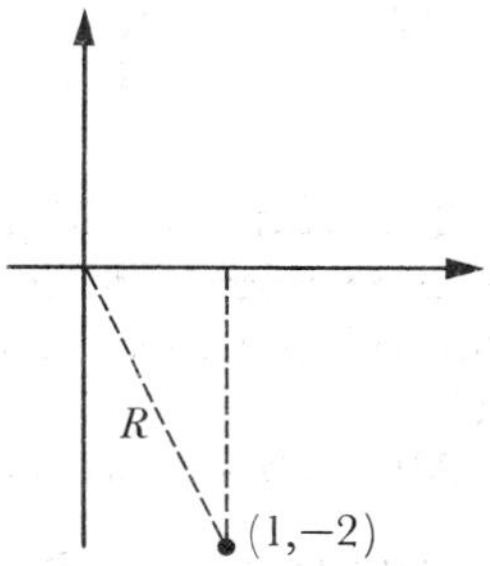

So (*i*) $f(x)$ has a *maximum value* $\sqrt{5}$

(when $x - 63° \, 26' = 90°, 450°$, etc.

i.e., $x = 153° \, 26', 513° \, 26' \ldots$)

and a *minimum value* $-\sqrt{5}$

(when $x = -26° \, 34', 333° \, 26'$, etc.)

and (*ii*) $\sin x - 2 \cos x = 1.5$

$\Rightarrow \quad \sqrt{5} \sin (x - 63° \, 26') = 1.5$

$$\Rightarrow \qquad \sin(x - 63°\ 26') = \frac{1.5}{2.236} = 0.670\,8$$

$$= \sin 42°\ 8',\ 137°\ 52',\ 402°\ 8', \text{ etc.}$$

$$\Rightarrow \qquad x - 63°\ 26' = 42°\ 8',\ 137°\ 52',\ 402°\ 8' \ldots$$

$$\Rightarrow \qquad x = 105°\ 34',\ 201°\ 18',\ 465°\ 34', \text{ etc.}$$

Alternatively, we can solve $\sin x - 2 \cos x = 1.5$

by letting $\quad t = \tan \frac{x}{2}$

Using the results of the last section, we obtain

$$\frac{2t}{1 + t^2} - \frac{2(1 - t^2)}{1 + t^2} = \frac{3}{2}$$

$$\Rightarrow \qquad 2t - 2 + 2t^2 = \frac{3}{2}(1 + t^2)$$

$$\Rightarrow \qquad t^2 + 4t - 7 = 0$$

$$\Rightarrow \quad \tan \frac{x}{2} = -2 \pm \sqrt{11} = 1.317 \text{ or } -5.317$$

$$\Rightarrow \qquad \frac{x}{2} = 52°\ 47', \quad 232°\ 47', \ldots$$

$$-79°\ 21', \quad 100°\ 39', \ldots$$

$$\Rightarrow \qquad x = 105°\ 34', \quad 201°\ 18',\ 465°\ 34' \ldots$$

So far we have confined our attention to the combination of functions which have the same period. But more interesting wave-forms arise from functions with different periods, as in the sound from a musical instrument. For a simple example, see No. 4 of the following exercise.

Exercise 5.7

1 Express the following functions in the form $A \sin(x + \alpha)$, and so find their maximum and minimum values:

(*i*) $\sin x + \cos x$; (*ii*) $2 \sin x + 3 \cos x$;
(*iii*) $4 \sin x - 3 \cos x$; (*iv*) $\cos x - 3 \sin x$.

2 Use a similar method to solve the equations for values between 0° and 360°:

(*i*) $\sin x + \cos x = 1$; (*ii*) $2 \sin x + 3 \cos x = -1$;
(*iii*) $4 \sin x - 3 \cos x = 2$; (*iv*) $\cos x - 3 \sin x = -2$.

3 Solve the equations of No. 2 by means of the substitution $t = \tan \frac{x}{2}$.

4 (Most suitable as a group exercise)

$f(x) = \sin x + \frac{1}{2}\sin 2x + \frac{1}{4}\sin 4x.$

Continue the following table at 10° intervals as far as $x = 370°$:

x	0°	10°	20°	30°
$\sin x$	0.0000	0.1736	0.3420	0.5000
$\sin 2x$	0.0000	0.3420	0.6428	0.8660
$\sin 4x$	0.0000	0.6428	0.9848	0.8660
$\sin x$	0.0000	0.1736	0.3420	0.5000
$\frac{1}{2}\sin 2x$	0.0000	0.1710	0.3214	0.4330
$\frac{1}{4}\sin 4x$	0.0000	0.1607	0.2462	0.2165
$f(x)$	0.0000	0.5053	0.9096	1.1495

Then plot on one graph the three component oscillations together with the curve of $f(x)$.

5.8 Factor formulae

There is another group of important formulae which are obtained from the equations of 5.5, simply by addition and subtraction:

$$\sin(\theta + \phi) + \sin(\theta - \phi) = 2\sin\theta\cos\phi$$

$$\sin(\theta + \phi) - \sin(\theta - \phi) = 2\cos\theta\sin\phi$$

$$\cos(\theta + \phi) + \cos(\theta - \phi) = 2\cos\theta\cos\phi$$

$$\cos(\theta + \phi) - \cos(\theta - \phi) = -2\sin\theta\sin\phi$$

If in these identities we put

$$\left.\begin{aligned}\theta + \phi &= A\\ \text{and}\quad \theta - \phi &= B\end{aligned}\right\}\quad\text{then}\quad\begin{aligned}\theta &= \tfrac{1}{2}(A + B)\\ \phi &= \tfrac{1}{2}(A - B)\end{aligned}$$

and we obtain:

$$\begin{aligned}\sin A + \sin B &= 2\sin\tfrac{1}{2}(A + B)\cos\tfrac{1}{2}(A - B)\\ \sin A - \sin B &= 2\cos\tfrac{1}{2}(A + B)\sin\tfrac{1}{2}(A - B)\\ \cos A + \cos B &= 2\cos\tfrac{1}{2}(A + B)\cos\tfrac{1}{2}(A - B)\\ \cos A - \cos B &= -2\sin\tfrac{1}{2}(A + B)\sin\tfrac{1}{2}(A - B)\\ &= 2\sin\tfrac{1}{2}(A + B)\sin\tfrac{1}{2}(B - A)\end{aligned}$$

As these enable us to factorise $\sin A \pm \sin B$ and $\cos A \pm \cos B$, they are sometimes called the *factor formulae.*

Example 1

Simplify $\dfrac{\sin A + \sin B}{\cos A - \cos B}$

$$\frac{\sin A + \sin B}{\cos A - \cos B} = \frac{2 \sin \frac{1}{2}(A + B) \cos \frac{1}{2}(A - B)}{-2 \sin \frac{1}{2}(A + B) \sin \frac{1}{2}(A - B)} = -\cot \tfrac{1}{2}(A - B)$$

Example 2

If A, B, C are the angles of a triangle, prove that

$$\sin A + \sin B + \sin C = 4 \cos \frac{A}{2} \cos \frac{B}{2} \cos \frac{C}{2}$$

$$\begin{aligned}
\sin A + \sin B + \sin C &= \sin A + 2 \sin \frac{B + C}{2} \cos \frac{B - C}{2} \\
&= 2 \sin \frac{A}{2} \cos \frac{A}{2} + 2 \sin \left(90^\circ - \frac{A}{2}\right) \cos \frac{B - C}{2} \\
&= 2 \sin \frac{A}{2} \cos \frac{A}{2} + 2 \cos \frac{A}{2} \cos \frac{B - C}{2} \\
&= 2 \cos \frac{A}{2} \left(\sin \frac{A}{2} + \cos \frac{B - C}{2}\right) \\
&= 2 \cos \frac{A}{2} \left(\cos \frac{B + C}{2} + \cos \frac{B - C}{2}\right) \\
&= 4 \cos \frac{A}{2} \cos \frac{B}{2} \cos \frac{C}{2}
\end{aligned}$$

Example 3

Solve, for values of θ between 0° and 360°, the equation

$$\cos x + \cos 2x = \sin 2x - \sin x$$

Now $\qquad \cos x + \cos 2x = \sin 2x - \sin x$

$$\Rightarrow \qquad 2 \cos \frac{3x}{2} \cos \frac{x}{2} = 2 \sin \frac{x}{2} \cos \frac{3x}{2}$$

$$\Rightarrow \quad 2 \cos \frac{3x}{2} \left(\sin \frac{x}{2} - \cos \frac{x}{2}\right) = 0$$

Case 1 $\quad \cos \dfrac{3x}{2} = 0 \quad \Rightarrow \quad \dfrac{3x}{2} = 90^\circ, 270^\circ, 450^\circ, \ldots$

$$\Rightarrow \qquad x = 60^\circ, 180^\circ, 300^\circ, \ldots$$

Case 2 $\sin\frac{x}{2} - \cos\frac{x}{2} = 0$

$$\Rightarrow \qquad \tan\frac{x}{2} = 1$$

$$\Rightarrow \qquad \frac{x}{2} = 45^\circ, 225^\circ, \ldots$$

$$\Rightarrow \qquad x = 90^\circ, 450^\circ, \ldots$$

So the required solutions are

$x = 60^\circ, 90^\circ, 180^\circ, 300^\circ$

Exercise 5.8

1 Factorise:

(*i*) $\sin 3\theta + \sin 5\theta$; (*ii*) $\sin 4A - \sin 2A$;
(*iii*) $\cos(x + 60^\circ) + \cos(x - 60^\circ)$; (*iv*) $\cos(x + 45^\circ) - \cos(x - 45^\circ)$.

2 Simplify:

(*i*) $\dfrac{\cos A + \cos B}{\sin A - \sin B}$; (*ii*) $\dfrac{\sin A + \sin B}{\cos A + \cos B}$;

(*iii*) $\dfrac{\sin 3A - \sin A}{\cos 3A - \cos A}$; (*iv*) $\dfrac{\sin 2A + \sin 3A}{\cos 2A - \cos 3A}$.

3 Prove:

(*i*) $\sin A + \sin 2A + \sin 3A = \sin 2A(2\cos A + 1)$;
(*ii*) $\cos A + 2\cos 3A + \cos 5A = 4\cos^2 A \cos 3A$;
(*iii*) $\cos A - 2\cos 3A + \cos 5A = 2\sin A(\sin 2A - \sin 4A)$;
(*iv*) $\dfrac{\sin A - \sin 2A + \sin 3A}{\cos A - \cos 2A + \cos 3A} = \tan 2A$.

4 If A, B, C are the angles of $\triangle ABC$, prove that

(*i*) $\sin 2A + \sin 2B + \sin 2C = 4\sin A \sin B \sin C$;
(*ii*) $\cos A + \cos B + \cos C = 1 + 4\sin\frac{A}{2}\sin\frac{B}{2}\sin\frac{C}{2}$;
(*iii*) $\sin A - \sin B + \sin C = 4\sin\frac{A}{2}\cos\frac{B}{2}\sin\frac{C}{2}$;
(*iv*) $\sin^2 A + \sin^2 B + \sin^2 C = 2 + 2\cos A \cos B \cos C$.

5 Solve the following equations, giving all solutions between 0° and 180°:

(*i*) $\sin x + \sin 5x = \sin 3x$;
(*ii*) $\cos x + \cos 2x + \cos 3x = 0$;
(*iii*) $\sin x - 2\sin 2x + \sin 3x = 0$;
(*iv*) $\cos x - \sin 2x = \cos 5x$.

5.9 Radians. Small angles

So far we have measured angles in degrees, using (like the Babylonians) a sub-division of the circle into 360 parts. There is, however, frequently great advantage if another unit is used, called the *radian*: this is the angle subtended at the centre of a circle by an arc whose length is equal to the radius.

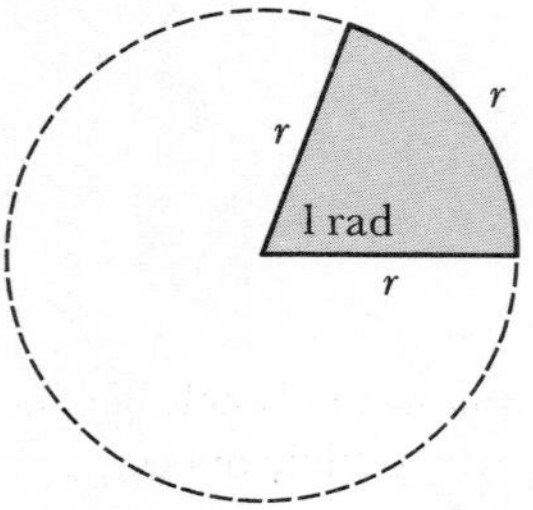

Now the total circumference of the circle is $2\pi \times$ radius.
So the total angle at the centre is $2\pi \times 1$ rad $= 2\pi$ rad.
Hence 2π rad $= 360°$

$$\Rightarrow \qquad 1 \text{ rad} = \frac{360°}{2\pi} = \frac{180°}{\pi} \approx 57° \, 18'$$

Conversely, $1° = \dfrac{\pi}{180} \approx 0.0175$ rad.

From now onwards all angles will be given in radians unless otherwise stated, so we simply say that

$$180° = \pi, \quad 90° = \pi/2, \quad 60° = \pi/3, \quad 45° = \pi/4, \quad 30° = \pi/6, \quad \text{etc.}$$

Length of arc and area of sector

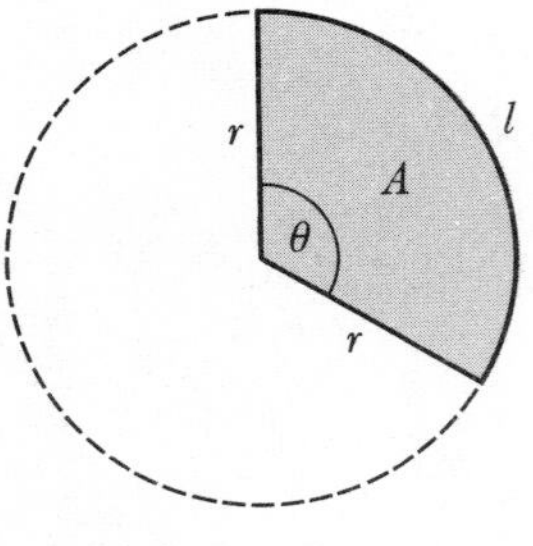

Suppose that a circular sector of radius r and angle θ has arc-length l and area A.

The whole angle 2π generates the complete circle, which has arc-length $2\pi r$ and area πr^2.

So the angle θ will generate a fraction $\dfrac{\theta}{2\pi}$ of the complete circle.

This sector has arc-length $= \dfrac{\theta}{2\pi} \times 2\pi r = r\theta$

and area $= \dfrac{\theta}{2\pi} \times \pi r^2 = \frac{1}{2}r^2\theta$

$$\Rightarrow \quad \boxed{\begin{aligned} l &= r\theta \\ A &= \tfrac{1}{2}r^2\theta \end{aligned}}$$

Example

Two circular discs of radius 2 cm are placed on a table so that each has its centre on the circumference of the other. Find the area of their overlap.

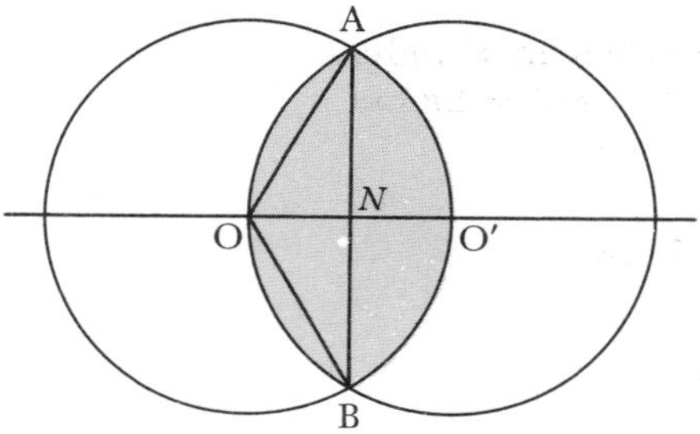

$\triangle AOO'$ is equilateral $\Rightarrow \angle AOB = 120° = \dfrac{2\pi}{3}$

Now ON $= 1$ and AN $= \sqrt{3}$

So area of sector of disc centre O between radii OA and OB

$$= \frac{1}{2}\,2^2\,\frac{2\pi}{3} = \frac{4\pi}{3} = 4.189 \text{ cm}^2$$

and area of $\triangle AOB = \sqrt{3} = 1.732 \text{ cm}^2$

$\Rightarrow$ area of minor segment AO′B $= 4.189 - 1.732 = 2.457 \text{ cm}^2$

$\Rightarrow$ area of overlap $\approx 4.91 \text{ cm}^2$

Exercise 5.9a

1 Express the following angles in radians:
(*i*) 10°; (*ii*) 36°; (*iii*) 120°; (*iv*) 180°;
(*v*) 270°; (*vi*) 135°; (*vii*) 1′; (*viii*) 1″.

2 Express in degrees:

(*i*) $\frac{\pi}{2}$ rad; (*ii*) $\frac{3\pi}{2}$ rad; (*iii*) $\frac{5\pi}{6}$ rad; (*iv*) $\frac{2\pi}{3}$ rad;

(*v*) $\frac{3\pi}{4}$ rad; (*vi*) $\frac{\pi}{12}$ rad; (*vii*) $\frac{\pi}{15}$ rad; (*viii*) $\frac{\pi}{10}$ rad.

3 What are the values of:

(*i*) $\sin\frac{\pi}{2}$; (*ii*) $\cos\frac{\pi}{4}$; (*iii*) $\tan\frac{\pi}{3}$; (*iv*) $\sin\frac{\pi}{6}$;

(*v*) $\cos\pi$; (*vi*) $\sin\frac{3\pi}{4}$; (*vii*) $\cos\frac{2\pi}{3}$; (*viii*) $\tan\pi$;

(*ix*) $\cos\frac{3\pi}{2}$; (*x*) $\sin\frac{7\pi}{6}$.

4 Simplify:

(*i*) $\sin\left(\frac{\pi}{2} - x\right)$; (*ii*) $\cos(\pi - x)$;

(*iii*) $\cos(\pi + x)$; (*iv*) $\sin(2\pi - x)$;

(*v*) $\tan\left(\frac{\pi}{2} - x\right)$; (*vi*) $\tan(\pi + x)$.

5 Solve, in radians (giving values of x between 0 and 2π):

(*i*) $\sin x = \frac{1}{2}$; (*ii*) $\sin x = \frac{1}{\sqrt{2}}$; (*iii*) $\cos x = -1$;

(*iv*) $\cos x = \frac{1}{2}$; (*v*) $\tan x = +1$; (*vi*) $\tan x = -\sqrt{3}$.

6 Evaluate in radians (to 3 decimal places):
(*i*) 40°; (*ii*) 46° 15′; (*iii*) 184° 5′; (*iv*) 394° 32′.

7 Express in degrees and minutes:
(*i*) 0.52 rad; (*ii*) 1.73 rad; (*iii*) 5.136 rad; (*iv*) 0.011 rad.

8 Find the length of arc and area of a circular sector whose radius and angle are:
(*i*) 3 m and 2 rad;
(*ii*) 2 m and $\frac{\pi}{4}$ rad;
(*iii*) 4 m and 25°.

9 A chord AB subtends 120° at the centre O of circle whose radius is 10 cm. Find:
(*i*) the length of the minor arc AB;
(*ii*) the area of ΔOAB;
(*iii*) the area of the major segment cut off by AB.

10 The section of a tunnel consists of the major segment of a circle standing on a chord of length 4 m. If the greatest height of the tunnel is 6 m, calculate the radius of the circle.

Prove that the angle subtended by the chord at the centre of the circle is approximately 1.287 rad, and hence find the area of the cross-section of the tunnel, correct to the nearest square metre. (O.C.)

11 A circular cone with base-radius r and slant-height l is unrolled into a circular sector. Find:
(*i*) the angle of this sector;
(*ii*) the curved surface area of the cone.

12 A nautical mile was defined as the distance between two points of equal longitude whose latitudes differed by 1 minute. Assuming the earth to be a perfect sphere of radius 6 380 km, find the number of kilometres in a nautical mile.

Small angles

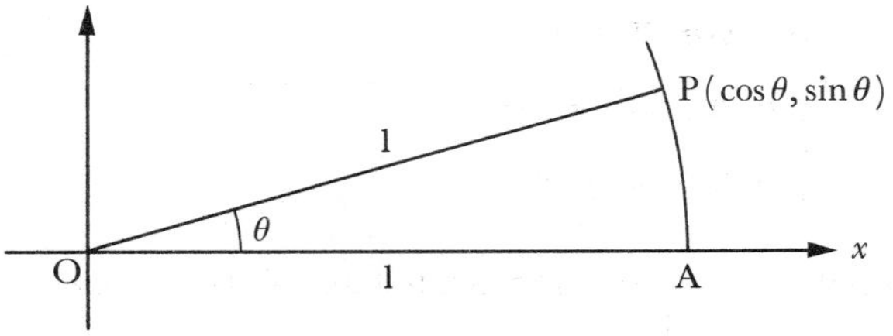

If θ is measured in radians and is small, it is seen from the figure that

$$\sin\theta \approx \widehat{AP} = \theta$$

and $$\cos\theta \approx OA = 1$$

Hence $$\tan\theta = \frac{\sin\theta}{\cos\theta} \approx \theta$$

So

$\sin\theta \approx \theta$	$\cos\theta \approx 1$	$\tan\theta \approx \theta$

Example

The distance of the Earth from the Sun is 1.5×10^8 km and this distance subtends an angle of 6″ at the nearest star. What is the distance of this star?

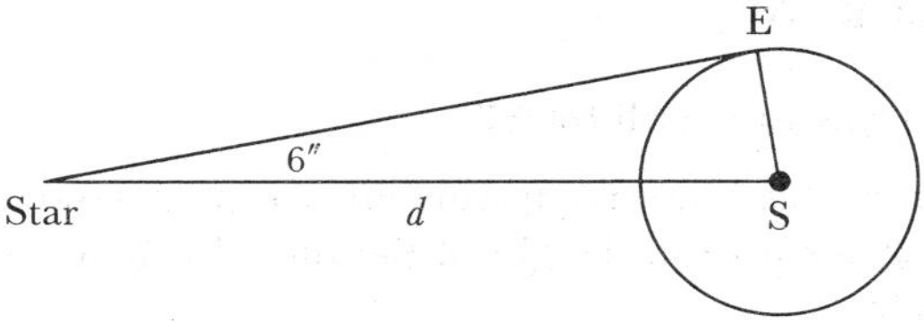

Now $$6'' = \frac{\pi}{180} \times \frac{1}{60} \times \frac{1}{10} \text{ radians}$$

So if the required distance is d km,

$$d \times \left(\frac{\pi}{180} \times \frac{1}{60} \times \frac{1}{10}\right) \approx 1.5 \times 10^8$$

$$\Rightarrow \qquad d \approx \frac{180 \times 60 \times 10 \times 1.5 \times 10^8}{\pi} = 1.6 \times 10^{13}$$

So the distance of the nearest star is approximately 1.6×10^{13} km ($\approx$ 1.7 light years).

Exercise 5.9b

1 The distances from the Earth of the Moon and the Sun are 3.84×10^5 km and 1.50×10^8 km respectively and their angular sizes for an observer on the Earth are 31.2′ and 31.9′ respectively. Calculate their diameters.

Even though the above approximations appear to be confirmed by a look at trigonometric tables, the meaning of the symbol $\approx$ has never been properly defined and has simply been taken to mean 'is approximately equal to'. How approximately? And what do we mean by θ being small?

In section 5.11 much is going to depend on a clear understanding of the behaviour of $\sin\theta$ as θ becomes small, so we must now investigate this in the more rigorous language of limits.

We shall consider the same section OAP and let AB be drawn perpendicular to OA.

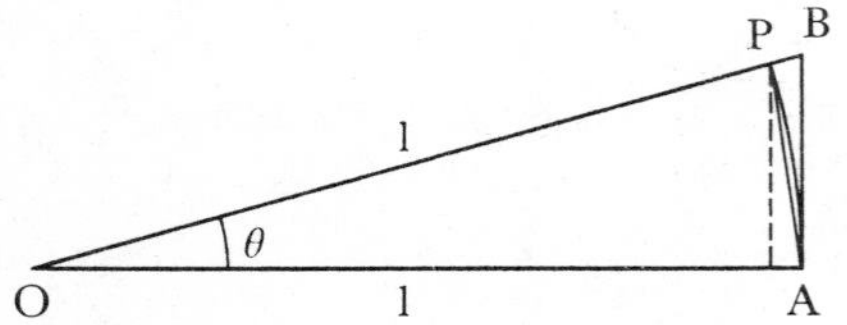

$$\triangle OAP < \text{Sector } OAP < \triangle OAB$$

$$\Rightarrow \quad \tfrac{1}{2}\times 1\times\sin\theta < \tfrac{1}{2}\times 1^2\times\theta < \tfrac{1}{2}\times 1\times\tan\theta$$

$$\Rightarrow \quad \sin\theta < \theta < \tan\theta$$

Dividing by $\sin\theta$, we obtain

$$1 < \frac{\theta}{\sin\theta} < \sec\theta$$

Now as $\theta \to 0$, $\sec\theta \to 1$

$$\Rightarrow \qquad \frac{\theta}{\sin\theta} \to 1$$

$$\Rightarrow \qquad \frac{\sin\theta}{\theta} \to 1$$

So $$\boxed{\lim_{\theta\to 0}\frac{\sin\theta}{\theta} = 1}$$

*5.10 Oscillations: amplitude, period, and frequency

We are now in a position to use sine and cosine functions to describe all kinds of waves, vibrations and oscillations.

Example

Two objects, A and B, are bobbing up and down in water and after t seconds their heights x m above their equilibrium positions are

(*a*) $x = 2\sin t$;   (*b*) $x = 3\sin 5\pi t$.

(*a*) *Object A:* $x = 2\sin t$

In this case the graph of x against t is

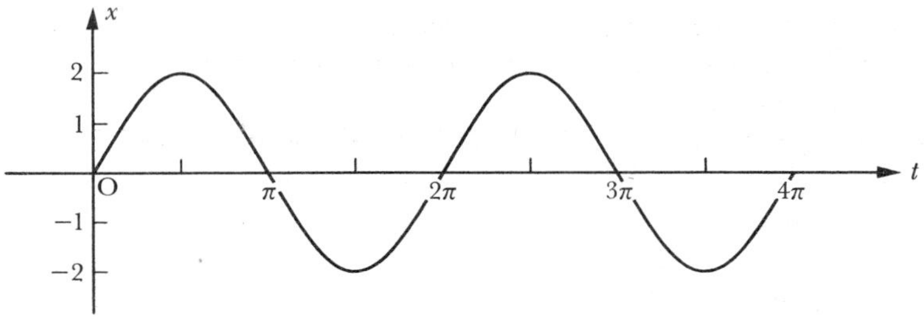

and

(*i*) the maximum displacement, or *amplitude*, of the oscillation is 2 m;
(*ii*) the *period* of the oscillation is clearly $2\pi \approx 6.28$ s. So the oscillations of this object are very slow (presumably because its density is nearly that of the water, or because it is badly water-logged);
(*iii*) the *frequency* of the oscillation is the number of oscillations (or *cycles*) per second, which is

$$\frac{1}{2\pi} \approx \frac{1}{6.28} \approx 0.159\ \text{s}^{-1}$$

(and in electrical situations 1 cycle per second is usually called 1 Hertz:

$1\ s^{-1} = 1$ Hz). Furthermore, the equation $x = 2 \sin t$ enables us to find the displacement at any moment of time: after 5 seconds, for instance,

$$x = 2 \sin 5$$
$$= 2 \sin 286^\circ\ 30' \quad (\text{since } 5 \text{ rad} = 286^\circ\ 30')$$
$$= 2 \times -0.9588 = -1.92$$

So the object is 1.92 m below its equilibrium position (and so only 0.08 m above its lowest point).

(*b*) *Object B:* $x = 3 \sin 5\pi t$

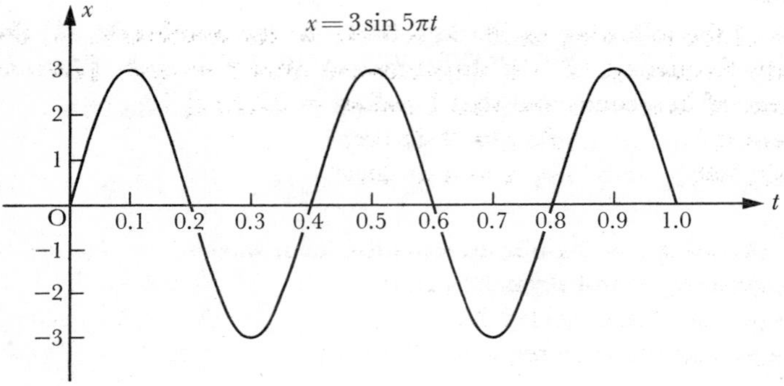

In this case the graph has
(*i*) a larger amplitude, 3 m;
(*ii*) successive cycles beginning when

$$5\pi t = 0, 2\pi, 4\pi, 6\pi$$

i.e., when $t = 0, 0.4, 0.8, 1.2\ldots.$

Hence the period $= 0.4$ s

and $\quad \text{frequency} = \dfrac{1}{0.4} = 2.5\ s^{-1}$

More generally, $x = a \sin nt$

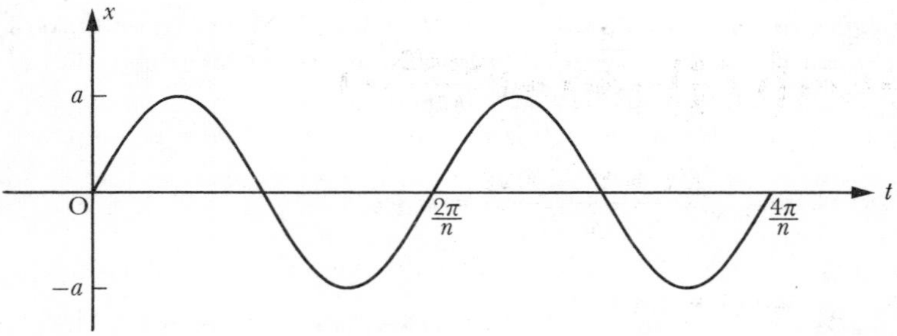

has (*i*) amplitude a
(*ii*) successive cycles beginning at

$$nt = 0, 2\pi, 4\pi, 6\pi \ldots$$

i.e., $$t = 0, \frac{2\pi}{n}, \frac{4\pi}{n}, \frac{6\pi}{n} \ldots$$

So its period is $\frac{2\pi}{n}$ and its frequency is $\frac{n}{2\pi}$.

Exercise 5.10

1 For each of the following oscillations find: (*a*) the amplitude, (*b*) the period, (*c*) the frequency, (*d*) the displacement after 2 seconds. [Assume that t is measured in seconds and that 1 radian = 57.29°.]
(*i*) $x = 4 \cos t$; (*ii*) $x = 2 \sin 10t$;
(*iii*) $x = 5 \sin 100t$; (*iv*) $x = a \sin 2\pi nt$.

2 Write in the form $x = a \sin nt$ an equation for a wave:
(*i*) with amplitude 10 and period 2π s;
(*ii*) with amplitude 3 and period 2 s;
(*iii*) with amplitude 0.5 and frequency 100 s^{-1}.

5.11 Differentiation of trigonometric functions

Let $$f(x) = \sin x$$

Then $$\frac{f(x+h) - f(x)}{h} = \frac{\sin(x+h) - \sin x}{h}$$

$$= \frac{2 \sin h/2 \cos (x + h/2)}{h}$$

$$= \cos\left(x + \frac{h}{2}\right) \frac{\sin h/2}{h/2}$$

As $h \to 0$, $\cos\left(x + \frac{h}{2}\right) \to \cos x$ and $\frac{\sin h/2}{h/2} \to 1$

So $f'(x) = \lim_{h \to 0} \frac{f(x+h) - f(x)}{h} = \cos x$

Hence: $\frac{d}{dx}(\sin x) = \cos x$

This can be illustrated:

$f(x) = \sin x$

$f'(x) = \cos x$

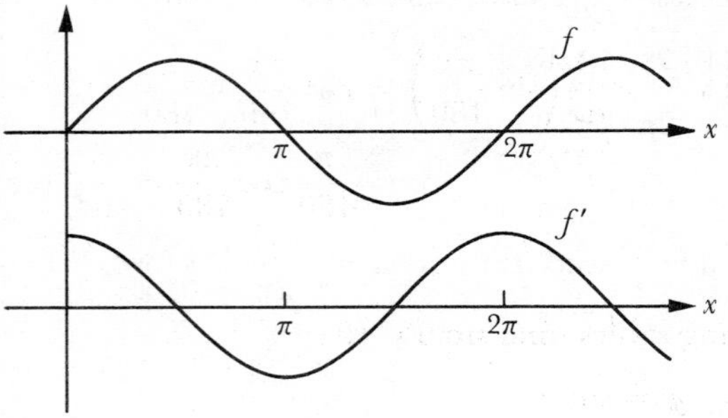

We see that the derived curve of a sine wave is an exactly similar curve, but with a phase difference of $\pi/2$. (Readers familiar with electrical induction will recognise that this is exactly what is happening in an A.C. transformer, where the current in the secondary circuit is proportional to the rate of change of that in the primary circuit.) We can therefore predict that the derivative of $\cos x$ will be yet another similar wave with a further phase-difference of $\pi/2$, i.e., be the function $-\sin x$.

For a formal proof, we can proceed as with $\sin x$ and consider the expression

$$\frac{\cos(x + h) - \cos x}{h}$$

Alternatively,

$$\frac{\mathrm{d}}{\mathrm{d}x}(\cos x) = \frac{\mathrm{d}}{\mathrm{d}x}\left\{\sin\left(x + \frac{\pi}{2}\right)\right\} = \cos\left(x + \frac{\pi}{2}\right) = -\sin x$$

So $$\boxed{\frac{\mathrm{d}}{\mathrm{d}x}(\sin x) = \cos x \quad \text{and} \quad \frac{\mathrm{d}}{\mathrm{d}x}(\cos x) = -\sin x}$$

The reader will have noted the tremendous benefit that has come from the use of radians, enabling us to say that

$$\frac{\sin h/2}{h/2} \rightarrow 1 \quad \text{as} \quad h \rightarrow 0.$$

Example 1

Differentiate:

(*i*) $\sin 3x$, (*ii*) $\cos^2 x$, (*iii*) $\sin x^\circ$.

(*i*) $$\frac{\mathrm{d}}{\mathrm{d}x}(\sin 3x) = (\cos 3x)\times 3 = 3\cos 3x;$$

(ii) $\frac{d}{dx}(\cos^2 x) = \frac{d}{dx}(\cos x)^2 = 2\cos x \times -\sin x = -2\cos x \sin x;$

(iii) $\frac{d}{dx}(\sin x^\circ) = \frac{d}{dx}\left(\sin\frac{\pi x}{180}\right) = \cos\frac{\pi x}{180} \times \frac{\pi}{180}$

$$= \frac{\pi}{180}\cos\frac{\pi x}{180} = \frac{\pi}{180}\cos x^\circ$$

Example 2

Without using tables, find sin 60° 10′.

Let $\quad y = \sin x$

Then $\quad x = 60^\circ \Rightarrow y = \sin 60^\circ = \frac{\sqrt{3}}{2} = 0.8660$

Also $\quad \delta y \approx (\cos x) \times \delta x \quad (x \text{ in radians})$

So $\quad \delta x = 10' = \frac{\pi}{1080}\text{ rad} \Rightarrow \delta y \approx \cos 60^\circ\left(\frac{\pi}{1080}\right)$

$$= \frac{\pi}{2160} \approx 0.0014$$

$\Rightarrow \quad \sin 60^\circ\, 10' = y + \delta y \approx 0.8660 + 0.0014$

$\Rightarrow \quad \sin 60^\circ\, 10' \approx 0.8674$

Exercise 5.11a

1 Prove from first principles that:

$\frac{d}{dx}(\cos x) = -\sin x$

2 Differentiate:

(i) $\sin 2x$; *(ii)* $\cos 3x$; *(iii)* $\sin 2\pi x$; *(iv)* $4\cos\frac{3x}{2}$;

(v) $\sin^2 x$; *(vi)* $\cos^3 x$; *(vii)* $x\sin 2x$; *(viii)* $\frac{\cos x}{x}$;

(ix) $(\sin x + \cos x)^2$; *(x)* $\sin^3 4x$; *(xi)* $\sqrt{\sin x}$; *(xii)* $\frac{x^2}{\sin x}$.

3 A curve is defined by the parametric equations:

$x = 3\cos\theta, \quad y = 2\sin\theta$

(i) Sketch the curve by plotting the position of this point for different values of θ;
(ii) eliminate θ to find its Cartesian equation;
(iii) find dy/dx in terms of θ, and illustrate its values on your sketch.

4 If $x = a\cos\theta, \quad y = b\sin\theta$,
(i) sketch this curve and find its Cartesian equation;
(ii) find dy/dx in terms of θ;

(*iii*) find the equations of the tangent and normal at the point whose parameter is θ.

5 A particle is moving along a straight line and its distance x m from a fixed point O in the line at time t s is given by

$$x = 2 + 3 \sin t + \cos t$$

Find:

(*i*) the velocity and acceleration at time t;

(*ii*) the value of t when the particle first comes to rest;

(*iii*) its acceleration and distance from O at this instant. (O.C.)

6 A particle is moving on a straight line, and its distance x from a fixed point O on the line at time t is given by

$$x = a(1 + \cos^2 t)$$

Show that the acceleration of the particle is $6a - 4x$.

Find the values of x at the points where the velocity of the particle is (*i*) zero, (*ii*) a maximum. (O.C.)

7 The horizontal displacement of a moving spot on a television screen after t milliseconds is x cm, where

$$x = 1 + 2 \cos t - 4 \cos 2t$$

Show that the spot is momentarily at rest when $t = 0$, and find the next two times when it is momentarily at rest.

How far does the spot travel between $t = \frac{1}{2}\pi$ and $t = \pi$? (S.M.P.)

8 The point P moves in a straight line so that after t seconds its distance x m from a fixed point O in the line is given by

$$x = \sin t + 2 \cos t$$

Its velocity is then v m s^{-1} and its acceleration a m s^{-2}.

(*i*) Show that $v^2 = 5 - x^2$ and $a = -x$;

(*ii*) find the greatest distance of P from O;

(*iii*) find the velocity of P after 2 second, taking 1 radian as 57° 18′.

9 Find the values of x for which the following functions have maxima and minima, distinguish between them and sketch the graphs of the functions:

(*i*) $2 \sin x - x \quad (-\pi \leqslant x \leqslant \pi)$;

(*ii*) $2 \sin x - \cos 2x \quad (0 \leqslant x \leqslant 2\pi)$.

10 Without using tables, find the values of

(*i*) cos 30° 30′, (*ii*) sin 45° 20′.

(Take: $\sqrt{2} = 1.414$, $\sqrt{3} = 1.732$, $\pi = 3.142$.)

11 Use Newton's method to solve the equation

$$\cos x = x$$

12 A girl is standing at the edge of a circular lake of radius 100 m. If she can walk at 2 m s^{-1} and swim at 1 m s^{-1}, what is the least time in which she can reach the point exactly opposite?

13 A cylinder is cut from a solid sphere of radius a so that its total surface area is as large as possible. What is the ratio of its height to its diameter?

Differentiation of other trigonometric functions

$$\frac{d}{dx}\left(\frac{\sin x}{\cos x}\right) = \frac{(\cos x)(\cos x) - (\sin x)(-\sin x)}{\cos^2 x}$$

$$= \frac{\cos^2 x + \sin^2 x}{\cos^2 x} = \frac{1}{\cos^2 x} = \sec^2 x$$

So $\dfrac{d}{dx}(\tan x) = \sec^2 x$

The reader can prove for himself that

$$\frac{d}{dx}(\cot x) = -\operatorname{cosec}^2 x; \quad \frac{d}{dx}(\sec x) = \sec x \tan x;$$

$$\frac{d}{dx}(\operatorname{cosec} x) = -\operatorname{cosec} x \cot x$$

So the derivatives of the six trigonometric functions can be summarised:

f	$\sin x$	$\cos x$	$\tan x$	$\cot x$	$\sec x$	$\operatorname{cosec} x$
f'	$\cos x$	$-\sin x$	$\sec^2 x$	$-\operatorname{cosec}^2 x$	$\sec x \tan x$	$-\operatorname{cosec} x \cot x$

Exercise 5.11b

1 Differentiate:

(*i*) $\tan 3x$; (*ii*) $\sec 4x$; (*iii*) $\cot \dfrac{x}{2}$;

(*iv*) $\operatorname{cosec} 2x$; (*v*) $x \tan x$; (*vi*) $\tan^2 x$;

(*vii*) $\sec^3 x$; (*viii*) $\dfrac{\sec x}{x}$; (*ix*) $\cot^2 3x$;

(*x*) $\sqrt{\tan x}$.

2 Show that

$$\frac{d}{dx}(\sec^2 x) = \frac{d}{dx}(\tan^2 x)$$

Why is this so?

3 (*i*) If $x = \sec\theta$, $y = \tan\theta$, sketch the curve described as θ varies, and find dy/dx in terms of θ. Also find the x, y equation of the curve and so check dy/dx by obtaining it in terms of x, y.

(*ii*) Repeat (*i*) if $x = a \operatorname{cosec}\theta$, $y = b \cot\theta$.

4 A boy who can run at 8 m s^{-1} and swim at 2 m s^{-1} wishes to cross a canal of width 40 m to a point 200 m along the other bank. If he sets off at an angle θ to the perpendicular width, find:
(*i*) the total time taken;
(*ii*) the value of θ which makes this a minimum;
(*iii*) the shortest possible time.

5.12 Integration of trigonometric functions

Integration of trigonometric functions is, as always, a speculative process, and depends on a thorough knowledge of differentiation, including the methods of substitution, and of the trigonometric identities. We can, of course, always check indefinite integrals by differentiation. In particular, we see that

$$\int \sin x \, dx = -\cos x + c \quad \text{and} \quad \int \cos x \, dx = \sin x + c$$

Example 1

$$\int \sin 2x \, dx$$

We first recall that $\frac{d}{dx}(\cos 2x) = -2 \sin 2x$

So $\int \sin 2x \, dx = -\frac{1}{2}\cos 2x + A$

Or we could say that

$$\int \sin 2x \, dx = \int 2 \sin x \cos x \, dx$$

$$= \sin^2 x + B$$

or $$- \cos^2 x + C$$

(Why are these three answers all correct?)

Example 2

$$\int \cos^2 x \, dx$$

Our first thought is probably $\frac{1}{3}\cos^3 x$.

But $\frac{d}{dx}(\frac{1}{3}\cos^3 x) = \frac{1}{3} 3 \cos^2 x (-\sin x) = -\cos^2 x \sin x$,

so we must think again.

Now $\cos 2x = 2\cos^2 x - 1$

$\Rightarrow \quad \cos^2 x = \frac{1}{2}(\cos 2x + 1)$

So
$$\int \cos^2 x \, dx = \int \tfrac{1}{2}(\cos 2x + 1)\, dx$$
$$= \tfrac{1}{4}\sin 2x + \frac{x}{2} + C$$
$$= \tfrac{1}{2}(\sin x \cos x + x) + C$$

Example 3

$$\int_0^{\pi/2} \sin^3 x \cos x \, dx$$

Here we can put $u = \sin x$

$$\Rightarrow \quad du = \cos x \, dx$$

So
$$\int_0^{\pi/2} \sin^3 x \cos x \, dx = \int_0^1 u^3 \, du = \left[\tfrac{1}{4}u^4\right]_0^1 = \tfrac{1}{4}$$

Example 4

$$\int \sin x \cos 2x \, dx$$

Here we recall that

$$\sin 3x - \sin x = 2\sin x \cos 2x$$

So
$$\int \sin x \cos 2x \, dx = \tfrac{1}{2}\int (\sin 3x - \sin x)\, dx$$
$$= -\tfrac{1}{6}\cos 3x + \tfrac{1}{2}\cos x + A$$
$$= \tfrac{1}{6}(3\cos x - \cos 3x) + A$$

Exercise 5.12

1 (*i*) $\int \sin 2x \, dx$; (*ii*) $\int \cos \frac{x}{2} \, dx$;

(*iii*) $\int_0^{\pi/2} \cos 3x \, dx$; (*iv*) $\int_0^{\pi/2} \sin \frac{x}{3} \, dx$.

2 Use the double-angle formulae to find:

(*i*) $\int \sin^2 x \, dx$; (*ii*) $\int \cos^2 \frac{x}{2} \, dx$;

(*iii*) $\int_0^{\pi} \sin^2 2x \, dx$; (*iv*) $\int_0^{\pi/2} \sin x \cos x \, dx$.

3 Use the factor formulae to find:

(*i*) $\int \sin 2x \cos x \, dx$; (*ii*) $\int \cos 2x \cos x \, dx$; (*iii*) $\int \sin 2x \sin x \, dx$.

4 Use the method of substitution to find the following integrals:

(*i*) $\int x \cos x^2 \, dx$; (*ii*) $\int x^2 \sin x^3 \, dx$;

(*iii*) $\int \sin^2 x \cos x \, dx$; (*iv*) $\int \cos^3 x \sin x \, dx$.

5 (*i*) $\int \sec^2 x \, dx$; (*ii*) $\int \sec x \tan x \, dx$;

(*iii*) $\int \tan^2 x \, dx$; (*iv*) $\int \cot^2 x \, dx$.

6 Sketch the graph of the curve $y = 2 + \sin x$ for values of x between 0 and 2π. Calculate:
(*i*) the area bounded by the curve, the x-axis and the lines $x = 0$, $x = 2\pi$;
(*ii*) the volume generated when this area makes a complete revolution about the x-axis. (L.)

7 A particle moves on the x-axis having an acceleration of $36 \cos 3t$, where t is the time. It is at the origin when $t = 0$ and its velocity then is 12. Prove that its position at time t is given by $x = 4(1 - \cos 3t) + 12t$.
How far from the origin is the particle when it first comes to rest. (O.C.)

8 What is the average value of $\sin^2 x$ between $x = 0$ and $x = \pi$?

9 Use the substitution $x = \tan \theta$ to find

$$\int_0^1 \frac{dx}{(x^2 + 1)^2}$$

10 (*i*) $\int_3^6 \frac{dx}{\sqrt{\{(x - 3)(6 - x)\}}}$ (put $x = 3\cos^2\theta + 6\sin^2\theta$);

(*ii*) $\int_2^3 \sqrt{\left(\frac{x - 2}{4 - x}\right)} dx$ (put $x = 2\cos^2\theta + 4\sin^2\theta$).

*5.13 Inverse trigonometric functions

If we are asked to find the angle whose sine is $\frac{1}{2}$, we have seen that there are any number of answers: $\pi/6$, $5\pi/6$, $13\pi/6$, $17\pi/6$, . . . (as well as $-7\pi/6$, $-11\pi/6$ . . .).

More generally, the angle whose sine is x does not define a function. Clearly the domain would have to be the set of numbers between -1 and $+1$, but even so there are any number of answers to the instruction, as can be seen from the graph:

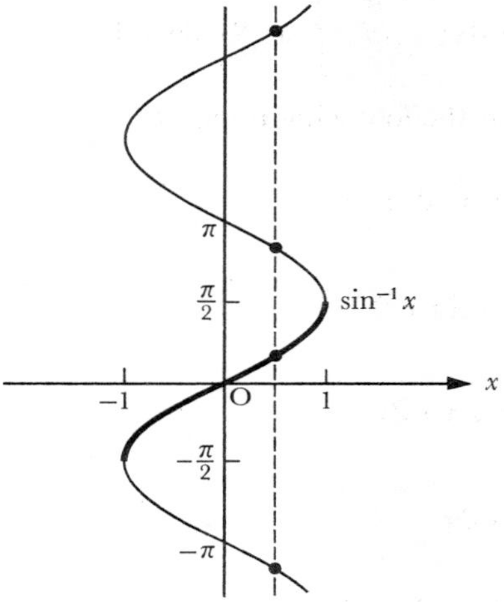

This is not the graph of a function, but it does show that we can obtain a function by making the more definite request for *the angle lying between* $-\pi/2$ *and* $+\pi/2$ *whose sine is* x. As we see from the graph, this has a unique answer, so does represent a function, called the inverse sine and written $\sin^{-1}$.

So $\sin^{-1}\frac{1}{2} = \dfrac{\pi}{6}$, $\sin^{-1} 0 = 0$, $\sin^{-1}(-1) = -\dfrac{\pi}{2}$, etc.

Similarly, $\cos^{-1} x$ is defined as *the angle between* 0 *and* π *whose cosine is* x. Hence

$\cos^{-1}\frac{1}{2} = \dfrac{\pi}{3}$, $\cos^{-1} 0 = \dfrac{\pi}{2}$, $\cos^{-1}(-1) = \pi$, etc.

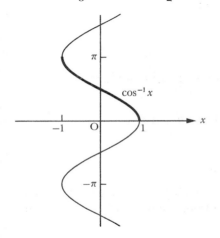

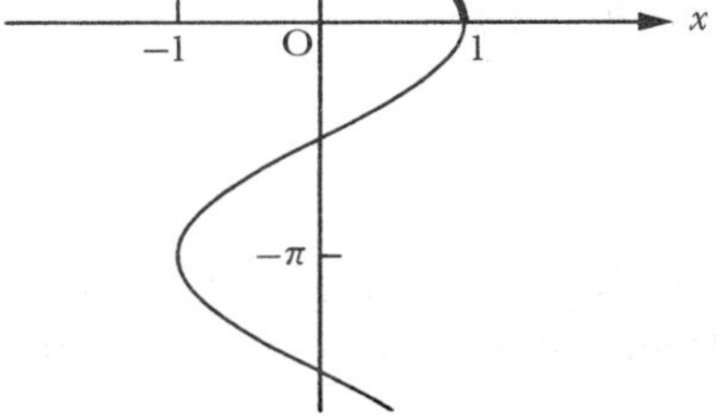

Finally, $\tan^{-1} x$ is defined as *the angle between* $-\pi/2$ *and* $+\pi/2$ *whose tangent is* x.

So $\tan^{-1} 1 = \dfrac{\pi}{4}$, $\tan^{-1}(-1) = -\dfrac{\pi}{4}$, etc.

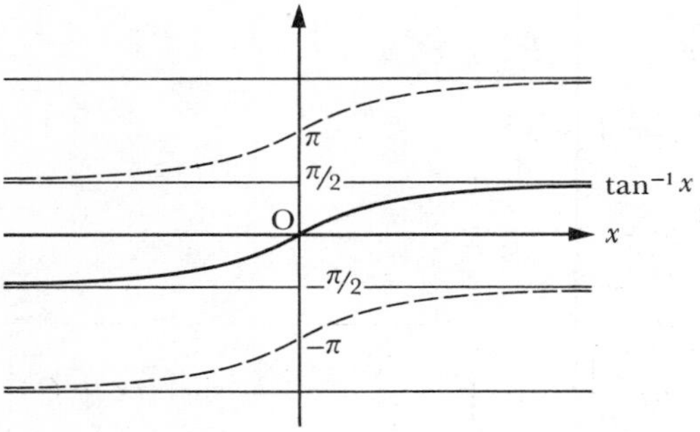

Exercise 5.13a

1 Evaluate:

(*i*) $\sin^{-1}\frac{1}{2}$, $\cos^{-1}\frac{1}{2}$, $\tan^{-1}\frac{1}{2}$;

(*ii*) $\sin^{-1} -1$, $\cos^{-1} -1$, $\tan^{-1} -1$;

(*iii*) $\sin^{-1} 0.7$, $\cos^{-1} 0.7$, $\tan^{-1} 0.7$.

2 Find, from question 1, a relationship between $\sin^{-1} x$ and $\cos^{-1} x$, and illustrate your answer by means of their graphs.

3 (*i*) Simplify: $\tan^{-1}\frac{1}{2} + \tan^{-1}\frac{1}{3}$

[Let $\theta = \tan^{-1}\frac{1}{2}$, $\phi = \tan^{-1}\frac{1}{3}$ and calculate $\tan(\theta + \phi)$.]

(*ii*) Simplify: $\tan^{-1} x + \tan^{-1} y$,

where $0 \leqslant x \leqslant 1$ and $0 \leqslant y \leqslant 1$.

Differentiation of inverse trigonometric functions

(*i*) $\dfrac{d}{dx}(\sin^{-1} x)$, $\dfrac{d}{dx}(\cos^{-1} x)$.

$$y = \sin^{-1} x \quad \Rightarrow \quad x = \sin y$$

$$\Rightarrow \quad \frac{dx}{dy} = \cos y = \pm\sqrt{(1 - x^2)}$$

$$\Rightarrow \quad \frac{dy}{dx} = \pm\frac{1}{\sqrt{(1 - x^2)}}$$

But $\sin^{-1} x$ is an increasing function,

so $$\frac{d}{dx}(\sin^{-1} x) = +\frac{1}{\sqrt{(1 - x^2)}}$$

By similar argument (or alternatively from $\sin^{-1} x + \cos^{-1} x = \frac{1}{2}\pi$)

we find that $\dfrac{d}{dx}(\cos^{-1} x) = -\dfrac{1}{\sqrt{(1 - x^2)}}$

(*ii*) $\dfrac{d}{dx}(\tan^{-1} x)$

$$y = \tan^{-1} x \quad \Rightarrow \quad x = \tan y$$

$$\Rightarrow \quad \frac{dx}{dy} = \sec^2 y = 1 + x^2$$

$$\Rightarrow \quad \frac{dy}{dx} = \frac{1}{1 + x^2}$$

So

$$\boxed{\begin{aligned} \frac{d}{dx}(\sin^{-1} x) &= \frac{1}{\sqrt{(1 - x^2)}} \\ \frac{d}{dx}(\tan^{-1} x) &= \frac{1}{1 + x^2} \end{aligned}} \quad \Rightarrow \quad \boxed{\begin{aligned} \int \frac{dx}{\sqrt{(1 - x^2)}} &= \sin^{-1} x + C \\ \int \frac{dx}{1 + x^2} &= \tan^{-1} x + C \end{aligned}}$$

The importance of the inverse trigonometric functions largely stems from their value in enabling us to integrate such functions as $\dfrac{1}{\sqrt{(1 - x^2)}}$ and $\dfrac{1}{1 + x^2}$. These results could equally well have been found using the method of substitution.

More generally,

$$\int \frac{dx}{\sqrt{(a^2 - x^2)}} = \int \frac{a \cos \theta \, d\theta}{\sqrt{(a^2 - a^2 \sin^2 \theta)}} \quad (x = a \sin \theta, \ dx = a \cos \theta \, d\theta)$$

$$= \int \frac{a \cos \theta \, d\theta}{a \cos \theta}$$

$$= \int d\theta = \theta + c = \sin^{-1} \frac{x}{a} + c$$

and

$$\int \frac{dx}{a^2 + x^2} = \int \frac{a \sec^2 \theta \, d\theta}{a^2 + a^2 \tan^2 \theta} \quad (x = a \tan \theta, \ dx = a \sec^2 \theta \, d\theta)$$

$$= \frac{1}{a} \int \frac{\sec^2 \theta \, d\theta}{\sec^2 \theta}$$

$$= \frac{1}{a} \int d\theta = \frac{\theta}{a} + c = \frac{1}{a} \tan^{-1} \frac{x}{a} + c$$

Exercise 5.13b

1 (i) $\int_0^1 \frac{dx}{\sqrt{(4 - x^2)}}$; (ii) $\int_0^3 \frac{dx}{x^2 + 9}$;

(iii) $\int_0^1 \sqrt{(1 - x^2)}\, dx$, $(x = \sin\theta)$.

2 (i) $\int \frac{dx}{x^2 + 2x + 2}$; (ii) $\int \frac{dx}{\sqrt{(2x - x^2)}}$;

(iii) $\int \sqrt{\left(\frac{x}{1 - x}\right)} dx$ $(x = \sin^2\theta)$.

3 (i) $\int \frac{dx}{\sqrt{(9 - 4x^2)}}$; (ii) $\int \frac{dx}{4x^2 + 1}$; (iii) $\int \sqrt{(a^2 - x^2)}\, dx$.

4 (i) $\int \frac{dx}{x\sqrt{(x^2 - 1)}}$ $(x = \sec\theta)$; (ii) $\int \frac{dx}{(1 + x^2)^2}$ $(x = \tan\theta)$.

*5.14 Polar coordinates

It is frequently convenient to identify a point P not by its Cartesian coordinates (x, y), but by r, its distance from a fixed point P (the origin, or *pole*) and θ, an angle through which a base-line Ox is rotated in order to line along OP. r, θ are then known as the *polar coordinates* of P, and clearly θ has a set of values, separated by multiples of 2π.

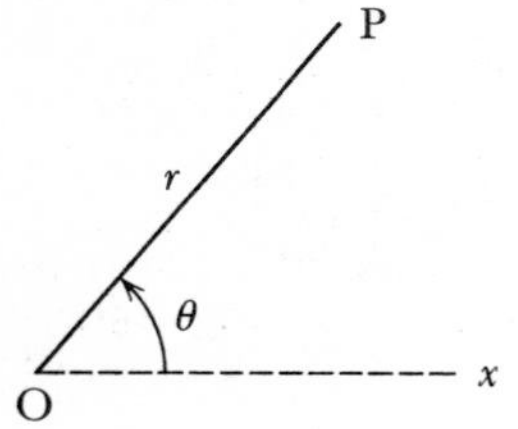

Furthermore, $r = f(\theta)$ usually denotes a curve and is called its *polar equation*.

Example 1

Sketch the curve

$r = 1 + \cos\theta$

Taking successive values of θ, we obtain:

$\theta = 0$	$\pm\pi/6$	$\pm\pi/3$	$\pm\pi/2$	$\pm 2\pi/3$	$\pm 5\pi/6$	$\pm\pi$
$r = 2$	1.87	1.50	1.00	0.50	0.13	0

from which we obtain the heart-shaped curve called the *cardioid*:

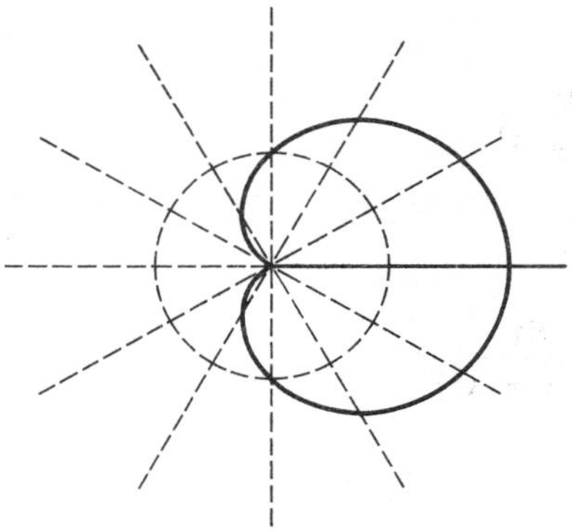

Example 2

$r = \frac{1}{2}\theta$ (where $\theta \geqslant 0$)

In this case,

$\theta = 0$	$\pi/4$	$\pi/2$	$3\pi/4$	π	$5\pi/4$	$3\pi/2$	$7\pi/4$	2π	$9\pi/4$
$r = 0$	0.39	0.79	1.18	1.57	1.96	2.36	2.75	3.14	3.53

from which we obtain the so-called Archimedean spiral:

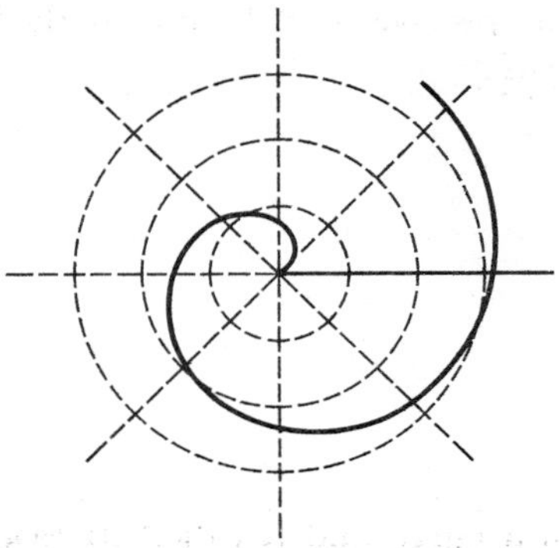

Exercise 5.14

Sketch the curves:

1 $r = a \sin \theta$, 2 $r = a \sin 2\theta$,

3 $r^2 = a^2 \cos 2\theta$ (a *lemniscate*),

4 $r = a \cos 3\theta$ (a *trifolium*),

5 Express x, y in terms of r, θ. Hence:

6 Find the polar equations of
(*i*) the circle $x^2 + y^2 = 2y$;
(*ii*) the parabola $y = x^2$;
(*iii*) the rectangular hyperbola $xy = 1$;
(*iv*) the straight line $x + y = 1$.

7 Sketch the following curves and find their Cartesian equations.
(*i*) $r = a \cos \theta$;
(*ii*) $r = a \operatorname{cosec} \theta$;
(*iii*) $r = a^2 \sin 2\theta$;
(*iv*) $r^2 \cos 2\theta = a^2$.

Miscellaneous problems 5

1 A wire of length 4 m is shaped into the arc and two radii of a circular sector. Find the maximum area it can enclose.

2 (*i*) Prove, without using tables, that

$$\tan^{-1} \tfrac{1}{2} + \tan^{-1} \tfrac{2}{3} + \tan^{-1} \tfrac{4}{7} = \tfrac{1}{2} \pi.$$

(*ii*) Find the values of θ between 0 and 360° which satisfy the equation $\sin 3\theta = \cos 2\theta$, and hence, or otherwise, prove without using tables that $\sin 18° = (\sqrt{5} - 1)/4$. (O.C.)

3 Without using tables, prove that

$$\tan \frac{\pi}{8} = \sqrt{2} - 1 \quad \text{and} \quad \tan \frac{\pi}{24} = \frac{1 + \sqrt{3} - \sqrt{6}}{\sqrt{3} + \sqrt{2} - 1}$$ (O.C.)

4 Show that $x \tan x = 1$ has an infinite number of real roots, and that if n is a large integer there is a root near $n\pi$.

Show that a better approximation is $n\pi + \dfrac{1}{n\pi}$.

5 A hollow right circular cone is of height h and semi-vertical angle θ. Show that the radius of the largest sphere which can be placed entirely within the cone is

$$\frac{h \sin \theta}{1 + \sin \theta}$$

Prove that the ratio of the volume of the sphere to that of the cone is greatest when $\sin \theta = \frac{1}{3}$, and is then equal to $\frac{1}{2}$.

6 Four towns are situated at the corners of a square of side 100 km. What is the shortest length of road which can be built to connect them?

7 Prove that $\tan^2 \dfrac{A}{2} = \dfrac{1 - \cos A}{1 + \cos A}$, and then use the cosine rule to show that in any $\triangle ABC$,

$$\tan\frac{A}{2} = \sqrt{\left\{\frac{(s-b)(s-c)}{s(s-a)}\right\}}$$

where s is the semi-perimeter.

Hence find A, B, C in exercise 5.4, 2(*i*).

8 Find all the values of x between 0 and 2π inclusive for which the function $3\cos x - \cos 3x$ has a turning value, distinguishing between maxima and minima.

How many inflexions has the function in the range $0 \leqslant x \leqslant 2\pi$?

Draw a rough graph of the function for this range. (O.C.)

9 A circle of radius a is rolling along the x-axis and after it has rotated through an angle θ the point P of the circle which was originally at O has coordinates x and y.

(*i*) Find x, y in terms of θ;

(*ii*) Find $\frac{dx}{d\theta}$, $\frac{dy}{d\theta}$ and $\frac{dy}{dx}$ in terms of θ;

(*iii*) Sketch the path of P.

10 A point P on an astroid has coordinates

$$x = a\cos^3 t, \quad y = a\sin^3 t$$

where t is a parameter.

Find:

(*i*) the gradient of the curve at P;

(*ii*) the equation of the tangent at P;

(*iii*) the coordinates of the points A, B where this tangent meets Ox, Oy;

(*iv*) the length AB.

Hence show how an astroid can be constructed, and sketch the curve. (C.)

11 The coordinates x, y of the spot of light on an oscilloscope are given by

$$x = a\sin mt, \quad y = b\sin(nt + \varepsilon)$$

and so are oscillating independently with amplitudes a, b and frequencies $m/2\pi$, $n/2\pi$ (ε being their initial phase difference). The paths traced by the spot as t varies for different values of a, b, m, n, ε, are known as Lissajou's Figures.

If $a = b = 1$, plot these when $\varepsilon = 0$, $\varepsilon = \pi/4$, $\varepsilon = \pi/2$, $\varepsilon = 3\pi/4$ for:

(*i*) $m = n = 1$;

(*ii*) $m = 2$, $n = 1$;

(*iii*) $m = 3$, $n = 1$;

(*iv*) other values of m and n.

How do the figures alter if a and b take other values?

6

Sequences and series

6.1 Sequences and series

(*i*) 1, 2, 3, 4, 5, 6, 7,. . .
(*ii*) 1, 2, 4, 7, 11, 16, 22,. . .
(*iii*) 1, 2, 3, 5, 8, 13, 21,. . .
(*iv*) 1, 2, 4, 8, 16, 32, 64,. . .
(*v*) 1, 2, 0, 3, −1, 4, −2, . . .
(*vi*) 1, 2, 6, 24, 120, 720, 5 040, . . .

The reader will find little difficulty in spotting how each row (or *sequence*) has been produced, and in writing down the next three members (or *terms*) of each sequence.

The question next arises whether we can state a general formula in each case for the *n*th term, which for convenience we shall call u_n. In some cases this is obvious. For example in

(*i*) $u_1 = 1, u_2 = 2, u_3 = 3, u_4 = 4$, and $u_n = n$,
(*iv*) $u_1 = 1, u_2 = 2, u_3 = 4, u_4 = 8$, and $u_n = 2^{n-1}$.

Some might be obtained by a certain amount of trial and error. For example in (*ii*),

$$u_n = \tfrac{1}{2}(n^2 - n + 2)$$

which should be checked for $n = 1, 2, 3, \ldots 9$.

But in other cases, it is doubtful if even the wildest speculation would lead to such a formula. In *(iii)*, for example,

$$u_n = \frac{1}{\sqrt{5}}\left\{\left(\frac{1+\sqrt{5}}{2}\right)^{n+1} - \left(\frac{1-\sqrt{5}}{2}\right)^{n+1}\right\}$$

as can be checked for particular values of n.

Factorials

The sequence *(vi)*:

1, 2, 6, 24, 120, 720, . . .

presents a special case.

These terms can be written

$u_1 = 1, u_2 = 2 \times 1, u_3 = 3 \times 2 \times 1, u_4 = 4 \times 3 \times 2 \times 1$, etc., so

$u_n = n \times (n-1) \times (n-2)\ldots 5 \times 4 \times 3 \times 2 \times 1$

This is an expression which occurs so frequently that we abbreviate it to $n!$, called 'n factorial' (or 'n shriek').

So	$1! = 1$	$5! = 120$
	$2! = 2$	$6! = 720$
	$3! = 6$	$7! = 5040$
	$4! = 24$	etc.

It is clear that the terms of this sequence are growing very rapidly and that the calculation of $n!$ when n is large (e.g., $100! = 100 \times 99 \times 98\ldots \ldots 4 \times 3 \times 2 \times 1$) would be very tedious.

There is, in fact, a formula, usually known as *Stirling's formula* (after the Scots mathematician, James Stirling, 1692–1770) which says, rather astonishingly, that:

$n! \approx \sqrt{(2\pi)}\, n^{n+1/2}\, e^{-n}$, where $e \approx 2.718$.

The percentage error of this approximation diminishes rapidly as n gets larger, and it is interesting to use the formula to compute known numbers like 10!, 12!, etc.

Exercise 6.1a

1 Write down the next three terms of the above sequences *(i)*, *(ii)*, *(iii)*, *(iv)*, *(v)*, *(vi)*.

2 Write down the values of u_1, u_2, u_3, u_4, where u_r is

(i) r^3; *(ii)* $\dfrac{r}{r+1}$; *(iii)* $1 + (-1)^r$; *(iv)* $(-1)^r r$.

3 Calculate:

(i) $\dfrac{7!}{6!}$; *(ii)* $\dfrac{12!}{10!}$; *(iii)* $\dfrac{(n+1)!}{n!}$; *(iv)* $\dfrac{n!}{(n-2)!}$.

Series

We frequently wish to find the sum of a number of terms of a sequence, such as

$$1 + 8 + 27 + 64$$

or $1 + 2 + 4 + 8 + 16 + 32$

or, more generally,

$$u_1 + u_2 + u_3 + u_4 + \cdots + u_n.$$

We call each of these a *series* and abbreviate the sum of n terms,

$$s_n = u_1 + u_2 + u_3 + u_4 + \cdots + u_n$$

as $\sum_{r=1}^{n} u_r$ (pronounced 'sigma' u_r from $r = 1$ to $r = n$),†

or more briefly as $\sum_{1}^{n} u_r$

So $\sum_{r=1}^{4} r^3 = 1^3 + 2^3 + 3^3 + 4^3 = 100$

and $\sum_{r=0}^{5} 2^r = 2^0 + 2^1 + 2^2 + 2^3 + 2^4 + 2^5 = 63.$

Exercise 6.1b

1 Write out fully and find the values of:

(*i*) $\sum_{r=1}^{4} r^2$; (*ii*) $\sum_{r=2}^{5} \frac{1}{r}$; (*iii*) $\sum_{r=0}^{7} 2^r$; (*iv*) $\sum_{r=1}^{6} (-1)^r$;

(*v*) $\sum_{r=0}^{6} \frac{1}{2^r}$; (*vi*) $\sum_{r=1}^{3} r(r+1)$; (*vii*) $\sum_{r=1}^{4} r2^r$; (*viii*) $\sum_{r=1}^{4} (-1)^{r-1}r^2$.

2 Use the $\sum$ notation to abbreviate (but do not evaluate):

(*i*) $1 + 4 + 9 + 16 + \cdots + 625$; (*ii*) $\frac{1}{2} + \frac{1}{3} + \frac{1}{4} \cdots + \frac{1}{100}$;
(*iii*) $1 + 8 + 27 + \cdots + 1\,000$; (*iv*) $1 + 3 + 9 + 27 + \cdots + 3^{50}$;
(*v*) $1 - \frac{1}{2} + \frac{1}{3} \cdots - \frac{1}{100}$; (*vi*) $1\times2 + 2\times3 + \cdots + n(n+1)$.

6.2 Arithmetic progressions

An *arithmetic progression* (a.p.) is a sequence which proceeds with constant difference, like

$1, 4, 7, 10 \ldots$ or $3, 2\frac{1}{2}, 2, 1\frac{1}{2} \ldots$

† This is, of course, a refinement of the notation already used in section 4.8.

Such a sequence is completely defined if we know its first term a and its constant difference d, the above two sequences being given by $a = 1, d = 3$ and by $a = 3$, $d = -\frac{1}{2}$.

The general a.p. can therefore be expressed as

$$a, a + d, a + 2d, a + 3d \ldots$$

and it is seen that the rth term is

$$\boxed{u_r = a + (r - 1)d.}$$

If we now concentrate on n terms of this sequence, stretching from the first term $a(=u_1)$ to the last term $l(=u_n)$, we see that their sum, S_n, is given by

$$S_n = a + (a + d) + (a + 2d) + \cdots + (l - d) + l$$

Writing this backwards,

$$S_n = l + (l - d) + (l - 2d) + \cdots + (a + d) + a$$

Adding, we obtain

$$\begin{aligned} 2S_n &= (a + l) + (a + l) + \cdots + (a + l) \\ &= n(a + l) \end{aligned}$$

$$\Rightarrow \quad S_n = \tfrac{1}{2}n(a + l) = n\frac{a + l}{2}.$$

Hence the sum of a sequence of terms of an a.p.

= number of terms × average of the first and last terms.

Further, since $u_n = a + (n - 1)\,d$,

$$S_n = \tfrac{1}{2}n\{a + a + (n - 1)d\}$$

$$\Rightarrow \quad \boxed{S_n = n\frac{a + l}{2} = \tfrac{1}{2}n\{2a + (n - 1)d\}.}$$

Example

A young man's initial annual salary was £875 and increased by £45 a year. How much did he expect to earn in his first six years?

Here we see that the constant increment is causing his salary to rise in an a.p., so we can say either

(*a*) Final term $= 875 + 5 \times 45 = 1\,100$

$$\Rightarrow \quad \text{Sum} = 6 \times \frac{875 + 1\,100}{2} = 5\,925$$

or (*b*) $S_n = \frac{n}{2}\{2a + (n - 1)\,d\}$,

where $a = 875$, $d = 45$, and $n = 6$

$$\Rightarrow \quad S_n = \frac{6}{2}\{1\,750 + 5 \times 45\} = 5\,925.$$

So, by either method, we see that his total income in 6 years was £5 925.

Exercise 6.2

1 Find the rth term and the sum of n terms of each of the following a.p.s:
(*i*) 2, 6, 10, 14, . . . ;
(*ii*) 1, $2\frac{1}{2}$, 4, $5\frac{1}{2}$, . . . ;
(*iii*) 3, 1, −1, −3, ...
Check your results for $r = 5$ and $n = 5$.

2 Find the number of terms and the sums of each of the following arithmetic series:
(*i*) $1 + 3 + 5 + \cdots + 99$;
(*ii*) $1 + 3\frac{1}{2} + 6 + \cdots + 101$;
(*iii*) $10 + 9 + 8 + \cdots - 18 - 19 - 20$.

3 Evaluate:

(*i*) $\sum_{r=1}^{25} (4r - 1)$; (*ii*) $\sum_{1}^{34} (3r - 2)$.

4 In a potato race the first and last potatoes are 5 m and 15 m respectively from the starting-line and the rest are equally spaced at intervals of 1 metre. What is the total distance travelled by a runner who brings them one at a time to his starting-line?

5 Find the sum of the numbers divisible by 3 which lie between 1 and 100. Find also the sum of the numbers from 1 to 100 inclusive which are *not* divisible by 3. (O.C.)

6 Find how many terms of the progression

$5 + 9 + 13 + 17 + \cdots$

have a sum of 2 414. (O.C.)

7 The first term of an a.p. is 3. Find the common difference if the sum of the first 8 terms is twice the sum of the first 5 terms. (O.C.)

8 The fifth term of an arithmetical progression is 24 and the sum of the first five terms is 80. Find the first term, the common difference and the sum of the first fifteen terms of the progression. (O.C.)

9 The sum to n terms of a certain a.p. is $2n\,(n + 5)$. What is:
(*i*) its nth term?
(*ii*) its constant difference?

6.3 Geometric progressions

A *geometric progression* (g.p.) is (by contrast with an a.p.) a sequence which proceeds with constant *ratio*, like 2, 6, 18, 54, . . ., or 8, −4, 2, −1, $\frac{1}{2}$, ...

Such a sequence is completely defined by its first term a and its constant ratio ρ (the Greek letter r, pronounced rō), the above two sequences being given by

$$a = 2, \rho = 3 \quad \text{and} \quad a = 8, \rho = -\tfrac{1}{2}.$$

The general g.p. can therefore be expressed as

$$a, a\rho, a\rho^2, a\rho^3 . . . . . .$$

so that $\boxed{u_r = a\rho^{r-1}}$

If we again concentrate on n terms of the sequence, from u_1 to u_n, we see that

$$S_n = a + a\rho + a\rho^2 + \cdots + a\rho^{n-1}$$

Multiplying by ρ, we obtain

$$\rho S_n = a\rho + a\rho^2 + \cdots + a\rho^{n-1} + a\rho^n$$

By subtraction,

$$(1 - \rho)\, S_n = a - a\rho^n = a(1 - \rho^n)$$

$$\Rightarrow \quad \boxed{S_n = \frac{a(1-\rho^n)}{1-\rho} = \frac{a(\rho^n - 1)}{\rho - 1}}$$

Example 1

On 1 January each year a man puts £50 in a bank which gives 4% interest each year. What will be the value of his investment on 31 December of the tenth year?

In one year £1 would grow to £1.04,
and so £50 would grow to £50 × 1.04.
Hence, by the end of the 10th year,

his 10th deposit will grow to £50 × 1.04,
his 9th deposit will grow to £50 × $(1.04)^2$,
his 8th deposit will grow to £50 × $(1.04)^3$,

his 1st deposit will grow to £50 × $(1.04)^{11}$.

So their total value (in £) is

$$50\,(1.04) + 50\,(1.04)^2 + \cdots + 50\,(1.04)^{11}$$

$$= \quad 50\,(1.04) \times \frac{(1.04)^{10} - 1}{1.04 - 1}$$

$$= \frac{52}{0.04} \times [(1.04)^{10} - 1]$$

$$= \quad 622.6,$$

and the final value of his investment is approximately £623.

Example 2

In the particular case when $a = 1$ and $\rho = \frac{1}{2}$ the g.p. becomes

$1, \frac{1}{2}, \frac{1}{4}, \frac{1}{8}, \frac{1}{16}, \cdots$

and the sum of its first n terms is

$$S_n = 1 + \tfrac{1}{2} + \tfrac{1}{4} + \cdots + \frac{1}{2^{n-1}}$$

So $S_1 = 1$, $S_2 = 1\frac{1}{2}$, $S_3 = 1\frac{3}{4}$, $S_4 = 1\frac{7}{8}$.

More generally,

$$S_n = \frac{1 - (\frac{1}{2})^n}{1 - \frac{1}{2}} = 2\{1 - (\tfrac{1}{2})^n\} = 2 - (\tfrac{1}{2})^{n-1}$$

As n increases, $(\frac{1}{2})^{n-1}$ gets smaller and smaller;

and as $n \to \infty$, $(\frac{1}{2})^{n-1} \to 0 \quad \Rightarrow \quad S_n \to 2$

In other words, as we take more and more terms of the series, their sum draws nearer and nearer to 2, approaching as closely as we wish.

We therefore say that the *infinite series*

$1 + \frac{1}{2} + \frac{1}{4} + \frac{1}{8} + \cdots$

converges to the sum 2.

More generally, consider the geometric series with first term a and common ratio ρ:

$a + a\rho + a\rho^2 + \cdots + a\rho^n + \cdots$

Again let $S_n = a + a\rho + \cdots + a\rho^{n-1}$.

Now if $|\rho| > 1$ (say $\rho = 2.5, -1.5$, etc.) successive terms are increasing in size and so the series cannot converge.

If $\rho = +1$ or -1, successive terms do not diminish in size and so again the series cannot converge.

But if $|\rho| < 1$ (e.g., $\rho = 0.8, -0.9$, etc.),

$\rho^n \to 0 \quad \text{as} \quad n \to \infty.$

So $S_n = \dfrac{a(1-\rho^n)}{1-\rho} = \dfrac{a - a\rho^n}{1-\rho} \to \dfrac{a}{1-\rho}$ as $n \to \infty$

> Hence $a + a\rho + a\rho^2 + a\rho^3 + \cdots$
>
> converges to the sum $\dfrac{a}{1-\rho}$, provided that $|\rho| < 1$.

Example 3

A snail moves 2 metres along a straight line in one hour and in each succeeding hour 10% less than in the preceding one. How far does it go?

Here the appropriate g.p. is

$$2,\quad 2\times\frac{9}{10},\quad 2\times\frac{9}{10}\times\frac{9}{10},\quad 2\times\frac{9}{10}\times\frac{9}{10}\times\frac{9}{10},\ldots$$

So the distance travelled in n hours is given by the sum of the first n terms of the series

$$2 + 2\left(\frac{9}{10}\right) + 2\left(\frac{9}{10}\right)^2 + 2\left(\frac{9}{10}\right)^3 + \cdots$$

But this infinite series is convergent to the sum

$$\frac{2}{1-\frac{9}{10}} = \frac{2}{\frac{1}{10}} = 20$$

So the snail gradually approaches a point on the line which is 20 metres from its starting-point.

Exercise 6.3

1 Find the last term and the sum of the following g.p.s:

(*i*) $2 + 6 + 18 + \cdots$ (7 terms);

(*ii*) $3 + 1 + \frac{1}{3} + \cdots$ (6 terms);

(*iii*) $2 - 4 + 8 - \cdots$ (8 terms);

(*iv*) $1 + x + x^2 + \cdots$ (10 terms);

(*v*) $x - x^2 + x^3 - \cdots$ (n terms).

2 Calculate the sum to infinity of the following g.p.s:

(*i*) $1 + \frac{2}{3} + \frac{4}{9} + \cdots$;

(*ii*) $9 - 6 + 4 - \cdots$;

(*iii*) $1 + 0.9 + 0.81 + \cdots$;

(*iv*) $1 - x + x^2 - \cdots$ ($|x| < 1$).

3 (*i*) The recurring decimal $0.\dot{1}\dot{2} = 0.121\,212\ldots$ can be written as

$$\frac{12}{100} + \frac{12}{10\,000} + \frac{12}{1\,000\,000} + \cdots$$

Hence express $0.\dot{1}\dot{2}$ as a fraction.

(*ii*) Express $0.\dot{1}2\dot{3}$ as a fraction.

4 A man offers to give to a charity by placing 1p on the first square of a chess-board, 2p on the second, 4p on the third, and so on until the board is covered. What is the value of his offer?

5 A man deposits £100 each year at 5% compound interest. What is the total value of his investment immediately after making his tenth deposit?

6 How much money should be deposited in order to yield £100 in a year's time? How much should be deposited in order to yield £100 at the end of each of the next five years and then leave the account empty?

7 A tennis ball is dropped from a height of 5 m and takes 1 s before hitting the ground. It then takes $1\frac{1}{2}$ s over its first bounce, and then bounces repeatedly so that the intervals between successive impacts are reduced in the ratio $\frac{3}{4}$ and successive heights in the ratio $\frac{9}{16}$.

Find the total time taken and the total distance travelled before the ball comes to rest.

8 The sum to infinity of a geometric progression is five times its first term. Find its common ratio.

Can the sum to infinity ever be $\frac{2}{3}$ of the first term?

Can it be $\frac{1}{3}$ of the first term?

*6.4 Finite series: the method of differences

In the last two sections we discovered results like

$$\sum_{r=1}^{n} r = 1 + 2 + 3 + \cdots + n = \tfrac{1}{2}n(n+1)$$

and $$\sum_{r=1}^{n} 2^r = 2 + 4 + 8 + \cdots + 2^n = 2\frac{2^n - 1}{2 - 1} = 2^{n+1} - 2$$

We now investigate the more general *method of differences.*

Example 1

Calculate:

$$\sum_{r=1}^{100} r^3 = 1^3 + 2^3 + 3^3 + \cdots + 100^3$$

The method depends on the fact that

$$r^3 \equiv \tfrac{1}{4}r^2[(r+1)^2 - (r-1)^2] \equiv \tfrac{1}{4}r^2(r+1)^2 - \tfrac{1}{4}(r-1)^2 r^2$$

So
$$1^3 = \tfrac{1}{4} \times 1^2 \times 2^2 - \tfrac{1}{4} \times 0^2 \times 1^2$$
$$2^3 = \tfrac{1}{4} \times 2^2 \times 3^2 - \tfrac{1}{4} \times 1^2 \times 2^2$$
$$\cdot \quad \cdot \quad \cdot \quad \cdot \quad \cdot \quad \cdot \quad \cdot \quad \cdot \quad \cdot \quad \cdot \quad \cdot$$
$$99^3 = \tfrac{1}{4} \times 99^2 \times 100^2 - \tfrac{1}{4} \times 98^2 \times 99^2$$
$$100^3 = \tfrac{1}{4} \times 100^2 \times 101^2 - \tfrac{1}{4} \times 99^2 \times 100^2$$

When we add these, we notice that all the terms of the right-hand side cancel out, except for

$\frac{1}{4} \times 100^2 \times 101^2$ and $\frac{1}{4} \times 0^2 \times 1^2$ (which is zero).

So $\sum_{r=1}^{100} r^3 = \frac{1}{4} \times 100^2 \times 101^2 = 25\,502\,500$

Example 2

Find the sum of the first n terms of the series

$1 \times 2 + 2 \times 3 + 3 \times 4 + \cdots,$

i.e., $\sum_{r=1}^{n} r(r+1)$

$$\text{Now}\quad r(r+1) \equiv \tfrac{1}{3}r(r+1)[(r+2)-(r-1)]$$
$$\equiv \tfrac{1}{3}r(r+1)(r+2) - \tfrac{1}{3}(r-1)r(r+1)$$

$$\text{So}\quad 1\times 2 = \tfrac{1}{3}\times 1\times 2\times 3 - \tfrac{1}{3}\times 0\times 1\times 2$$
$$2\times 3 = \tfrac{1}{3}\times 2\times 3\times 4 - \tfrac{1}{3}\times 1\times 2\times 3$$
$$3\times 4 = \tfrac{1}{3}\times 3\times 4\times 5 - \tfrac{1}{3}\times 2\times 3\times 4$$

. . . .

and $n(n+1) = \frac{1}{3}n(n+1)(n+2) - \frac{1}{3}(n-1)n(n+1)$

Adding, we obtain

$$\sum_{r=1}^{n} r(r+1) = \tfrac{1}{3}n(n+1)(n+2)$$

Exercise 6.4a

1 Use the identity

$$r(r+1)(r+2) \equiv \tfrac{1}{4}r(r+1)(r+2)[(r+3)-(r-1)]$$

in order to find a formula for

$$\sum_{r=1}^{n} r(r+1)(r+2)$$

Check your result when $n = 4$.

2 Use $\dfrac{1}{r(r+1)} \equiv \dfrac{1}{r} - \dfrac{1}{r+1}$ in order to find $\displaystyle\sum_{r=1}^{n} \frac{1}{r(r+1)}$

Check your result when $n = 4$.

Find the sum to infinity of $\dfrac{1}{1\times 2} + \dfrac{1}{2\times 3} + \dfrac{1}{3\times 4} + \cdots$

Standard Series

It is clear that $\sum_{1}^{n} 1 = n$

and we now know that $\sum_{1}^{n} r = \frac{1}{2}n(n+1)$

and $\sum_{1}^{n} r(r+1) = \frac{1}{3}n(n+1)(n+2)$

and $\sum_{1}^{n} r(r+1)(r+2) = \frac{1}{4}n(n+1)(n+2)(n+3)$

These *standard series* can often be used for the calculation of other sums.

Example 3

$$\begin{aligned}\sum_{1}^{n} r^2 &= \sum_{1}^{n}\{r(r+1) - r\}\\ &= \sum_{1}^{n} r(r+1) - \sum_{1}^{n} r\\ &= \tfrac{1}{3}n(n+1)(n+2) - \tfrac{1}{2}n(n+1)\\ &= \tfrac{1}{6}n(n+1)[2(n+2) - 3]\\ &= \tfrac{1}{6}n(n+1)(2n+1)\end{aligned}$$

Example 4

Find the sum of the first 50 terms of

$1^2 \times 2 + 2^2 \times 3 + 3^2 \times 4 + \cdots$

This can be written $\sum_{1}^{50} r^2(r+1)$

$$\begin{aligned}\sum_{1}^{n} r^2(r+1) &= \sum_{1}^{n}\{r(r+1)(r+2) - 2r(r+1)\}\\ &= \tfrac{1}{4}n(n+1)(n+2)(n+3) - \tfrac{2}{3}n(n+1)(n+2)\\ &= \tfrac{1}{12}n(n+1)(n+2)[3(n+3) - 8]\\ &= \tfrac{1}{12}n(n+1)(n+2)(3n+1)\end{aligned}$$

So $\sum_{1}^{50} r^2(r+1) = \frac{1}{12} \times 50 \times 51 \times 52 \times 151 = 1\,668\,550$

Exercise 6.4b

1 Show that $\sum_{r=1}^{n} (2r - 1)^2 \equiv 4 \sum_{r=1}^{n} r(r + 1) - 8 \sum_{r=1}^{n} r + \sum_{r=1}^{n} 1.$

Hence find the sum of the squares of the first n odd integers, and calculate $1^2 + 3^2 + 5^2 + \cdots + 99^2$

2 Find the rth term and the sum to n terms of
(*i*) $1\times3 + 2\times4 + 3\times5 + \cdots;$
(*ii*) $1\times2 + 3\times4 + 5\times6 + \cdots;$
(*iii*) $1\times3 + 4\times6 + 7\times9 + \cdots.$

*6.5 Mathematical induction

We now come to an exceedingly general method of proof, once a possible result has been suspected. This we shall illustrate by a number of examples.

Example 1

To find the sum of the cubes of the first n integers (see 4.2), i.e., to find a formula for

$$\sum_{1}^{n} r^3 = 1^3 + 2^3 + \cdots + n^3$$

After a certain amount of trial and error with the simplest cases, it may be noticed that

$$1^3 = 1 = 1^2 = (\tfrac{1}{2} \times 1 \times 2)^2$$

$$1^3 + 2^3 = 9 = 3^2 = (\tfrac{1}{2} \times 2 \times 3)^2$$

$$1^3 + 2^3 + 3^3 = 36 = 6^2 = (\tfrac{1}{2} \times 3 \times 4)^2$$

$$1^3 + 2^3 + 3^3 + 4^3 = 100 = 10^2 = (\tfrac{1}{2} \times 4 \times 5)^2$$

and we are led to suspect that

$$\sum_{1}^{n} r^3 = 1^3 + 2^3 + 3^3 + \cdots + n^3 = [\tfrac{1}{2}n(n + 1)]^2 = \tfrac{1}{4}n^2(n + 1)^2$$

But the fact that this result has been shown to be true for $n = 1, 2, 3, 4$ and can be verified for any other particular value of n does not constitute a general proof. For, however often our hopes are confirmed, there will remain a fear that for some value of n the result will be untrue. That this is not so can be shown by the method of *mathematical induction*:

To prove that

$$\sum_{1}^{n} r^3 = 1^3 + 2^3 + 3^3 + \cdots + n^3 = \tfrac{1}{4}n^2(n + 1)^2, \quad (n \in Z^+)$$

First of all, let us suppose that this is true for a particular value of n, which we shall call k. We shall try to deduce from this hypothesis that it is then automatically true for the next value, $n = k + 1$.

If the result is true for $n = k$,

$$\text{then} \quad \sum_1^k r^3 = \tfrac{1}{4}k^2(k+1)^2$$

$$\text{But} \quad \sum_1^{k+1} r^3 = \sum_1^k r^3 + (k+1)^3$$

$$= \tfrac{1}{4}k^2(k+1)^2 + (k+1)^3$$

$$= \tfrac{1}{4}(k+1)^2[k^2 + 4(k+1)]$$

$$= \tfrac{1}{4}(k+1)^2(k+2)^2$$

which has established its truth for the next value, $n = k + 1$.

So *if* it is true for $n = k$, it is also true for $n = k + 1$.

$$\text{But} \quad 1^3 = \tfrac{1}{4} \times 1^2 \times 2^2$$

so we know that the result *is* true for $n = 1$.

Hence, from what we have just shown, it is true for the next value, $n = 2$; hence for the next value, $n = 3$; and so on, for all values of n.

So by establishing

(*i*) that its truth for $n = k$ *would imply* its truth for $n = k + 1$; and
(*ii*) that it *is* true for the first value $n = 1$,

we have proved that

$$\sum_1^n r^3 = \tfrac{1}{4}n^2(n+1)^2$$

for all positive integral values of n.

Example 2

Prove by induction that

$$\sum_1^n r(r+1) = \tfrac{1}{3}n(n+1)(n+2)$$

$$\text{i.e.,} \quad 1\times2 + 2\times3 + 3\times4 + \cdots + n(n+1) = \tfrac{1}{3}n(n+1)(n+2)$$

(We have already proved this by the method of differences.)

Suppose that the result is true for $n = k$,

$$\text{i.e., that} \quad \sum_1^k r(r+1) = \tfrac{1}{3}k(k+1)(k+2)$$

$$\text{Then} \quad \sum_1^{k+1} r(r+1) = \sum_1^k r(r+1) + (k+1)(k+2)$$

$$= \tfrac{1}{3}k(k+1)(k+2) + (k+1)(k+2)$$

$$= \tfrac{1}{3}(k+1)(k+2)(k+3)$$

So *if* the result is true for $n = k$, it is also true for $n = k + 1$. But we also see that,

$1 \times 2 = \frac{1}{3} \times 2 \times 3$, so the result *is* true for $n = 1$

Hence, *by induction*, it is true for all positive integral n that:

$$\sum_{1}^{n} r(r+1) = \tfrac{1}{3}n(n+1)(n+2)$$

Example 3

Prove, for all positive integral values of n, that

$$\frac{d}{dx}(x^n) = nx^{n-1}$$

Suppose this is true for $n = k$,

i.e., that $\dfrac{d}{dx}(x^k) = kx^{k-1}$

$$\text{Then} \quad \frac{d}{dx}(x^{k+1}) = \frac{d}{dx}(x \times x^k)$$

$$= x \times \frac{d}{dx}(x^k) + \frac{d}{dx}(x) \times x^k$$

$$= x \times kx^{k-1} + x^k$$

$$= kx^k + x^k = (k+1)x^k$$

So *if* the result is true for $n = k$, it will also be true for $n = k + 1$.

But $\dfrac{d}{dx}(x^1) = 1 = 1 \times x^0$

so the result *is* true for $n = 1$.

Hence, by induction, it is true for all positive integral n

that $\dfrac{d}{dx}(x^n) = nx^{n-1}$

Example 4

Show that, for all positive integers n, $u_n = 9^n + 7$ is divisible by 8.

Suppose that the result is true if $n = k$, i.e., that $u_k = 9^k + 7$ is divisible by 8.

$$\text{Now} \quad u_{k+1} - u_k = (9^{k+1} + 7) - (9^k + 7)$$

$$= 8 \times 9^k, \text{ which is divisible by 8}$$

$$\text{So} \quad u_{k+1} = u_k + 8 \times 9^k$$

and if u_k is divisible by 8, so is u_{k+1}.

So *if* the result is true for $n = k$, it must also be true for $n = k + 1$.

But $u_1 = 9^1 + 7 = 16$,

and so it *is* true for $n = 1$.

Hence, by induction, it is true for all positive integral n that $u_n = 9^n + 7$ is divisible by 8.

Exercise 6.5

Check the following conjectures when $n = 1, 2, 3, 4$ and then either prove them for $n \in Z^+$ by the method of induction or show that they are false.

1 $1 + 2 + 3 + \cdots + n = \frac{1}{2}n(n + 1)$.

2 $1^2 + 2^2 + 3^2 \cdots + n^2 = \frac{1}{6}n(n + 1)(2n + 1)$.

3 $1 + 3 + 5 + \cdots + (2n - 1) = n^2$.

4 $\dfrac{1}{1\times2} + \dfrac{1}{2\times3} + \cdots + \dfrac{1}{n(n + 1)} = \dfrac{n}{n + 1}$.

5 $1 + \dfrac{1}{2^2} + \dfrac{1}{3^2} + \cdots + \dfrac{1}{n^2} = \dfrac{2n + 1}{n + 2}$.

6 $1 + x + x^2 + \cdots + x^{n-1} = \dfrac{1 - x^n}{1 - x} \; (x \neq 1)$.

7 $n^3 - n$ is divisible by 6.

8 $8^n + 6$ is divisible by 14.

9 n straight coplanar lines, no two of which are parallel and no three concurrent, meet in $\frac{1}{2}n(n - 1)$ points.

10 The above lines divide the plane into $\frac{1}{2}(n^2 + n + 2)$ regions.

*6.6 Iteration: recurrence relations

It is frequently very easy, particularly when using a computer, to generate the terms of a sequence by considering the *process* by which the next term u_{n+1} is produced from its predecessor (or predecessors). This is called *iteration*, simply because it is repeated again and again, and is usually defined by a *recurrence relation* which expresses u_n in terms of its predecessor(s).

Example 1

Find the sequence defined by the recurrence relation $u_{n+1} = 2u_n + 1$ and the initial condition $u_1 = 1$.

We can see that

$$u_2 = 2u_1 + 1 = 2\times1 + 1 = 3$$

$$u_3 = 2u_2 + 1 = 2\times3 + 1 = 7$$

$$u_4 = 2u_3 + 1 = 2\times7 + 1 = 15$$

etc.

and so obtain the sequence 1, 3, 7, 15, ...

It is then quickly conjectured that the nth term of the sequence is

$$u_n = 2^n - 1.$$

This can easily be proved by the method of induction:

For $\quad u_k = 2^k - 1$

$$\Rightarrow \quad u_{k+1} = 2u_k + 1$$

$$= 2(2^k - 1) + 1$$

$$= 2^{k+1} - 1$$

So if the conjecture is true for $n = k$, it is also true for $n = k + 1$.

But $\quad u_1 = 1 = 2^1 - 1$

So the conjecture is true for $n = 1$.

Hence, by induction, it is true for every positive integer n.

Example 2

Find the sequence defined by

$$u_{n+1} = \tfrac{1}{2}\left(u_n + \frac{3}{u_n}\right); \quad u_1 = 1$$

We see that

$$u_2 = \frac{1}{2}\left(1 + \frac{3}{1}\right) = 2$$

$$u_3 = \frac{1}{2}\left(2 + \frac{3}{2}\right) = \frac{7}{4} = 1.75$$

$$u_4 = \frac{1}{2}\left(\frac{7}{4} + \frac{12}{7}\right) = \frac{97}{56} \approx 1.732$$

$$u_5 = \frac{1}{2}\left(\frac{97}{56} + \frac{168}{97}\right) = \frac{18\,817}{10\,864} \approx 1.732$$

Clearly the sequences *seem* to be tending to a limit, and we can see that *if* this is so and the limit is l, then

$$l = \frac{1}{2}\left(l + \frac{3}{l}\right)$$

$$\Rightarrow \quad l = \frac{3}{l} \quad \Rightarrow \quad l = \sqrt{3} \quad (\text{since } u_n > 0)$$

So if the series is tending to a limit, this limit must be $\sqrt{3}$, and this is in fact a very quick means (already employed in **1.5**) of calculating $\sqrt{3}$ to as high a degree of accuracy as we wish.

Exercise 6.6

1 Find the first seven terms of the sequence u_n defined by:

(*i*) $u_{n+1} = u_n + 2$ and $u_1 = 1$;
(*ii*) $u_{n+1} = u_n + n$ and $u_1 = 2$;
(*iii*) $u_{n+1} = 3u_n + 1$ and $u_1 = 4$;
(*iv*) $u_{n+1} = -2u_n$ and $u_1 = -1$.

2 What are the first seven terms of the sequence u_n defined by:

$u_{n+1} = (n + 1)\, u_n$ and $u_1 = 1$

Find a formula for u_n.

3 Find the first five terms of the sequence defined by:

$u_{n+1} = 5u_n - 6u_{n-1}$; $u_1 = 5$, $u_2 = 13$

and prove that

$u_n = 3^n + 2^n$

4 Calculate $\sqrt{10}$ to six places of decimals by means of the recurrence relation:

$$u_{n+1} = \frac{1}{2}\left(u_n + \frac{10}{u_n}\right)$$

5 $u_{n+1} = \frac{1}{2}\left(u_n + \frac{10}{u_n^2}\right)$; $u_1 = 2$

If $u_n \to l$, what is the value of l? Find it, correct to 2 decimal places.

*6.7 Interlude: Fibonacci numbers and prime numbers

Fibonacci numbers

Examine a pineapple. It will be seen that its units are arranged in 5 parallel rows rising gently to the right, 8 steeper rows rising to the left and 13 very steep rows rising to the right (though some pineapples are left-handed!). Why 5 and 8 and 13?

Examine the ancestry of a male bee (or *drone*). It has a mother (a *queen*) but no father, whilst a queen has both a father and a mother. So the genealogy of a drone can be represented:

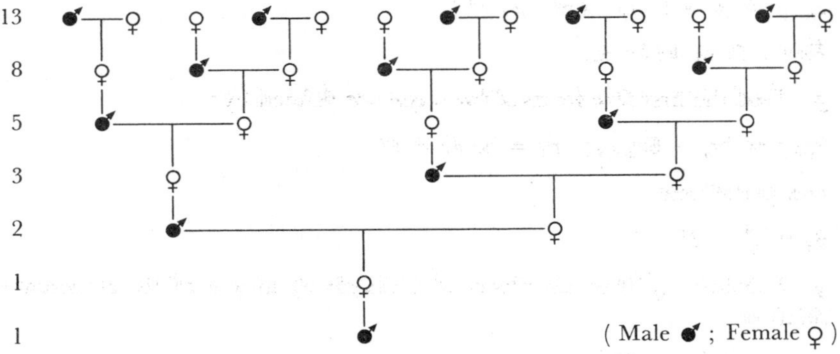

Going backwards in time, the numbers in the different generations are

1, 1, 2, 3, 5, 8, 13, . . .,

and again we notice the numbers 5, 8, and 13.

This sequence of *Fibonacci*† *numbers* is evidently generated by the recurrence relation

$$u_{n+1} = u_n + u_{n-1}$$

Exercise 6.7

1 What are the values of u_8, u_9, u_{10}, u_{11} in the Fibonacci sequence?

† The nickname of Leonardo of Pisa, a thirteenth century mathematician.

2 Show that:

(*i*) $\dfrac{u_{n+1}}{u_n} = 1 + \dfrac{u_{n-1}}{u_n}$;

(*ii*) if $\dfrac{u_{n+1}}{u_n} \to l$, then $l = 1 + \dfrac{1}{l}$.

Hence show that a distant generation of ancestors of a particular drone had $\dfrac{\sqrt{5}+1}{2}$ as many parents. What is the ratio of the sexes in this generation?

3 Check, for $n = 1, 2, 3, 4$, that

$$u_n = \frac{1}{\sqrt{5}}\left\{\left(\frac{1+\sqrt{5}}{2}\right)^n - \left(\frac{1-\sqrt{5}}{2}\right)^n\right\}$$

Prime numbers

A prime number is any positive integer larger than 1 which has no factors apart from 1 and itself.

So the sequence of prime numbers is 2, 3, 5, 7, 11, 13, 17, 19, 23. . . .

Unlike Fibonacci numbers, these are highly irregular and there is no iterative method for producing the next of the sequence. Clearly 2 is the only even prime, and as we move through the positive integers there is an increasing number of potential factors and so a diminishing likelihood that a number is prime. Nevertheless, consecutive pairs of prime numbers like 11, 13; 17, 19; 29, 31 keep persisting (the largest known pair at the time of writing is 1 000 000 009 649 and 1 000 000 009 651), though it is not yet certain whether there is an infinite supply of such pairs.†

A simpler question, however, is whether or not there is an infinite supply of the prime numbers themselves? Certainly the further we proceed, the scarcer they become. Are they ever exhausted? This question, 'Is this sequence of prime numbers infinite?', or 'Is there always a next prime number?', was first faced by Euclid and his proof is one of the most famous of mathematics:

Suppose we know that there are n prime numbers, which we shall call $p_1, p_2, p_3 \ldots p_n$ (so $p_1 = 2, p_2 = 3, p_3 = 5$, etc.). We now wish to try and establish that there is another prime number larger than p_n.

Let us consider the number

$$(p_1 p_2 p_3 \ldots p_n) + 1,$$

which we shall call N.

[So if $n = 4$, $N = 2\times3\times5\times7 + 1 = 211$.]

When p_1 is divided into N, there is a remainder 1.

† Similarly it appears that every even number can be expressed as the sum of two prime numbers, e.g., $8 = 3 + 5$, $28 = 11 + 17$, $100 = 47 + 53$. But, although there is no known exception, this conjecture also has never been proved.

When p_2 is divided into N, there is a remainder 1;
and so on.

So N is not divisible by any of the prime numbers $p_1, p_2 \ldots p_n$.

So *either* N is divisible by a prime number larger than p_n *or* N is itself a prime number, and in either case we have established the existence of a prime number larger than p_n; there is always a *next* prime number and their supply is never exhausted.

6.8 Permutations and combinations

Before we leave the subject of sequences and series, we shall investigate a particularly important two-way sequence, known as *Pascal's triangle*, and the closely related *binomial theorem*. But first we must consider the number of ways of *arranging* a number of objects from a given set, called their *permutations*, and also the number of ways of *choosing* them, called their *combinations*. We shall assume, unless otherwise stated, that no two objects are alike.

Permutations

The number of ways of arranging 6 boys in order is found by choosing one of the 6 for the first position (6 ways); then one of the remaining 5 fi r the next position (5 ways); then one of the remaining 4 for the third position (4 ways), etc. So the total number of permutations is

$$6 \times 5 \times 4 \times 3 \times 2 \times 1 = 6! = 720$$

More generally,

> The number of permutations of n different objects is
>
> $$n(n-1)(n-2)\ldots 4 \times 3 \times 2 \times 1 = n!$$

If we now ask in how many ways we can arrange (or *permute*) 4 out of a group of 6 boys, which we shall call the number of *permutations of 4 out of 6*, and write ${}^6P_4$, we see that

$${}^6P_4 = 6 \times 5 \times 4 \times 3 = 360$$

and it is also clear that this can be written as

$$\frac{6 \times 5 \times 4 \times 3 \times 2 \times 1}{2 \times 1} = \frac{6!}{2!}$$

So $${}^6P_4 = \frac{6!}{2!}$$

More generally,

$$^nP_r = n(n-1)(n-2)\dots(n-r+1)$$

$$= \frac{n(n-1)\dots(n-r+1)(n-r)(n-r-1)\dots 4\times 3\times 2\times 1}{(n-r)(n-r-1)\dots 4\times 3\times 2\times 1}$$

$$= \frac{n!}{(n-r)!}$$

When $r = n$, this becomes $\frac{n!}{0!}$, which has no meaning as $0!$ is undefined. But $^nP_n = n!$, so it is highly convenient if we now *define* $0!$ to be 1.

So $\boxed{^nP_r = \frac{n!}{(n-r)!}}$ (even when $r = n$)

Example 1

In how many ways can three prizes be awarded to a class of ten boys, one for English, one for French, and one for mathematics:

(*a*) if there is a rule that no boy may win more than one prize,
(*b*) with no such rule.

(*a*) There are 10 possible winners of the English prize, then 9 for French, then 8 for mathematics

So Number of ways $= 10 \times 9 \times 8 = 720$.

(*b*) There are 10 possible winners for English, then 10 for French, then 10 for mathematics,

So Number of ways $= 10 \times 10 \times 10 = 1\,000$.

Example 2

In how many ways can the letters L E E D S be arranged?

The difficulty here clearly lies with the repeated E. We can, however, mark them with suffixes, L E_1 E_2 D S, and we then see that re-arrangements occur in pairs which are distinguishable only by their suffixes,

e.g., E_1 L D E_2 S and E_2 L D E_1 S.

So the number of different re-arrangements $= \frac{5!}{2!} = 60$

Combinations

If we have a set of n different objects, in how many ways can we choose (*irrespective of order*) r out of them?

We shall call this the number of *combinations* of r objects out of n, written ${}^nC_r$.

For particular values of n and r, we can find ${}^nC_r$ simply by enumeration. For instance, three letters can be chosen out of the five letters ABCDE in the following ways:

ABC, ABD, ABE, ACD, ACE, ADE, BCD, BCE, BDE, CDE.

So ${}^5C_3 = 10$

But we could also look at the problem of choosing 3 letters out of 5 rather differently, and note that

the number of ways of *arranging* 3 letters out of 5

= (the number of ways of *choosing* 3 out of 5)
× (the number of ways of *arranging those chosen*).

So ${}^5P_3 = {}^5C_3 \times 3!$

$$\Rightarrow \quad {}^5C_3 = \frac{{}^5P_3}{3!} = \frac{5!}{3!2!}$$

More generally,

the number of ways of *arranging* r objects out of n

= (the number of ways of *choosing* r objects out of n)
× (the number of ways of *arranging those chosen*).

So ${}^nP_r = {}^nC_r \times r!$

$$\Rightarrow \quad {}^nC_r = \frac{{}^nP_r}{r!} = \frac{n!}{r!(n-r)!}$$

We can check this result in particular cases, such as

$${}^5C_3 = \frac{5!}{3!2!} = \frac{5\times4\times3\times2\times1}{3\times2\times1\times2\times1} = 10$$

If, however, $r = 0$, there is no meaning for ${}^nC_0$.

But $\dfrac{n!}{0!n!} = 1$, so we *define* ${}^nC_0$ to be 1.

Hence $$\boxed{{}^nC_r = \frac{n!}{r!(n-r)!}}$$ (even when $r = 0$)

Example 1

If a Council consists of 6 Conservative, 5 Labour and 2 Independent members, in how many ways can a committee be formed with 3 Conservative, 3 Labour, and 1 Independent members?

$$\text{Number of ways} = {}^6C_3 \times {}^5C_3 \times {}^2C_1$$
$$= 20 \times 10 \times 2 = 400$$

Example 2

Prove that ${}^{n+1}C_r = {}^nC_{r-1} + {}^nC_r$

To choose r objects out of $n + 1$, one can

either choose the first and then the other $r - 1$ from the remaining n,

or not choose the first, but choose all r from the remaining n.

So ${}^{n+1}C_r = {}^nC_{r-1} + {}^nC_r$

Exercise 6.8

1 In how many ways can the captain of a cricket team arrange his batting order?

2 In a newspaper competition eight desirable qualities of a good family car (roominess, reliability, etc.) have to be put in order of importance. How many coupons must be completed in order to be sure of obtaining the winning order?

3 If 6 couples go to a party, in how many ways can they pair off to dance?

4 (*i*) If 10 horses run in a race, in how many different ways can the first three places be filled?
(*ii*) If 10 greyhounds race against each other on three successive nights, in how many different ways can the winning places be filled?

5 In how many different orders can colours appear when 5 balls are drawn from a bag containing:
(*i*) 1 red ball, 1 yellow, 1 blue, and 2 white;
(*ii*) 1 red ball, 1 yellow, and 3 white;
(*iii*) 2 red balls and 3 white.

6 Repeat the last question, supposing that each ball drawn is immediately replaced.

7 A cricket touring party consists of 16 members. In how many ways could eleven players be selected for the first match?

8 A man tries to forecast the result (win, lose, or draw) of eleven first-division football matches. In how many ways could he be correct:
(*i*) in exactly 9 of the matches;
(*ii*) in at least 9 matches?

9 If n different points are given in space:
(*i*) how many lines are there connecting them in pairs;
(*ii*) how many triangles are there with three of the points as vertices;
(*iii*) how many tetrahedra are there with four of the points as vertices?

10 In how many ways can 12 golfers be divided into three groups of four?

11 By carefully considering their meaning, simplify:

(*i*) ${}^nC_{n-r}$; (*ii*) $\sum_{r=0}^{n} {}^nC_r$.

12 Use the formula for ${}^nC_r$ in order to simplify:
(*i*) ${}^nC_{n-r}$; (*ii*) ${}^nC_{r-1} + {}^nC_r$.

13 Six similar discs are coloured in pairs: red, green, and blue. In how many different ways can these discs be placed in a row, and of these how many show all three pairs of discs of the same colour lying next to each other? (M.E.I.)

14 How many 5-figure numbers can be constructed from the nine digits 1, 2, . . . 9:
(*i*) if no digit is repeated;
(*ii*) if repetitions are allowed?

15 (*i*) In how many different arrangements, relative to each other, can 6 people be seated at a circular table?
(*ii*) In how many orders, relative to each other, can 6 different-coloured beads be strung on a circular thread?

6.9 Pascal's Triangle: the binomial theorem

We now take a different starting point and look at the two-way sequence which arises from the expansions of $(a + b)^n$. For successive values of n, we see that:

$$(a + b)^0 = 1$$

$$(a + b)^1 = 1a + 1b$$

$$(a + b)^2 = 1a^2 + 2ab + 1b^2$$

$$(a + b)^3 = 1a^3 + 3a^2b + 3ab^2 + 1b^3$$

$$(a + b)^4 = 1a^4 + 4a^3b + 6a^2b^2 + 4ab^3 + 1b^4$$

$$(a + b)^5 = 1a^5 + 5a^4b + 10a^3b^2 + 10a^2b^3 + 5ab^4 + 1b^5$$

Picking out numerical coefficients, we obtain the array known as Pascal's triangle:†

1

1 1

1 2 1

1 3 3 1

1 4 6 4 1

1 5 10 10 5 1

† Blaise Pascal (1623–62), French philosopher and mathematician.

and a moment's glance shows how it can be extended for as many rows as we wish. We also notice that it is made up of values of ${}^nC_r$ (e.g., $4 = {}^4C_1$, $6 = {}^4C_2$, $10 = {}^5C_2$, $10 = {}^5C_3$, etc.), and that if we number its rows and columns from zero, the entry in the nth row and rth column is precisely ${}^nC_r$

n \ r	0	1	2	3	4	5
0	1					
1	1	1				
2	1	2	1			
3	1	3	3	1		
4	1	4	6	4	1	
5	1	5	10	10	5	1

If we now look again at the expansion of

$$(a + b)^n = (a + b)\,(a + b)\ldots(a + b)$$

we see that this product of n brackets has terms

$$\begin{array}{ll} a^n, & \\ a^{n-1}b, & ({}^nC_1 \text{ terms}) \\ a^{n-2}b^2, & ({}^nC_2 \text{ terms}) \\ \vdots & \\ a^{n-r}b^r, & ({}^nC_r \text{ terms}) \\ \vdots & \\ ab^{n-1}, & ({}^nC_{n-1} \text{ terms}) \\ b^n. & \end{array}$$

There are clearly ${}^nC_r$ terms of the type $a^{n-r}b^r$, as these are formed by taking

b from r of the brackets (which can be done in ${}^nC_r$ ways)

and a from the remaining $n - r$.

Hence we obtain the *binomial theorem*:

$$(a + b)^n = a^n + {}^nC_1a^{n-1}b + \cdots + {}^nC_ra^{n-r}b^r + \cdots + {}^nC_{n-1}ab^{n-1} + b^n$$

$$(a + b)^n = \sum_{r=0}^{n} {}^nC_ra^{n-r}b^r$$

Example 1

Use the binomial theorem to expand $(3x - 2y)^4$

$$(3x - 2y)^4 = (3x)^4 + {}^4C_1(3x)^3(-2y) + {}^4C_2(3x)^2(-2y)^2 + {}^4C_3(3x)(-2y)^3 + (-2y)^4$$

$$= (3x)^4 + 4(3x)^3(-2y) + 6(3x)^2(-2y)^2 + 4(3x)(-2y)^3 + (-2y)^4$$

$$= 81x^4 - 216x^3y + 216x^2y^2 - 96xy^3 + 16y^4$$

Example 2

Find the first three terms when $(1 - 2x)^{10}$ is expanded in ascending powers of x, and so evaluate $(0.998)^{10}$.

$$(1 - 2x)^{10} = 1^{10} + {}^{10}C_1 1^9(-2x) + {}^{10}C_2 1^9(-2x)^2 \cdots$$

$$= 1 + 10(-2x) + 45(-2x)^2 \ldots$$

$$= 1 - 20x + 180x^2 \ldots$$

Now put $x = 0.001$

$$(0.998)^{10} = 1 - 20(0.001) + 180(0.001)^2 \ldots$$

$$\approx 1 - 0.02 + 0.000\,180$$

$$\approx 0.9802$$

Exercise 6.9

1 Write down the expansions of:

(*i*) $(a + b)^4$; (*ii*) $(2x + y)^3$;

(*iii*) $(p - 2q)^5$; (*iv*) $\left(x - \frac{1}{x}\right)^4$;

(*v*) $(2x - y)^6$; (*vi*) $(3x^2 - 2y^2)^4$.

2 Write down the first three terms when each of the following is expanded in ascending powers of x:

(*i*) $(1 + x)^8$; (*ii*) $(1 + 2x)^6$;

(*iii*) $\left(1 - \frac{x}{2}\right)^7$; (*iv*) $(2 - 3x)^5$.

3 Find the given term in the following expansions:

(*i*) x^2 in expansion of $(1 + x)^8$;

(*ii*) x^3 in expansion of $(3 - 2x)^4$;

(*iii*) x^4 in expansion of $\left(x^2 + \frac{2}{x}\right)^5$;

(*iv*) constant term in expansion of $\left(2x + \frac{1}{x^2}\right)^3$;

(*v*) constant term in expansion of $\left(x^2 - \frac{1}{x}\right)^6$.

4 Use the binomial theorem to evaluate:
(i) $(1.01)^{10}$ to 4 significant figures;
(ii) $(0.99)^6$ to 5 significant figures;
(iii) $(2.001)^8$ to 6 significant figures;
(iv) $(0.998)^5$ to 5 significant figures.

5 Expand as far as the term in x^2:
(i) $(1 - x - x^2)^5$;
(ii) $(1 + 2x + 3x^2)^6$;
(iii) $(2 - x - x^2)^4$.

Miscellaneous problems 6

1 Investigate whether the sum of the cubes of three consecutive integers is always divisible by 9. Try to prove your conjecture.

2 n points are taken on the circumference of a circle. Investigate u_n, the number of regions into which the chords joining them divide the circle, if no three chords are concurrent. Show that $u_2 = 2$, $u_3 = 4$, $u_4 = 8$ and calculate u_5. Hence try to find a formula for u_n and check when $n = 6$.

3 A magic square of order n consists of the integers $1, 2, 3 \ldots n^2$ arranged in a square in such a way that the numbers in each row, in each column, and in each diagonal have the same sum. What is this sum?

4 Find the sum of all integers less than $10n$ which are not multiples of 2 or 5. (C.S.)

5 A company borrows £100 000 at 10% compound interest, repaying £15 000 at the end of each year. How much remains unpaid after the tenth repayment?

6 *(i)* A snowflake is made by taking an equilateral triangle of area 1 cm^2 and adding smaller equilateral triangles based on the middle thirds of the sides. This is repeated on the middle thirds of the segments of the new perimeter, and so on ad infinitum.

What is the area of such a snowflake, and what is its perimeter?

(ii) An anti-snowflake is made in a similar way, but by subtracting the equilateral triangles. Calculate its area and perimeter.

7 Find the sum of n terms of the series

$$x + x^2 + x^3 + \cdots + x^n$$

Hence, by differentiation, find the sum of the first n terms of

$$x + 2x^2 + 3x^3 + \cdots$$

and of $x + 2^2x^2 + 3^2x^3 + \cdots$

8 $f(x) = \sin x + \sin 2x + \cdots + \sin nx$

By considering the product

$$f(x) \sin \frac{x}{2}$$

and using the factor formulae, find the sum of this series.

9 ${}^nC_r = c_r$, so that

$$(1 + x)^n = c_0 + c_1x + c_2x^2 + \cdots + c_nx^n$$

Hence find:
(*i*) $c_0 + c_1 + c_2 + \cdots + c_n$;
(*ii*) $c_0 - c_1 + c_2 + \cdots + (-1)^n c_n$;
(*iii*) $c_1 + 2c_2 + 3c_3 + \cdots + nc_n$;
(*iv*) $c_0^2 + c_1^2 + c_2^2 + \cdots + c_n^2$.

10 An absent-minded mathematician writes n letters and addresses n envelopes for them. u_n is the number of ways in which the letters could be put in the envelopes so that each is in a wrong envelope.
(*i*) Find u_n for $n = 1, 2, 3, 4, 5$.
(*ii*) Try to find a recurrence relation for u_{n+1} in terms of u_n, u_{n-1} and hence calculate u_6, u_7.
(*iii*) Try to find u_n in terms of u_{n-1}.

7

Probability and statistics

7.1 Trials, events, and probabilities

If a coin is spun and is just as likely to turn up heads as tails we call it *fair*, or *unbiased*.

Such a spin is usually called a *trial* and the appearance of a head is an *event*. As this event occurs in only one of the two possible *outcomes*, and these are equally likely, we say that the probability of a head is $\frac{1}{2}$:

$$p\ (\text{head}) = \tfrac{1}{2}$$

This definition is fortified by our belief that, in a large number of similar trials, the proportion of heads would gradually approach $\frac{1}{2}$.†

Similarly, if a playing card is chosen from a well-shuffled pack, the probability of its being an ace is $\frac{4}{52}$, as the event of choosing an ace takes place in 4 of the 52 equally possible outcomes:

$$p\ (\text{ace}) = \frac{4}{52} = \frac{1}{13}$$

and likewise

$$p\ (\text{heart}) = \frac{13}{52} = \frac{1}{4}$$

More generally, let us consider the case of a trial that has N equally likely outcomes, in some of which a particular event $\boldsymbol{E}$ takes place.

We shall use $\mathscr{E}$ to denote the set of all N outcomes, and E to denote the set of outcomes in which $\boldsymbol{E}$ takes place.

† Indeed, there are many probabilities (e.g., that of a drawing-pin landing point upwards when dropped) which can be assessed only from such a large number of trials.

Then the probability of $\boldsymbol{E}$ is defined as

$$p(\boldsymbol{E}) = \frac{\text{number of elements of } E}{\text{number of elements of } \mathscr{E}} = \frac{n(E)}{N}$$

Example 1

Trial: The toss of a fair coin

$\mathscr{E} = \{\text{head, tail}\}$ and $n(\mathscr{E}) = 2$

If $\boldsymbol{E}$ is the event of the coin turning up heads then

$E = \{\text{head}\}$ and $n(E) = 1$,

So $p(\boldsymbol{E}) = \dfrac{n(E)}{n(\mathscr{E})} = \dfrac{1}{2}.$

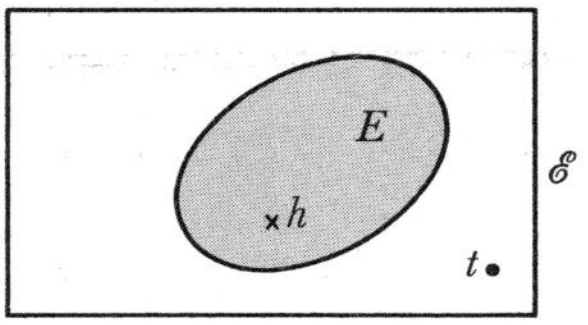

Example 2

Trial: the choice of a card from a well-shuffled pack.

$\mathscr{E} = \{\text{the 52 cards of the pack}\}$ and $n(\mathscr{E}) = 52$

If $\boldsymbol{A}$ is the event of choosing an ace, then

$A = \{\text{the four aces}\}$ and $n(A) = 4$

So $p(A) = \dfrac{n(A)}{n(\mathscr{E})} = \dfrac{4}{52} = \dfrac{1}{13}$

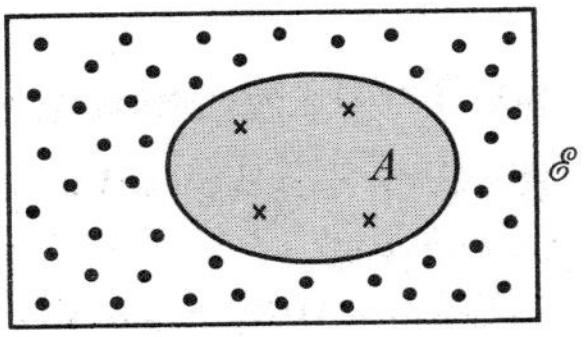

If $\boldsymbol{H}$ is the event of choosing a heart, then

$H = \{\text{the thirteen hearts}\}$ and $n(H) = 13$.

So $p(\boldsymbol{H}) = \dfrac{n(H)}{n(\mathscr{E})} = \dfrac{13}{52} = \dfrac{1}{4}$

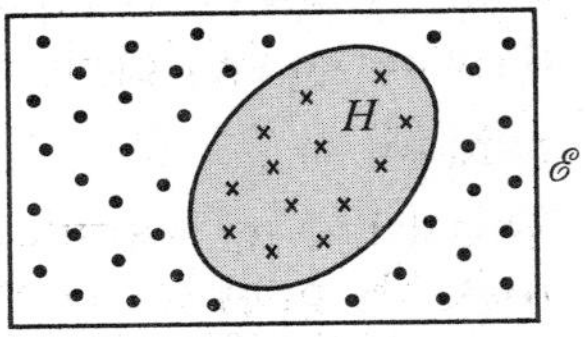

If we take particular cases:

(*i*) when $\boldsymbol{E}$ never happens, i.e., is impossible.

The set E has no members, and so

$$p(\boldsymbol{E}) = \frac{n(E)}{N} = \frac{0}{N} = 0$$

(*ii*) when $\boldsymbol{E}$ always happens, i.e., is certain.

The set E is identical with $\mathscr{E}$, so

$$p(\boldsymbol{E}) = \frac{n(E)}{N} = \frac{n(\mathscr{E})}{N} = \frac{N}{N} = 1$$

Moreover, if $\boldsymbol{E}'$ is the *event of* $\boldsymbol{E}$ *not occurring* (the *negation* of $\boldsymbol{E}$), we can use E' to represent the set of outcomes corresponding to the event $\boldsymbol{E}'$.

As every outcome belongs either to E or to E', but never to both:

$$n(E) + n(E') = N$$

$$\Rightarrow \quad \frac{n(E)}{N} + \frac{n(E')}{N} = 1$$

$$\Rightarrow \quad p(\boldsymbol{E}) + p(\boldsymbol{E}') = 1$$

Example 3

A fair six-sided die is thrown and $\boldsymbol{E}$ is the event of it showing a 5 or a 6.

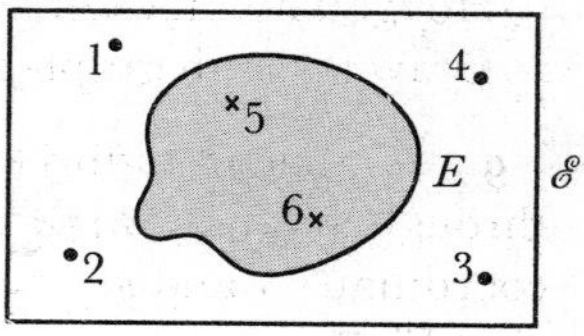

Here $\mathscr{E} = \{1, 2, 3, 4, 5, 6\}$

$E = \{5, 6\}$

$E' = \{1, 2, 3, 4\}$

So $p(\boldsymbol{E}) = \dfrac{n(E)}{n(\mathscr{E})} = \dfrac{2}{6} = \dfrac{1}{3}$

and $p(\boldsymbol{E}') = \dfrac{n(E')}{n(\mathscr{E})} = \dfrac{4}{6} = \dfrac{2}{3}$

Example 4

Two six-sided dice are thrown. What is the chance of scoring more than 10?

$\mathscr{E} = \{\text{All possible pairs of scores}\}$

$= \{(1, 1), (1, 2), (1, 3) \ldots (6, 4), (6, 5), (6, 6)\}$

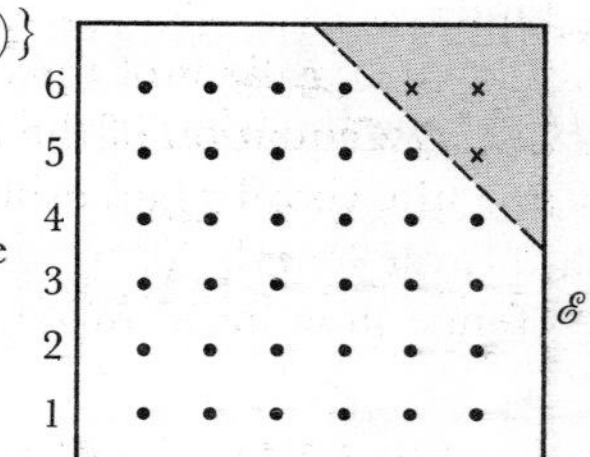

So $n(\mathscr{E}) = 36$

If $\boldsymbol{E}$ is the event of scoring more than 10,

then $E = \{(5, 6), (6, 5), (6, 6)\}$

and $n(E) = 3$

Hence $p(\boldsymbol{E}) = \dfrac{n(E)}{n(\mathscr{E})} = \dfrac{3}{36} = \dfrac{1}{12}$

It might alternatively be argued that the total score on two dice could be

2, 3, 4, 5, . . ., 12

and that of these eleven possible totals exactly two are greater than 10.

So is it true to say that the probability of scoring more than 10 is $\dfrac{2}{11}$?

Why not? Are the eleven possible occurrences all equally likely?

Exercise 7.1

In the following trials state the universal set $\mathscr{E}$ corresponding to all possible occurrences and the sets corresponding to the stated events. Hence calculate the probabilities of these events.

1 A card is chosen from a complete pack:
(*i*) ***A*:** the choice of an ace;
(*ii*) ***B*:** the choice of a black card;
(*iii*) ***C*:** the choice of a court card (ace, king, queen, or jack).

2 A dice is thrown:
(*i*) ***S*:** scoring a six;
(*ii*) ***E*:** scoring an even number;
(*iii*) ***Z*:** scoring less than 4 or more than 5.

3 Two coins are spun:
(*i*) ***H*:** two heads;
(*ii*) ***T*:** two tails;
(*iii*) ***E*:** a head and a tail.

4 Three coins are spun:
(*i*) ***H*:** three heads;
(*ii*) ***M*:** more heads than tails;
(*iii*) ***S*:** the same number of heads and tails.

5 Two poker dice are thrown:
(*i*) $\boldsymbol{E_2}$**:** two aces;
(*ii*) $\boldsymbol{E_1}$**:** one ace;
(*iii*) $\boldsymbol{E_0}$**:** no aces.

6 Two boys each choose a positive integer less than 10:
(*i*) ***A*:** they choose the same number;
(*ii*) ***B*:** they choose different numbers;
(*iii*) ***C*:** their total is less than 5.

7 I meet a friend who has four children in his family:
(*i*) ***A*:** it will contain 2 boys and 2 girls;
(*ii*) ***B*:** it will contain more boys than girls;
(*iii*) ***C*:** the eldest and youngest will be boys.

8 Five tulips, of which two are red, are planted in a row:
(*i*) ***E*:** the red tulips are at the two ends;
(*ii*) ***N*:** the red tulips are next to each other.

9 Two boys have their birthdays in the same week:
(*i*) ***M*:** both birthdays are on the Monday;
(*ii*) ***S*:** both birthdays are on the same day;
(*iii*) ***C*:** the birthdays are on consecutive days.

7.2 Compound events

If ***A*** and ***B*** are two separate events, then

(*i*) ***AB*** represents the event of both ***A*** and ***B*** taking place; and
(*ii*) $\boldsymbol{A + B}$ represents the event of either ***A*** or ***B*** (or both) taking place.

Suppose for example, that in the trial of drawing a card from a pack:

A is the event of choosing an ace: $A = \{\text{the 4 aces}\}$; and

B is the event of choosing a black card: $B = \{\text{the 26 spades and clubs}\}$.

Then $\quad AB = \{\text{ace of spades, ace of clubs}\}$

and $\quad A + B = \{\text{all 4 aces together with all the remaining black cards}\}$.

Further, we see that:

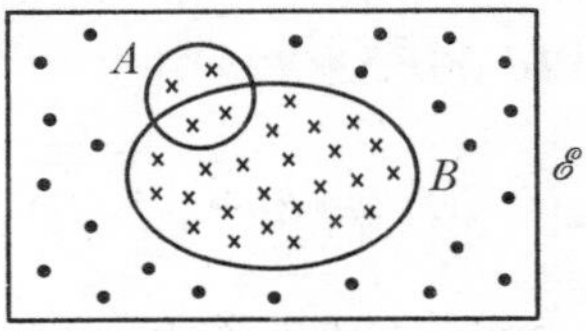

$n(A) = 4, \qquad n(B) = 26$

$n(AB) = 2, \quad n(A + B) = 28$

These are connected by the equation:

$$n(A + B) + n(AB) = n(A) + n(B)$$

and inspection of other Venn diagrams will satisfy the reader that this must always be so.

Hence, dividing by N:

$$\frac{n(A + B)}{N} + \frac{n(AB)}{N} = \frac{n(A)}{N} + \frac{n(B)}{N}$$

$$\Rightarrow \quad \boxed{p(\boldsymbol{A} + \boldsymbol{B}) + p(\boldsymbol{AB}) = p(\boldsymbol{A}) + p(\boldsymbol{B})}$$

Exclusive events

If ***A*** and ***B*** are two events which can never happen together, they are called *exclusive* events.

In such a case, $\quad AB = \phi \quad \Rightarrow \quad p(\boldsymbol{AB}) = 0$

$$\Rightarrow \quad p(\boldsymbol{A} + \boldsymbol{B}) = p(\boldsymbol{A}) + p(\boldsymbol{B})$$

For example, in drawing a card from a pack

$$\left.\begin{aligned} &\boldsymbol{A} = \text{the choice of an ace} \\ \text{and} \quad &\boldsymbol{K} = \text{the choice of a king} \end{aligned}\right\} \text{ are exclusive events}$$

In this case,

$$p(\boldsymbol{A}) = \frac{1}{13}, \quad p(\boldsymbol{K}) = \frac{1}{13}, \quad \text{and} \quad p(\boldsymbol{A} + \boldsymbol{K}) = \frac{2}{13}$$

Similarly,

$$\left.\begin{aligned} &\boldsymbol{C} = \text{the choice of a club} \\ \text{and} \quad &\boldsymbol{R} = \text{the choice of a red card} \end{aligned}\right\} \text{ are exclusive}$$

and in this case

$p(C) = \frac{1}{4}, \quad p(R) = \frac{1}{2}, \quad p(C + R) = \frac{3}{4}$

By contrast,

$\left.\begin{array}{ll} & A = \text{the choice of an ace} \\ \text{and} & B = \text{the choice of a black card} \end{array}\right\}$ are *not* exclusive

In this case,

$$p(A) = \frac{1}{13}, \quad p(B) = \frac{1}{4}, \quad p(A + B) = \frac{28}{52} = \frac{7}{13}$$

and $p(A + B) \neq p(A) + p(B)$

Exhaustive events

If A and B are two events at least one of which must happen, they are called *exhaustive*. In such a case, $p(A + B) = 1$

$\Rightarrow \quad 1 + p(AB) = p(A) + p(B)$

Exercise 7.2a

In nos. 1–6 an ordinary six-sided die is thrown:

A is the event of scoring a 6;
B is the event of scoring an even number;
C is the event of scoring less than 3.

1 Find (*i*) $p(A)$, $p(B)$, $p(C)$;
(*ii*) $p(BC)$, $p(CA)$, $p(AB)$;
(*iii*) $p(B + C)$, $p(C + A)$, $p(A + B)$.

2 Check that:

$p(A + B) + p(AB) = p(A) + p(B)$

and the two similar results.

3 Which pairs of events are:
(*i*) exclusive; (*ii*) exhaustive?

4–6 Repeat questions 1–3 for the events A', B', and C'.

7 The probability of a boy being selected for the Cricket XI is $\frac{1}{10}$, for the Football XI is $\frac{1}{20}$, and for both is $\frac{1}{40}$. What is the chance of his being selected for:
(*i*) at least one of the teams;
(*ii*) just one of the teams?

8 An integer is chosen at random. Calculate the probability that it is not divisible by 3 or 7. (M.E.I.)

Independent events

Suppose that a person is chosen at random from a football crowd, perhaps by selecting the ten thousandth to be admitted to the ground.

Let $\boldsymbol{A}$ be the event of the person having a surname beginning with the letter A;
and $\boldsymbol{B}$ be the event of the person being bald.

Then it may happen that

$$p(\boldsymbol{A}) = \frac{1}{10} \quad \text{and} \quad p(\boldsymbol{B}) = \frac{1}{20}$$

What would be the probability of the person having a surname starting with A *and* being bald?

As there is no likely connection between baldness and the first letter of one's surname, we should expect this probability to be

$$\frac{1}{10} \times \frac{1}{20} = \frac{1}{200}$$

so that $p(\boldsymbol{AB}) = p(\boldsymbol{A})p(\boldsymbol{B})$

If, however,

$\boldsymbol{A}$ is the event of the person being an adult,
and $\boldsymbol{B}$ is the event of being bald,

we should expect a very different answer. It is almost certain that a bald person will be adult, so $p(\boldsymbol{AB})$ is not very different from $p(\boldsymbol{B})$ and in this case

$$p(\boldsymbol{AB}) \neq p(\boldsymbol{A})\,p(\boldsymbol{B})$$

Whenever events $\boldsymbol{A}$ and $\boldsymbol{B}$ are such that

$$p(\boldsymbol{AB}) = p(\boldsymbol{A})\,p(\boldsymbol{B})$$

we call them *independent*.

So having a surname beginning with A and being bald are independent events, but being adult and being bald are not independent.

If $\boldsymbol{A}$ and $\boldsymbol{B}$ are independent, it can easily be shown that $\boldsymbol{A}'$ and $\boldsymbol{B}$ are also independent.

For $\boldsymbol{A}$ and $\boldsymbol{B}$ independent $\Leftrightarrow p(\boldsymbol{AB}) = p(\boldsymbol{A})\,p(\boldsymbol{B})$

$$\begin{aligned} \text{But} \quad p(\boldsymbol{A'B}) &= p(\boldsymbol{B}) - p(\boldsymbol{AB}) \\ &= p(\boldsymbol{B}) - p(\boldsymbol{A})\,p(\boldsymbol{B}) \\ &= (1 - p(\boldsymbol{A}))\,p(\boldsymbol{B}) \\ &= p(\boldsymbol{A}')\,p(\boldsymbol{B}) \end{aligned}$$

So $\boldsymbol{A}'$ and $\boldsymbol{B}$ are independent.

Similarly it can be shown that $\boldsymbol{A}$ and $\boldsymbol{B}'$ are independent and that $\boldsymbol{A}'$ and $\boldsymbol{B}'$ are independent.

Contingency tables

Compound events, whether independent or not, can be illustrated by means of probability, or *contingency*, tables.

Example 1

A card is drawn from a pack:

$\boldsymbol{A}$: The choice of an ace.
and $\boldsymbol{B}$: The choice of a black card.

These are independent events and the various probabilities are as follows:

	$\boldsymbol{B}$	$\boldsymbol{B}'$	
$\boldsymbol{A}$	$p(\boldsymbol{AB}) = \frac{1}{26}$	$p(\boldsymbol{AB}') = \frac{1}{26}$	$p(\boldsymbol{A}) = \frac{1}{13}$
$\boldsymbol{A}'$	$p(\boldsymbol{A}'\boldsymbol{B}) = \frac{6}{13}$	$p(\boldsymbol{A}'\boldsymbol{B}') = \frac{6}{13}$	$p(\boldsymbol{A}') = \frac{12}{13}$
	$p(\boldsymbol{B}) = \frac{1}{2}$	$p(\boldsymbol{B}') = \frac{1}{2}$	1

Example 2

A card is drawn from a pack:

$\boldsymbol{B}$: The choice of a black card.
and $\boldsymbol{C}$: The choice of a club.

As a club must be a black card, these are *not* independent events and the probabilities are:

	$\boldsymbol{B}$	$\boldsymbol{B}'$	
$\boldsymbol{C}$	$\frac{1}{4}$	0	$\frac{1}{4}$
$\boldsymbol{C}'$	$\frac{1}{4}$	$\frac{1}{2}$	$\frac{3}{4}$
	$\frac{1}{2}$	$\frac{1}{2}$	1

Example 3

10% of a large consignment of apples is known to be bad. If three are chosen at random, what is the chance that:

(*i*) all will be bad;
(*ii*) none will be bad;
(*iii*) at least one will be bad?

As the consignment is large, the chance that an apple will be bad is not affected by those previously chosen.

Now $$p(\boldsymbol{B}) = \frac{1}{10}$$

So $$p(\boldsymbol{B}, \boldsymbol{B}, \boldsymbol{B}) = \frac{1}{10} \times \frac{1}{10} \times \frac{1}{10} = \frac{1}{1\,000} = 0.001$$

Also $$p(\boldsymbol{B}') = \frac{9}{10}$$

So $$p(\boldsymbol{B}', \boldsymbol{B}', \boldsymbol{B}') = \frac{9}{10} \times \frac{9}{10} \times \frac{9}{10} = \frac{729}{1\,000} = 0.729$$

$\Rightarrow$ Chance of having at least one bad $= 1 - 0.729 = 0.271$.

Exercise 7.2b

In the following trials calculate $p(\boldsymbol{A})$, $p(\boldsymbol{B})$, and $p(\boldsymbol{AB})$. In each case state whether events $\boldsymbol{A}$ and $\boldsymbol{B}$ are independent.

1 Andrew and Barbara each throw a die:
$\boldsymbol{A}$: Andrew throws a 6;
$\boldsymbol{B}$: Barbara throws a 6.

2 Andrew and Barbara each has a pack of cards and picks a card:
$\boldsymbol{A}$: Andrew picks a heart;
$\boldsymbol{B}$: Barbara picks a heart.

3 Andrew and Barbara each take a card from the same pack at the same time:
$\boldsymbol{A}$: Andrew picks a heart;
$\boldsymbol{B}$: Barbara picks a heart.

4 Andrew chooses at random a number between 1 and 200:
$\boldsymbol{A}$: it is divisible by 4;
$\boldsymbol{B}$: it is divisible by 5.

5 Barbara chooses at random a number between 1 and 200:
$\boldsymbol{A}$: it is divisible by 4;
$\boldsymbol{B}$: it is divisible by 10.

6 A pack of 52 ordinary playing cards is shuffled and one card is withdrawn. $\boldsymbol{D}$ denotes the event that the card is a diamond, $\boldsymbol{K}$ that it is a king,

R that it is red. Prove by calculating the appropriate probabilities that D and K are independent, that K and R are independent, but that D and R are not independent.

What is the value of $p(D'K + DK')$? (M.E.I.)

7 How many times should an unbiassed die be thrown if the probability that a six should appear at least once is to be greater than $\frac{9}{10}$?

8 The chance of hitting a target with a single shot is $\frac{1}{3}$. How many shots must be fired to make the chance of at least one hit on the target more than 90 per cent?

What is the chance that the first hit scored will be with the third shot? (O.C.)

* 7.3 Conditional probability

We frequently wish to consider the probability of an event A amongst occurrences of another event B. This we call the *conditional probability of* A *upon* B, written $p(A/B)$, or the probability of A given B.

For example, in drawing a card from an ordinary pack,

p(ace/club) = probability of a card being an ace if it is known to be a club $= \frac{1}{13}$

p(black card/club) = probability of a card being black if it is known to be a club = 1

and similarly p(club/black card) $= \frac{1}{2}$,

p(heart/black card) = 0

Now $p(A/B)$ is the probability of event A amongst occurrences of B, and so is the probability of event AB amongst occurrences of B.

So $$p(A/B) = \frac{n(AB)}{n(B)} = \frac{\{n(AB)/N\}}{\{n(B)/N\}}$$

$$\Rightarrow \quad \boxed{p(A/B) = \frac{p(AB)}{p(B)}}$$

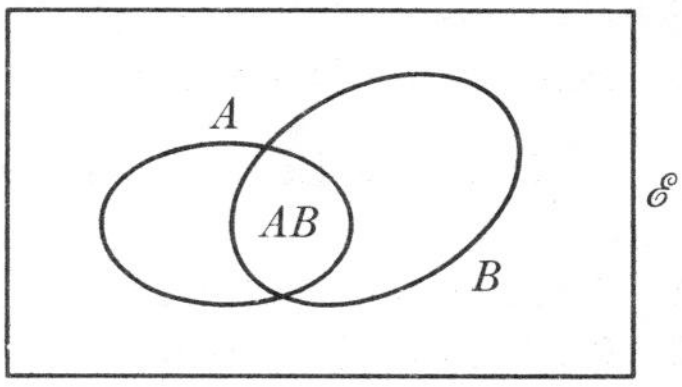

and this can quickly be verified in the above examples.

In the particular case when A and B are independent

$$p(AB) = p(A)\,p(B)$$

$$\Rightarrow \quad p(A/B) = \frac{p(AB)}{p(B)} = \frac{p(A)\,p(B)}{p(B)} = p(A)$$

$$\text{and} \quad p(B/A) = \frac{p(BA)}{p(A)} = \frac{p(B)\,p(A)}{p(A)} = p(B)$$

So if two events are independent the probability of either is (as we would expect) always the same, whether or not the other is known to occur.

Exercise 7.3a

1 When a card is chosen from a pack $\boldsymbol{A}$ is the event of obtaining an ace and $\boldsymbol{B}$ is the event of obtaining a black card. Calculate:

(*i*) $p(\boldsymbol{A})$, $p(\boldsymbol{A'})$, $p(\boldsymbol{B})$, $p(\boldsymbol{B'})$
(*ii*) $p(\boldsymbol{AB})$, $p(\boldsymbol{AB'})$, $p(\boldsymbol{A'B})$, $p(\boldsymbol{A'B'})$
(*iii*) $p(\boldsymbol{A}/\boldsymbol{B})$, $p(\boldsymbol{A}/\boldsymbol{B'})$, $p(\boldsymbol{A'}/\boldsymbol{B})$, $p(\boldsymbol{A'}/\boldsymbol{B'})$
(*iv*) $p(\boldsymbol{B}/\boldsymbol{A})$, $p(\boldsymbol{B}/\boldsymbol{A'})$, $p(\boldsymbol{B'}/\boldsymbol{A})$, $p(\boldsymbol{B'}/\boldsymbol{A'})$

Which of the following pairs of events are independent: $\boldsymbol{A}, \boldsymbol{B}$; $\boldsymbol{A}, \boldsymbol{B'}$; $\boldsymbol{A'}, \boldsymbol{B}$; $\boldsymbol{A'}, \boldsymbol{B'}$?

2 Repeat no. **1** if a die is thrown and

(*i*) $\boldsymbol{A}$ is the event of obtaining a six,
(*ii*) $\boldsymbol{B}$ is the event of obtaining an even number.

3 Prove that if two events $\boldsymbol{A}, \boldsymbol{B}$ are independent, so are $\boldsymbol{A}, \boldsymbol{B'}$; $\boldsymbol{A'}, \boldsymbol{B}$; $\boldsymbol{A'}, \boldsymbol{B'}$.

Conditional probabilities are very conveniently illustrated by means of *probability trees*:

Example

The probabilities that a man makes a certain dangerous journey by car ($\boldsymbol{C}$), motor-bike ($\boldsymbol{M}$), or on foot ($\boldsymbol{F}$) are $\frac{1}{2}$, $\frac{1}{6}$, and $\frac{1}{3}$ respectively. If the probabilities of an accident ($\boldsymbol{A}$) when he uses these means of transport are $\frac{1}{5}$, $\frac{3}{5}$, and $\frac{1}{10}$ respectively, what is the chance of an accident? Here the situation can best be summed up in a figure:

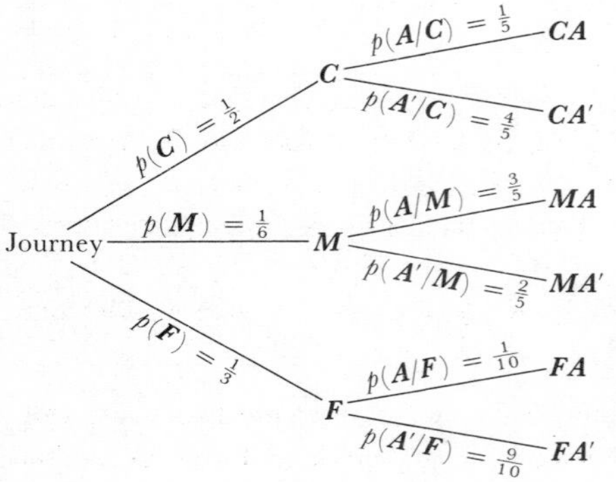

$p(\boldsymbol{CA}) = \frac{1}{2} \times \frac{1}{5} = \frac{1}{10}$

$p(\boldsymbol{CA'}) = \frac{1}{2} \times \frac{4}{5} = \frac{2}{5}$

$p(\boldsymbol{MA}) = \frac{1}{6} \times \frac{3}{5} = \frac{1}{10}$

$p(\boldsymbol{MA'}) = \frac{1}{6} \times \frac{2}{5} = \frac{1}{15}$

$p(\boldsymbol{FA}) = \frac{1}{3} \times \frac{1}{10} = \frac{1}{30}$

$p(\boldsymbol{FA'}) = \frac{1}{3} \times \frac{9}{10} = \frac{3}{10}$

So $p(A) = p(CA) + p(MA) + p(FA)$

$$= p(C)\,p(A/C) + p(M)\,p(A/M) + p(F)\,p(A/F)$$

$$= \tfrac{1}{2}\times\tfrac{1}{5} + \tfrac{1}{6}\times\tfrac{3}{5} + \tfrac{1}{3}\times\tfrac{1}{10}$$

$$= \tfrac{1}{10} + \tfrac{1}{10} + \tfrac{1}{30}$$

$$= \tfrac{7}{30}$$

So the chance of an accident $= \frac{7}{30}$.

It is also interesting to ask the reverse question: if an accident is known to have happened, what is the probability that the man was travelling by car, motor-bike, or on foot?

Using the above notation, these probabilities can be written

$$p(C/A) = \frac{p(CA)}{p(A)} = \frac{\tfrac{1}{10}}{\tfrac{7}{30}} = \frac{3}{7}$$

$$p(M/A) = \frac{p(MA)}{p(A)} = \frac{\tfrac{1}{10}}{\tfrac{7}{30}} = \frac{3}{7}$$

and $$p(F/A) = \frac{p(FA)}{p(A)} = \frac{\tfrac{1}{30}}{\tfrac{7}{20}} = \frac{1}{7}$$

So if it is known that the man had an accident, the probabilities of his having travelled by car, motor-bike, or on foot are $\frac{3}{7}$, $\frac{3}{7}$, $\frac{1}{7}$ respectively.

If he is known to have arrived safely, show that the corresponding probabilities are $\frac{12}{23}$, $\frac{2}{23}$, $\frac{9}{23}$.

Exercise 7.3b

1 A coin is spun twice. Draw a probability tree showing the various possible combinations of heads and tails.

2 Repeat no. 1 when the coin is spun three times. What are the probabilities of 0, 1, 2, 3 heads?

3 The chance of tomorrow being fine is $\frac{3}{4}$. If it is fine, our football team's chance of victory is $\frac{4}{5}$; but otherwise it is only $\frac{1}{2}$. What is our chance of victory.

4 Ten counters are in a bag, 7 being white and 3 black. They are taken out at random one by one without replacement. Calculate the probabilities that (*i*) the first taken is black; (*ii*) the second is black; (*iii*) the first two are black; (*iv*) exactly one of the first two is black; (*v*) at least one of the first two is black. (O.C.)

[This question should be answered both by use of combinations and by a probability tree.]

5 The probability that a candidate attempts a certain question is $\frac{9}{10}$ and, having done so, the probability of success is $\frac{2}{3}$. Find the probability

that the examiner will find at least one correct solution in the first three scripts which he marks. (S.M.P.)

6 A garden has 3 flower beds. The first bed has 50 flowers of which 10 are red, the second bed has 30 flowers of which 10 are red and the third bed has 20 flowers of which 10 are red. One of the beds is chosen at random and a flower is selected at random from that bed and removed from the garden. Find:
(*i*) the probability that a red flower will be taken from the first bed;
(*ii*) the probability that a red flower will be taken from the garden.
If a flower is chosen at random from the 100 flowers in the garden, find the probability that the flower will be red. (M.E.I.)

7 John and Peter each have two pennies. John tosses each of his coins once and gives to Peter those which fall heads. Peter now tosses once each of the coins in his possession, giving to John those which fall heads. Using a probability tree, or otherwise, determine the probability that:
(*i*) Peter now has all the coins.
(*ii*) John now has all the coins.
(*iii*) John and Peter each now have two coins.

8 Events $\boldsymbol{A}$ and $\boldsymbol{B}$ are such that

$$p(\boldsymbol{A}) = \tfrac{2}{3}, \quad p(\boldsymbol{B}/\boldsymbol{A}) = \tfrac{1}{2}, \quad p(\boldsymbol{B}/\boldsymbol{A}') = \tfrac{1}{4}$$

(*i*) Draw the corresponding probability tree with $\boldsymbol{A}$ before $\boldsymbol{B}$ and all probabilities marked.
(*ii*) Calculate $p(\boldsymbol{AB})$, $p(\boldsymbol{AB}')$, $p(\boldsymbol{A}'\boldsymbol{B})$, $p(\boldsymbol{A}'\boldsymbol{B}')$.
(*iii*) Calculate $p(\boldsymbol{B})$, $p(\boldsymbol{B}')$.
(*iv*) Calculate $p(\boldsymbol{A}/\boldsymbol{B})$, $p(\boldsymbol{A}'/\boldsymbol{B})$, $p(\boldsymbol{A}/\boldsymbol{B}')$, $p(\boldsymbol{A}'/\boldsymbol{B}')$.
(*v*) Draw the probability tree with $\boldsymbol{B}$ before $\boldsymbol{A}$, again marking all the probabilities.

9 Box A contains two white balls and Box B a white ball and a black ball. A man spins a coin in order to decide the box from which to choose a ball. If we are told that the chosen ball is white, what is the chance of its having come from Box A?

10 20% of the inhabitants of a city have been inoculated against a certain disease. In an epidemic the chance of infection amongst those inoculated is $\frac{1}{10}$, but amongst the rest is $\frac{3}{4}$.
(*i*) What proportion are infected?
(*ii*) If a man is chosen at random and found to be infected, what is the chance of his having been inoculated?
(*iii*) What is the corresponding chance if he is not infected?

11 A bowler delivers off-breaks, leg-breaks and straight balls with probabilities $\frac{1}{3}$, $\frac{1}{6}$, and $\frac{1}{2}$, and the batsman's chances of being bowled by these balls are $\frac{1}{10}$, $\frac{1}{10}$, and $\frac{1}{20}$ respectively.
(*i*) What is the chance of his being bowled by a particular ball?

(*ii*) If he is bowled, what is the chance of the ball having been straight?
(*iii*) If he is not bowled, what is the chance of it having been a leg-break?

12 A company advertises a professional vacancy in three national daily newspapers A, B, and C which have readerships in the proportions 2, 3, 1 respectively. From a survey of the occupations of the readers of these papers it is thought that the probabilities of an individual reader replying are 0.002, 0.001, 0.005 respectively.
(*i*) If the company receives one reply, what are the probabilities that the applicant is a reader of papers A, B, and C respectively?
(*ii*) If two replies are received, what is the probability that both applicants are readers of paper A?

[You may assume that each reader sees only one paper.] (M.E.I.)

*7.4 Probability distributions: three standard types

We shall now investigate three different trials that can be carried out with dice, calculating the probabilities of the various events that can occur, and investigating the way in which these are distributed. The reader may like to carry out these trials a number of times and find the discrepancies between his distributions and the ones we obtain. A further investigation of these discrepancies is one of the major tasks of statistics. If for instance, we find that the discrepancy is considerable and that the chance of such a large discrepancy is exceedingly small, we begin to suspect the basis of our calculation and are forced to doubt that the dice were unbiased. But for the moment we shall concentrate on calculating the probabilities of the various events which are possible.

The rectangular distribution

Suppose that a single die is thrown just once. Then the probabilities of scoring 1, 2, 3, 4, 5, 6 are all $\frac{1}{6}$ and we obtain a *rectangular distribution*:

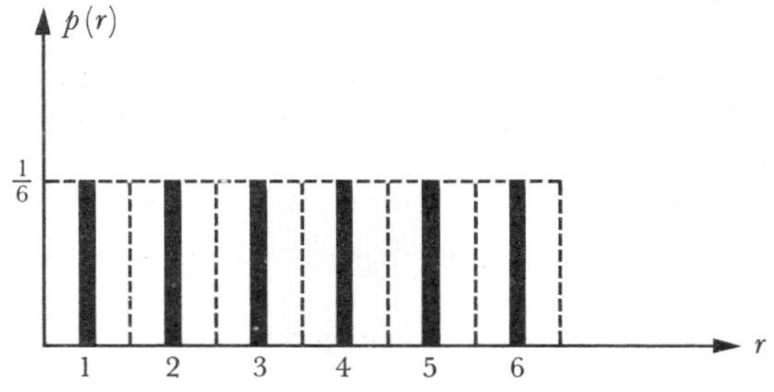

Exercise 7.4a

1 Two dice are thrown and their total score r recorded. Calculate the probabilities $p(r)$ $(2 \leqslant r \leqslant 12)$ and sketch the corresponding probability distribution.

2 Repeat no. 1 when r is the numerical difference between their scores.

The binomial distribution

If four dice are now thrown, what are the probabilities of obtaining

0, 1, 2, 3, 4 sixes?

With a single die, the chance:

of throwing a six is $\frac{1}{6}$;
of not throwing a six is $\frac{5}{6}$.

So when four dice are thrown,

$$p(0 \text{ sixes}) = \left(\frac{5}{6}\right)^4$$

If 1 six is thrown, this can be in any one of ${}^4C_1$ $(= 4)$ ways.

So $$p(1 \text{ six}) = {}^4C_1 \left(\frac{5}{6}\right)^3 \left(\frac{1}{6}\right).$$

Now 2 sixes can occur in any one of ${}^4C_2$ $(= 6)$ ways.

So $$p(2 \text{ sixes}) = {}^4C_2 \left(\frac{5}{6}\right)^2 \left(\frac{1}{6}\right)^2$$

Similarly $$p(3 \text{ sixes}) = {}^4C_3 \left(\frac{5}{6}\right) \left(\frac{1}{6}\right)^3$$

and $$p(4 \text{ sixes}) = \left(\frac{1}{6}\right)^4$$

So we see that the probabilities of 0, 1, 2, 3, 4 sixes are given by the terms of the binomial expansion

$$\left(\frac{5}{6} + \frac{1}{6}\right)^4$$

More generally, if in a single trial the probability of a certain event is p and $q = 1 - p$, then in n independent trials the probabilities of

0, 1, 2, . . ., n events are given by the terms of $(q + p)^n$

This is therefore called a *binomial distribution.*

The individual probabilities can easily be calculated.

In the above case, for instance, $n = 4$, $p = \frac{5}{6}$, and $q = \frac{1}{6}$ and the chances of 0, 1, 2, 3, 4 sixes were given by the terms of

$$\left(\frac{5}{6} + \frac{1}{6}\right)^4$$

and so are

$$\left(\frac{5}{6}\right)^4, \quad 4\left(\frac{5}{6}\right)^3\left(\frac{1}{6}\right), \quad 6\left(\frac{5}{6}\right)^2\left(\frac{1}{6}\right)^2, \quad 4\left(\frac{5}{6}\right)\left(\frac{1}{6}\right)^3, \quad \left(\frac{1}{6}\right)^4$$

which can easily be shown to be approximately:

0.482, 0.386, 0.116, 0.015, 0.001

These can be most conveniently illustrated

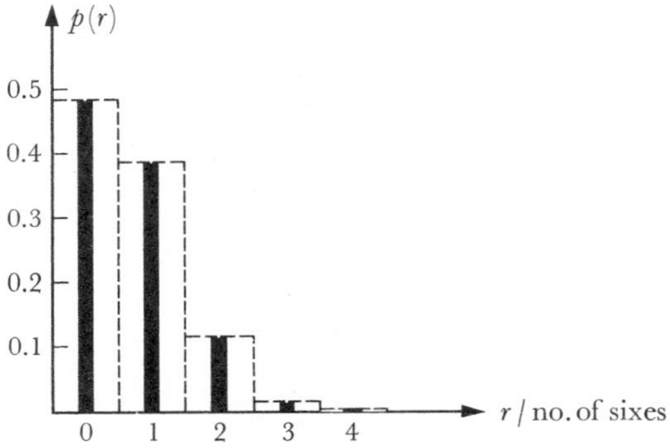

If, instead of concentrating on the number of sixes, we had counted the number of dice which showed either a five or a six, then the probability for a single die would be $\frac{1}{3}$.

So $n = 4, \quad p = \frac{1}{3}, \quad q = \frac{2}{3}$

and the corresponding probabilities of a five or a six being shown by 0, 1, 2, 3, 4 dice are given by the terms of

$$\left(\frac{2}{3} + \frac{1}{3}\right)^4$$

These can easily be shown to be approximately 0.198, 0.395, 0.296, 0.099, 0.012:

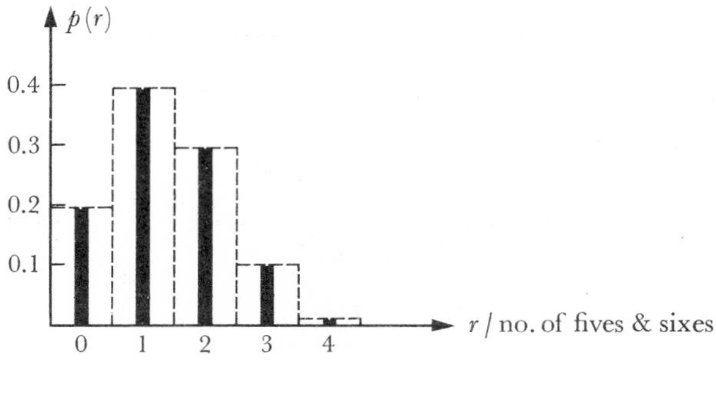

Exercise 7.4b

1 Find the probabilities of the various possible number of heads when a boy spins
(*i*) 1 coin; (*ii*) 2 coins; (*iii*) 3 coins;
(*iv*) 4 coins; (*v*) 5 coins; (*vi*) 6 coins.
Illustrate each answer by a probability diagram.

2 A marksman has an 80% chance of scoring a bull. When he fires five shots at a target calculate the probabilities of his scoring 0, 1, 2, 3, 4, 5 bulls.

3 A machine is faulty and 10% of the articles it produces are defective. In a batch of ten articles, what is the chance of more than one being defective?

4 A question paper contains six questions, whose answers are either 'yes' or 'no'. If a boy guesses at all the answers, what is the chance that more than half will be correct?

5 The probability of germination of a certain type of seed is $\frac{4}{5}$. If the seeds are planted out in rows of six, find the probability that, in any one row:
(*a*) all the seeds will germinate;
(*b*) five or six seeds will germinate. (M.E.I.)

6 A test is applied to a certain type of component produced in a factory. If the probability that a component chosen at random fails is $\frac{1}{5}$, calculate the following probabilities for a test on 10 components so chosen: (*i*) that none will fail; (*ii*) that less than two will fail; (*iii*) that exactly two will fail; (*iv*) that more than two will fail. (M.E.I.)

7 The probability that a colony of bees will survive a particularly hard winter is 0.35. If I have four colonies what is the probability that exactly one of these colonies will survive? I have a friend who has twice as many colonies (eight). Write down an expression for the probability that exactly one of his colonies will survive. Is the probability greater or less than in my case? (S.M.P.)

8 A gun is engaging a target and it is desired to make at least two direct hits. It is estimated that the chance of a direct hit with a single round is $\frac{1}{4}$ and that this chance remains constant throughout the firing. A burst of five rounds is fired and if at least two direct hits are scored firing ceases. Otherwise a second burst of five rounds is fired. Find the chance that at least two direct hits will be scored:
(*i*) only five rounds being fired;
(*ii*) ten rounds having to be fired. (O.C.)

9 The probability of a bowler taking a wicket with a particular delivery is $\frac{1}{4}$. What is the chance in a six-ball over of his taking:
(*i*) 2 wickets;

(*ii*) at least 2 wickets;
(*iii*) a hat-trick (and no other wickets).

10 A coin, which is thought to be unbiassed, is spun 10 times and only once does it come down heads. What is the probability:
(*i*) of obtaining so few heads;
(*ii*) of obtaining such an extreme result, either heads or tails?

The geometric distribution

A simple die is thrown repeatedly until a 6 occurs. The chance of this happening

at the 1st throw is $\frac{1}{6} \approx 0.167$,

at the 2nd throw is $\frac{5}{6} \times \frac{1}{6} \approx 0.138$,

at the 3rd throw is $\left(\frac{5}{6}\right)^2 \times \frac{1}{6} \approx 0.116$,

and as these form the terms of an infinite g.p., we call it a *geometric distribution*.

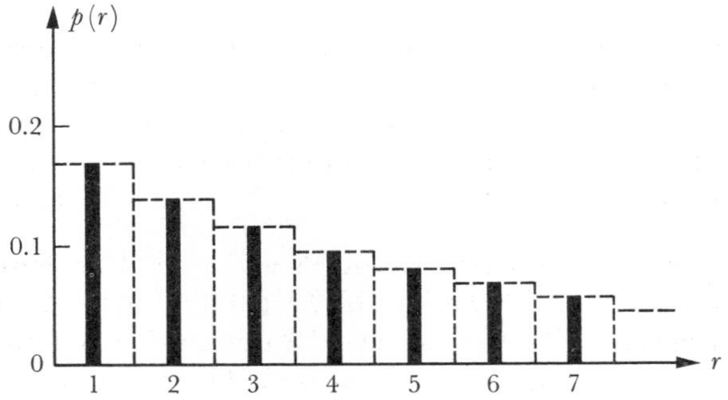

Exercise 7.4c

1 The chance of a girl passing her driving test is 2/3, and does not alter if she has to take it again. What is the probability that she will have to take it at least three times?

2 The last pair of our side is batting and is just about to face a six-ball over.
(*i*) If the chance of a wicket is 1/4 for each successive ball, what is the chance that they will not survive the over?
(*ii*) If the probability is the same for the other bowler, what is the chance that the innings will end during the following over?

3 Two guns are firing together at a fortified emplacement. Their chances of hitting the target are 1/3 and 1/4 respectively. What is the chance that it will survive three simultaneous rounds:

(*i*) if simultaneous hits are needed to destroy it;

(*ii*) if one hit is sufficient?

4 A rifleman is firing at a distant target and has only a 10% chance of hitting it. How many rounds must he fire in order to have a 50% chance of hitting it at least once?

*7.5 Expectation

Suppose that I throw a child's die and am promised:

if it turns up 6 ($p = \frac{1}{6}$), a reward of £30;

if it turns up 5, 4 or 3 ($p = \frac{1}{2}$), a reward of £20;

if it turns up 2 or 1 ($p = \frac{1}{3}$), no reward.

In the course of n such throws I should expect these events to take place roughly $\frac{n}{6}, \frac{n}{2}, \frac{n}{3}$ times respectively, and the total gain to be roughly

$$£\left(\frac{n}{6}\times 30 + \frac{n}{2}\times 20 + \frac{n}{3}\times 0\right)$$

So, for a single throw, on the average we expect to gain

$$£\left(\tfrac{1}{6}\times 30 + \tfrac{1}{2}\times 20 + \tfrac{1}{3}\times 0\right)$$

$$= £15$$

This amount we call the *expectation*. It would be a fair price to pay for the privilege of throwing the die under such rules: at a higher price we should expect, in the long run, to make a loss, and at a lower price a profit.

More generally, suppose that, in a certain trial n possible events occur with probabilities

$$p_1, p_2, p_3, \ldots p_n$$

and that associated with them are the values

$$x_1, x_2, x_3, \ldots x_n$$

(known as values of the *random variable*, x).

Then the *expectation*, or *expected value*, of x is defined as

$$E[x] = p_1x_1 + p_2x_2 + \cdots + p_nx_n$$

In the above example:

Event:	6	5, 4, 3	2, 1
p:	$\frac{1}{6}$	$\frac{1}{2}$	$\frac{1}{3}$
x:	30	20	0

$$E[x] = \tfrac{1}{6} \times 30 + \tfrac{1}{2} \times 20 + \tfrac{1}{3} \times 0 = 15$$

It must be emphasised that $E[x]$ is a function not of x, but of its probability distribution; i.e., a value which depends on the set of values $x_1, x_2, \ldots, x_n$ and the probabilities with which they occur.

Example

When two dice are thrown what is the expected number of sixes?

In this case the random variable x is simply the number of sixes:

Event	0 six	1 six	2 sixes
p	$\frac{25}{36}$	$\frac{10}{36}$	$\frac{1}{36}$
x	0	1	2

So $E[x] = \frac{25}{36}\times 0 + \frac{10}{36}\times 1 + \frac{1}{36}\times 2$

$\Rightarrow \quad E[x] = \frac{1}{3}$

Clearly 'one third of a six' has no meaning, but this is simply a way of saying that in N trials the expected number of sixes is $\frac{1}{3}N$.

Exercise 7·5

In nos. 1–6 of this exercise there is ample opportunity for experiment, and the reader can readily compare his expectations with observations from a large number of trials.

1 What is the expected value of:
(*i*) the score when a die is thrown;
(*ii*) the total score when two dice are thrown?

2 If a boy throws a die and is promised a reward in pence equal to the square of its score, what is his expectation?

3 What is the expected number of heads when three coins are spun?

4 When four coins are spun what numerical difference between the number of heads and the number of tails is to be expected? Repeat this question for five coins and for six coins.

5 If two dice are thrown, what is the expected value of the higher score (or the score on each, if both are the same)?

6 A metre rule is marked at intervals of one decimetre, including its ends. If two of these eleven points are chosen at random, what is the expected distance between them?

7 A child should not start at Ludo until he has thrown a six with his die, but even if he is unsuccessful his father allows him to begin at his third turn. What is the expected number of turns before the child can start?

8 Jones and Smith each cut a pack of cards. If both their cards are of the same suit Smith pays Jones £5 unless they are both hearts, when he pays £10. If, however, the suits are different Jones pays Smith £2. Whom does this favour, and by how much?

9 Two players stake £1 each in a game in which eight counters are tossed on the ground. The counters are painted black on one side and white on the other. If the number of blacks showing is odd the thrower wins the other stake, if all counters are of the same colour he wins a double stake, and otherwise he loses his stake. Calculate the expected gain of the thrower and state what assumptions you make. (M.E.I.)

Statistics

Whilst the calculation of probabilities concerns the uncertainties of the future, the study of statistics appears at first sight to be about the certainties of the past. But our predictions of the future usually depend on the evidence of the past; and our study of the past frequently involves investigation of the probability, on certain assumptions, that events should have happened as they did. So the two subjects are very closely interwoven.

Statistics, therefore, is largely concerned with assessing the strength of numerical evidence. But first must come the humbler task of recording, classifying and describing this evidence as clearly as possible, and this will be our objective in these final sections.

7.6 Location and spread

One of the commonest pieces of evidence is a set of values, such as the results of a short mathematical test given to a class of 20 pupils:

13, 12, 9, 14, 17, 12, 13, 13, 7, 10, 9, 11, 15, 13, 14, 9, 13, 12, 12, 10

which can be put in ascending order:

7, 9, 9, 9, 10, 10, 11, 12, 12, 12, 12, 13, 13, 13, 13, 13, 14, 14, 15, 17

These can be conveniently expressed in a frequency table, and the cumulative frequencies up to (and including) a particular mark can also be recorded:

Mark	6	7	8	9	10	11	12	13	14	15	16	17
Frequency	0	1	0	3	2	1	4	5	2	1	0	1
Cumulative frequency	0	1	1	4	6	7	11	16	18	19	19	20

Such figures can be illustrated by a *frequency diagram* and a *cumulative frequency diagram*.†

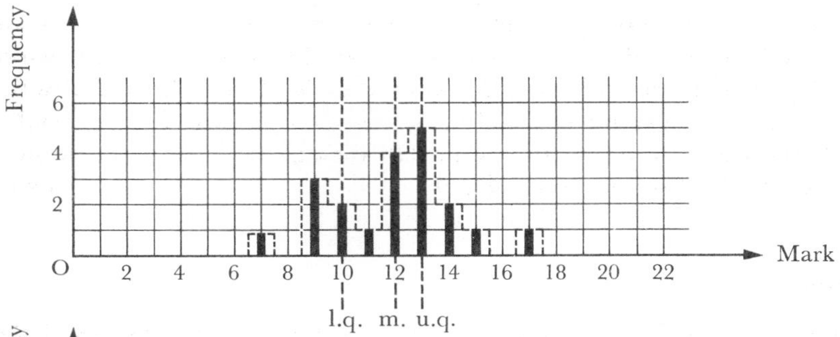

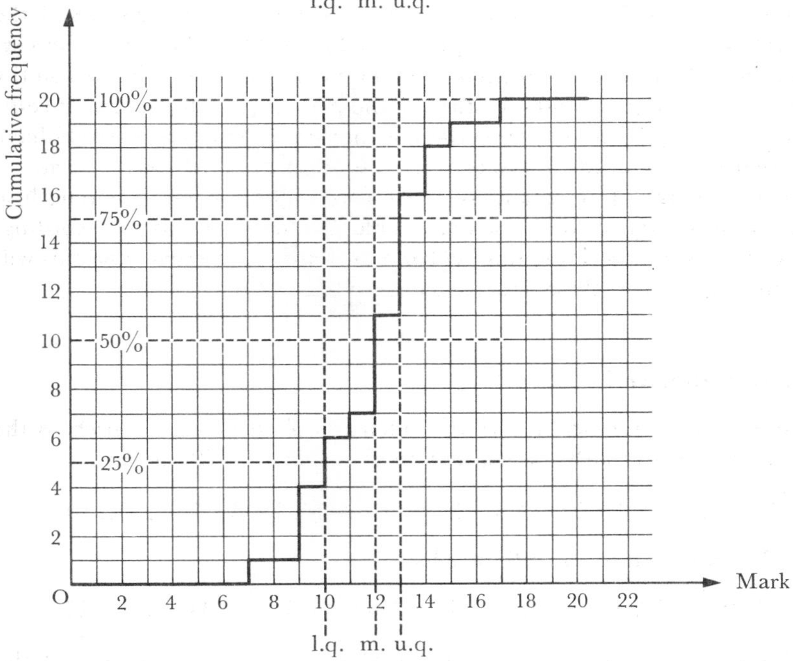

l.q. = Lower quartile = 10

m. = Median = 12

u.q. = Upper quartile = 13

† Though if the class intervals are unequal, a *histogram* is drawn with frequencies represented not by heights but by areas.

Location

If we wish to describe the general level, or *location*, of such marks we can use any one of three measures:

(*i*) The *mode* is the mark which occurs most frequently. There might possibly be two or more such marks, but in our example 13 is the most frequent mark, so the mode is 13.
(*ii*) The *median* is the central mark (or average of two central marks) as we go through them in order of magnitude. In our example there are two central marks (the 10th and 11th): as these are both 12, the median is 12.

(*iii*) The *mean*, or average of the marks

$$\tfrac{1}{20}(7 + 9 + 9 + 9 + 11 + \cdots + 14 + 14 + 15 + 17) = \tfrac{1}{20} \times 238 = 11.90$$

More generally, the mean of n values $x_1, x_2, x_3, \ldots, x_n$ can be written

$$m = \frac{x_1 + x_2 + \cdots + x_n}{n} = \frac{1}{n}\sum_{r=1}^{n} x_r$$

Or, more briefly: $$\boxed{m = \frac{1}{n}\sum x_r}$$

m is sometimes called the *arithmetic mean* A, to distinguish it from the *geometric mean*,

$$G = \sqrt[n]{(x_1 x_2 x_3 \ldots x_n)}$$

and the *harmonic mean*, H, where

$$\frac{1}{H} = \frac{1}{n}\left(\frac{1}{x_1} + \frac{1}{x_2} + \cdots \frac{1}{x_n}\right)$$

So, for the pair of numbers 1 and 2

$$A = \tfrac{1}{2}(1 + 2) = 1.500$$

$$G = \sqrt{1 \times 2} \approx 1.414$$

and $$\frac{1}{H} = \frac{1}{2}\left(\frac{1}{1} + \frac{1}{2}\right) = \frac{3}{4}$$

$$\Rightarrow \quad H \approx 1.333$$

More generally, it is a useful exercise to show that, for any two positive numbers, $H \leqslant G \leqslant A$ (which is also true for any set of positive numbers). Both the geometric and (less frequently) the harmonic means are sometimes used in statistics, but we shall restrict ourselves to the overwhelmingly more important arithmetic mean.

Spread

We frequently wish to indicate not only the general level of a set of values, but also the way in which they are spread about this level. The first such measure is clearly the *range* of values, or difference between extremes, and in our example is 17 − 7 = 10.

This, however, is entirely dependent on extreme values. When we wish to look at the way in which the generality of marks are spread, we more frequently investigate the position of the *quartiles*, or values which are one-quarter and three-quarters of the way through the set, when arranged in order. In our example, for instance, the lower and upper quartiles can be taken as the 5th and 15th values as we go up the scale, and so are 10 and 13 respectively. (Or should we have taken the 16th? But such doubts become unimportant when there is a large set of values.)

A useful measure of spread is one-half of the separation of the quartiles, or *quartile deviation*, q.d. (also called the *semi-interquartile range*). In our example,

$$\text{q.d.} = \tfrac{1}{2}(13 - 10) = 1.5$$

When there is a large number of values, we see that the median and the quartiles are the values which divide the area of the frequency diagram (and so the ordinate of the cumulative frequency diagram) into four equal parts. More generally, we can obtain *percentiles* by similar subdivision into a hundred parts.

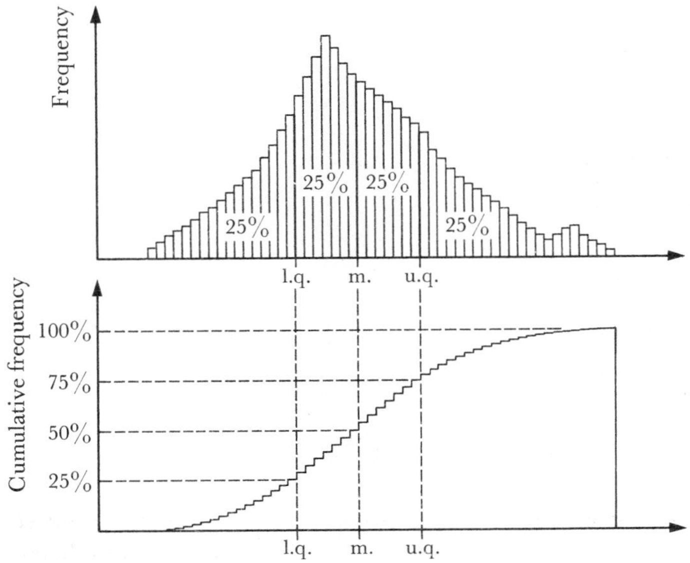

We can investigate the spread of a set of values by calculating their average deviation from the mean:

For the given set of marks

7, 9, 9,..., 14, 15, 17,

the deviations from the mean (11.9) are

$-4.9, \ -2.9, \ -2.9, \cdots +2.1, \ +3.1, \ +5.1$

and their average is found to be zero. This is hardly a surprise, since

$$\begin{aligned}\frac{1}{n}\sum (x_r - m) &= \frac{1}{n}\sum x_r - \frac{1}{n}\sum m \\ &= m - \frac{1}{n}nm = 0\end{aligned}$$

So the average of such deviations is always zero.

Nevertheless, we could use the *numerical* values of such deviations in order to calculate the *mean absolute deviation*:

$\frac{1}{20}(+4.9 + 2.9 + \cdots + 2.1 + 3.1 + 5.1) = 1.83$

More generally,

$$\text{mean absolute deviation} = \frac{1}{n}\sum |x_r - m|.$$

In practice, however, it is so inconvenient to deal with such numerical values that we almost always avoid this by squaring deviations from the mean and calling their average value the *variance*. This is usually written s^2, and s is called the *standard deviation*.

In our example,

$$\begin{aligned} s^2 &= \tfrac{1}{20}\{(-4.9)^2 + (-2.9)^2 + \cdots + (3.1)^2 + (5.1)^2\} \\ &= 5.39 \end{aligned}$$

$$\Rightarrow \quad s \approx 2.32$$

More generally, $$\boxed{s^2 = \frac{1}{n}\sum (x_r - m)^2}$$

Now $\quad m = \dfrac{1}{n}\sum x_r$

So
$$\begin{aligned} s^2 &= \frac{1}{n}\sum (x_r - m)^2 \\ &= \frac{1}{n}\left\{\sum x_r^2 - 2m\sum x_r + \sum m^2\right\} \\ &= \frac{1}{n}\left\{\sum x_r^2 - 2mnm + nm^2\right\} \\ &= \frac{1}{n}\sum x_r^2 - m^2 \end{aligned}$$

This can be checked in our example:

$$s^2 = \tfrac{1}{20}\{7^2 + 9^2 + 9^2 + \cdots + 15^2 + 17^2\} - (11.90)^2 = 5.39$$

Summarising $\quad m = \dfrac{1}{n}\sum x_r$

and $\quad s^2 = \dfrac{1}{n}\sum (x_r - m)^2 = \dfrac{1}{n}\sum x_r^2 - m^2$†

Change of variable

For purposes of calculation it is often more convenient to use a different origin for values of x. In the above example, the calculation can be simplified if we use 12 as our working origin, i.e., if we investigate the mean and standard deviation of the values of $x - 12$. Suppose, more generally, that the values of x undergo the transformation

$$y = \lambda x + a$$

Let m_x, s_x and m_y, s_y be the means and standard deviations of the values of x and y.

The transformation $y = \lambda x + a$ can be represented

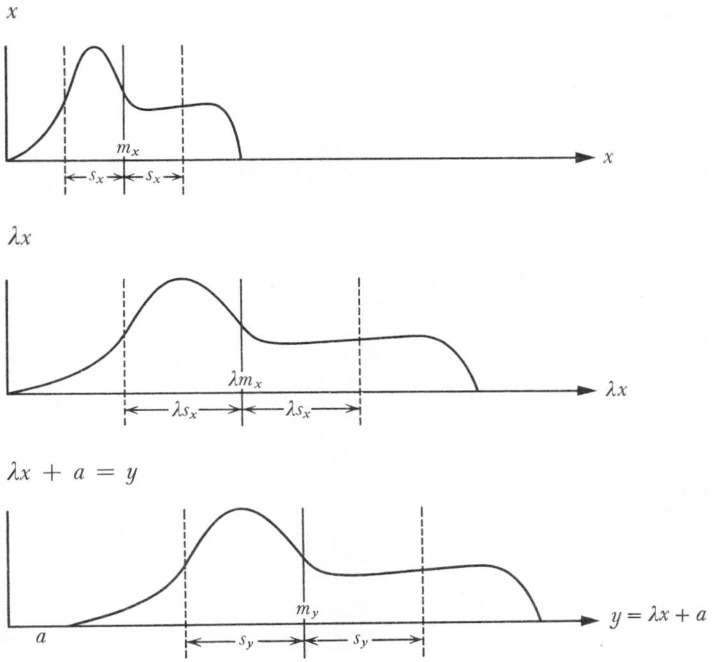

† Conveniently remembered as *the mean of the squares minus the square of the mean.*

So the transformation corresponds to a horizontal stretch with scale factor λ followed by a translation through a distance a, and from the diagrams we should expect m_x to undergo the same transformation, and s_x to be multiplied by the scale factor λ.

This is easily confirmed, since

$$
\begin{aligned}
m_y = \frac{1}{n}\sum y_r &= \frac{1}{n}\sum (\lambda x_r + a) \\
&= \frac{1}{n}\lambda \sum x_r + \frac{1}{n}\sum a \\
&= \frac{1}{n}\lambda n m_x + \frac{1}{n} na \\
&= \lambda m_x + a
\end{aligned}
$$

$$
\begin{aligned}
\text{and} \quad s_y^2 &= \frac{1}{n}\sum (y_r - m_y)^2 \\
&= \frac{1}{n}\sum \{(\lambda x_r + a) - (\lambda m_x + a)\}^2 \\
&= \frac{1}{n}\sum (\lambda x_r - \lambda m_x)^2 \\
&= \lambda^2 \frac{1}{n}\sum (x_r - m_x)^2 \\
&= \lambda^2 s_x^2
\end{aligned}
$$

$$\text{So} \quad y = \lambda x + a \Rightarrow \boxed{\begin{aligned} m_y &= \lambda m_x + a \\ s_y &= \lambda s_x \end{aligned}}$$

In our example, the marks can all be referred to a new origin 12 by means of the transformation

$$y = x - 12$$

The values of y, therefore, are

$-5, -3, -3, -3, -2, -2, -1, 0, 0, 0, 0, 1, 1, 1, 1, 1, 2, 2, 3, 5$

$$\text{So} \quad m_y = \frac{1}{20}(-5 - 3 - 3 \cdots + 2 + 2 + 3 + 5)$$

$$\Rightarrow \quad m_y = -0.1$$

$$\text{and}\quad s_y^2 = \frac{1}{20}(25 + 9 + 9 + \cdots + 4 + 9 + 25) - (-0.1)^2$$

$$= \frac{1}{20} \times 108 - 0.01$$

$$= 5.40 - 0.01 = 5.39$$

$$\Rightarrow \quad s_y = 2.32$$

But $\quad y = x - 12$

$\Rightarrow \quad m_y = m_x - 12 \quad \text{and} \quad s_y = s_x$

So $\quad -0.1 = m_x - 12 \quad \text{and} \quad 2.32 = s_x$

Hence $\quad m_x = 11.90 \quad \text{and} \quad s_x = 2.32$

Exercise 7.6

More realistic calculations will be set in the next section, but in the meantime it is important to become familiar with the terms which have been introduced.

The following data refers to a class of 10 students who have been given short tests (each marked out of 10) in a number of subjects:

1	Mathematics	5,	4,	6,	6,	8,	6,	7,	5,	6,	7
2	Physics	7,	7,	8,	6,	9,	5,	8,	7,	6,	7
3	Chemistry	4,	10,	9,	8,	6,	7,	8,	6,	5,	7
4	Biology	4,	10,	5,	9,	9,	8,	7,	8,	6,	9
5	Economics	10,	5,	5,	4,	9,	8,	8,	5,	9,	5
6	French	5,	6,	9,	6,	10,	6,	9,	8,	9,	8

For the marks of each subject separately:

(*i*) Construct tables of frequencies and cumulative frequencies and draw the corresponding frequency and cumulative frequency diagrams.

(*ii*) Calculate the mode(s) (if any), median and mean.

(*iii*) Calculate the quartiles and quartile deviation (and note that they are satisfactorily defined only with a larger population).

(*iv*) Calculate the variance and the standard deviation:

(*a*) working from the mean;

(*b*) working from the origin;

(*c*) after first making a convenient transformation.

7 Calculate the mean and standard deviation of the marks

66, 36, 72, 46, 87, 56, 79, 94

(Use the transformation $y = x - 60$ to produce a working origin at 60.)

8 In an experiment to find the acceleration due to gravity the following results were obtained:

$x\,(\mathrm{m\,s^{-2}})$: 9.813, 9.803, 9.811, 9.814, 9.809, 9.815, 9.816, 9.815

(*i*) Use the transformation $y = \dfrac{x - 9.810}{0.001}$ to convert them to a set with origin at 9.810, and 0.001 as unit.
(*ii*) Find m_y and s_y.
(*iii*) Hence find the mean and standard deviation of the original results.

9 Five estimates for the radius of the Sun (in km) are:

695 400; 695 800; 695 300; 695 900; 695 100

Find their mean and standard deviation.

10 In no. 7 use the results obtained and find a linear transformation which converts these marks to a set with mean 50 and standard deviation 15. (The use of a calculating machine is desirable.)

*7.7 Frequency distributions

It is clear that the calculations of the last section involved unnecessary repetition when a number of pupils had the same mark. In our first example, for instance,

$$m = \frac{1}{20}(7 + 9 + 9 + 9 + 10 + 10 + 11 + 12 + 12 + 12 + 12 + 13 + 13 + 13 + 13 + 13 + 14 + 14 + 15 + 17)$$

is needlessly repetitive, and can be simplified to

$$m = \frac{1}{20}(1\times7 + 3\times9 + 2\times10 + 1\times11 + 4\times12 + 5\times13 + 2\times14 + 1\times15 + 1\times17)$$

More generally, if the value x_r occurs with frequency f_r, we can express the formulae for the mean m and variance s^2 in the form:

$$m = \frac{1}{n}\sum f_r x_r$$

$$s^2 = \frac{1}{n}\sum f_r (x_r - m)^2 = \frac{1}{n}\sum f_r x_r^2 - m^2$$

(where the summations are taken over all the different values of x_r).

The second of the formulae for s^2 is usually the more convenient to use, and the full calculation is best done in tabular form:

x_r	f_r	$f_r x_r$	$f_r x_r^2$
7	1	7	49
8	0	0	0
9	3	27	243
10	2	20	200
11	1	11	121
12	4	48	576
13	5	65	845
14	2	28	392
15	1	15	225
16	0	0	0
17	1	17	289

$$n = \sum f_r = 20; \ \sum f_r x_r = 238; \ \sum f_r x_r^2 = 2\,940$$

$$m = \frac{1}{20} \times 238 = 11.90$$

$$s^2 = \frac{1}{20} \times 2\,940 - (11.9)^2$$

$$= 147 - 141.61$$

$$= 5.39$$

$$\Rightarrow \quad s \approx 2.32$$

So the mean is 11.90 and the standard deviation is 2.32.

Alternatively, the calculation can be performed more easily if (as before) we first use the transformation $y = x - 12$

x_r	y_r	f_r	$f_r y_r$	$f_r y_r^2$
7	−5	1	−5	25
8	−4	0	0	0
9	−3	3	−9	27
10	−2	2	−4	8
11	−1	1	−1	1
12	0	4	0	0
13	1	5	5	5
14	2	2	4	8
15	3	1	3	9
16	4	0	0	0
17	5	1	5	25

$$\sum f_r = 20; \ \sum f_r y_r = -2; \ \sum f_r y_r^2 = 108$$

So $$m_y = \frac{1}{20} \times -2 = -0.10$$

$$s_y^2 = \frac{1}{20} \times 108 - (-0.1)^2 = 5.39$$

$$\Rightarrow \quad s_y = 2.32$$

But $$y = x - 12$$

$$\Rightarrow \quad m_y = m_x - 12 \quad \text{and} \quad s_y = s_x$$

$$\Rightarrow \quad -0.10 = m_x - 12 \quad \text{and} \quad 2.32 = s_x$$

So the mean is 11.90 and the standard deviation is 2.32.

Grouped data

When presented with a large number of marks or readings we usually like to arrange them in groups according to size, and preferably with equal intervals. Marks out of 100, for instance, might be divided into the groups 0–4, 5–9, 10–14, . . ., 90–94, 95–100 (all groups except the last consisting of a 5-mark band); and in the following example the group 23.27–23.31 mm will include any bolt whose head diameter lies between 23.265 mm and 23.315 mm, and so will be centred upon 23.29 mm. There is, of course, a very slight distortion of the data if all bolts belonging to this group are regarded as having a head diameter of exactly 23.29 mm and it is possible to offset this distortion (particularly when calculating the standard deviation) by use of adjustments known as Sheppard's corrections. For the moment, however, we shall analyse grouped data without such adjustment.

Example

The frequency distribution of a random sample of 250 steel bolts according to their head diameter, measured to the nearest 0.01 mm, is shown in the following table:

Diameter (mm)	*Number of bolts*
23.07–23.11	10
23.12–23.16	20
23.17–23.21	28
23.22–23.26	36
23.27–23.31	52
23.32–23.36	38
23.37–23.41	32
23.42–23.46	21
23.47–23.51	13

In order to calculate the mean and standard deviation we make use of the fact that these intervals are all equal and label them with their mean diameter (x) and their position (y) relative to the central interval. It is immediately clear that

$$x = 23.29 + 0.05y,$$

and the calculation can then be set out:

x	y	f	fy	fy^2
23.09	−4	10	−40	160
23.14	−3	20	−60	180
23.19	−2	28	−56	112
23.24	−1	36	−36	36
23.29	0	52	0	0
23.34	1	38	38	38
23.39	2	32	64	128
23.44	3	21	63	189
23.49	4	13	52	208

$$n = \sum f = 250; \quad \sum fy = 25; \quad \sum fy^2 = 1\,051$$

$$\text{So} \quad m_y = \frac{1}{n}\sum fy = \frac{25}{250} = 0.1$$

$$s_y^2 = \frac{1}{n}\sum fy^2 - m_y^2$$

$$= \frac{1\,051}{250} - (0.1)^2$$

$$= 4.204 - 0.01 = 4.194$$

$$\Rightarrow \quad s_y \approx 2.05$$

$$\text{But} \quad x = 23.29 + 0.05y$$

$$\Rightarrow \quad m_x = 23.29 + 0.05m_y \quad \text{and} \quad s_x \approx 0.05s_y$$

$$= 23.29 + 0.005 \qquad\qquad \approx 0.102(5)$$

$$= 23.295$$

Hence the head diameters of the bolts have mean 23.29(5) mm and standard deviation 0.10 mm.

Exercise 7.7

1 The marks of 492 candidates who entered for an examination are shown:

marks		*marks*	
0 to 4	1	5 to 9	2
10 to 14	5	15 to 19	6
20 to 24	10	25 to 29	11
30 to 34	21	35 to 39	26
40 to 44	44	45 to 49	53
50 to 54	63	55 to 59	66
60 to 64	58	65 to 69	42
70 to 74	30	75 to 79	21
80 to 84	15	85 to 89	9
90 to 94	5	95 to 99	4

The maximum mark was 99.

(*i*) Draw the cumulative frequency curve and estimate the median and inter-quartile range.
(*ii*) What should be the pass mark if at least 80% of the candidates are to pass?
(*iii*) If the pass mark is 47, estimate how many candidates fail by no more than 5 marks. (S.M.P.)

2 The marks of 1 000 candidates in an examination were distributed as follows:

Mark	0–9	10–19	20–29	30–39	40–49
No. of candidates	26	64	112	170	212

Mark	50–59	60–69	70–79	80–89	90–99
No. of candidates	192	130	62	22	10

Construct the cumulative frequency curve for these marks. It is required to pass 60% of the candidates, to give a grade A to the top 10%, and to give a grade B to the next 20% of the candidates. It being understood that all the marks given are whole numbers, obtain from your cumulative frequency curve an estimate of the least mark which a candidate must obtain to be given (*a*) a pass, (*b*) a grade A, (*c*) a grade B. (L)

3 The number of bargains marked in certain ordinary shares on the London Stock Exchange on 50 business days were:

66	133	89	97	112	104	96	105	117	88
77	89	91	113	109	102	94	107	119	87
71	127	125	109	99	103	96	108	117	88
81	91	124	98	109	101	105	93	119	86
84	131	92	97	100	101	106	93	121	87

Draw up a frequency table showing the numbers in the intervals 60 to 69, 70 to 79, etc., and from this table calculate the mean of the distribution.

Draw a histogram to illustrate the data.

Draw a cumulative frequency curve to fit the distribution and deduce the values of the median and the semi-inter-quartile range. (O.C.)

4 Two dice were thrown together 200 times and for each throw the total score was recorded. The frequencies with which the scores occurred were as follows:

Score	12	11	10	9	8	7
Frequency	7	10	20	20	30	34
Score	6	5	4	3	2	
Frequency	27	21	14	11	6	

By taking the score of 7 as the base of your calculations, find the mean and the standard deviation of the scores. (L)

5 The contents of 50 boxes of matches of the same make were found to be as follows:

Number of matches	42	43	44	45	46	47	48
Number of boxes	10	6	20	6	5	2	1

Find the mean number of matches in a box and the standard deviation. (L)

6 The following data give the percentage ash in the bones of chicks which have been fed on a diet containing supplementary vitamin D. Form a frequency tabulation and plot a histogram.

Calculate *from the grouped data* the mean and standard deviation of the percentage.

Percentage ash in chick bones

34.68	33.57	36.17	37.53	37.17	38.51	42.13
38.70	37.67	37.35	34.64	39.19	38.88	43.44
38.30	40.23	37.14	37.58	36.39	37.17	39.38
35.74	34.92	35.37	41.85	34.64	36.69	35.57
34.28	36.41	35.98	34.99	34.15	35.30	37.87
34.49	40.44	36.47	30.41	35.00	41.09	41.20
35.17	38.66	41.39	37.27	35.30	33.32	37.41
34.96	40.65	37.19	36.41	39.72	29.11	

(M.E.I.)

7, 8, 9 Calculate the means and standard deviations of the grouped data in nos. 1, 2, 3.

10 In 1970 it was estimated that the age-distribution of the population (in thousands) of England and Wales in the year 2000 would be as follows:

Age	*Male*	*Female*	*Total*
0–4	2 528	2 394	4 922
5–9	2 438	2 318	4 756
10–14	2 345	2 238	4 583
15–19	2 225	2 150	4 375
20–24	2 168	2 118	4 286
25–29	2 072	2 022	4 094
30–34	2 035	1 978	4 013
35–39	2 088	2 021	4 109
40–44	1 794	1 741	3 535
45–49	1 640	1 601	3 241
50–54	1 803	1 775	3 578
55–59	1 434	1 443	2 877
60–64	1 235	1 283	2 518
65–69	1 045	1 175	2 220
70–74	815	1 097	1 912
75–79	594	1 003	1 597
80–84	287	626	913
85–	158	510	668

[Population Projections, 1970–2010: H.M.S.O. 1971]

(*i*) Plot frequency and cumulative frequency diagrams for the total population.
(*ii*) Calculate the median age of the population and its lower and upper quartiles.
(*iii*) Calculate the mean and standard deviation.

11 Repeat no. 10 for the male population.

12 Repeat no. 10 for the female population.

13 Comment on your results for nos. 10, 11, 12.

Miscellaneous problems 7

1 An electrically heated hot water service is operated by two relay switches, A and B, whose probabilities of failure, at the moment the service is switched on, are

$$p(\text{A}) = 0.1, \qquad p(\text{B}) = 0.05$$

Assuming the switches operate independently of each other, calculate the probability that at least one of the switches will fail at this moment.

In practice it has been observed that the failure of switch A depends on switch B and that the conditional probability of A failing given that B has not failed is 0.105. Calculate the probability that both of the switches will fail, at the moment the service is switched on, if the individual probabilities of failure are as before. (M.E.I.)

2 A random-number table consists of a succession of digits each chosen at random and independently from the set $\{0, 1, 2, 3, \ldots, 9\}$. Two successive digits are taken from such a table: show that the probability that their sum is 9 is $\frac{1}{10}$.

Four successive digits are taken from the table. Calculate the probability that the sum of the first two equals the sum of the third and fourth. (M.E.I.)

3 Electric light bulbs are chosen at random and tested until a dud is found. It was thought that 10% were faulty, but the thirtieth bulb is the first dud. How surprising is it that so many have been tested successfully?

4 Three boxes A, B, C contain respectively 2 white and 2 black balls, 3 white and 2 black balls, and 4 white and 2 black balls. A ball is drawn unseen from A and placed in B and then a ball is drawn from B and placed in C. If a ball is now drawn from C, find the probability that it will be black. (C)

5 A large batch of manufactured articles is accepted if either a random sample of 10 articles contains no defective article or, failing this, a second random sample of ten contains no defective article. Otherwise it is rejected.

If, in fact, 5% of the articles in a batch to be examined are defective, what is the chance of its being accepted? (O.C.)

6 Ignoring February 29, what is the probability of at least two members of a class having the same birthday if it contains (*i*) 20 members, (*ii*) 25 members? How large must it be in order for this event to be more likely than not?

7 If $|x| < 1, \quad 1 + x + x^2 + x^3 \cdots \equiv \dfrac{1}{1 - x}$

Differentiate this in order to find the sum of $1 + 2x + 3x^2 + 4x^3 + \cdots$.

The chance of a friend being out when I telephone is 4/5. Use your result to find the number of telephone calls I expect to make in order to speak with him.

8 In a game of tennis one point is scored either by A or his opponent B. The winner of the game is the player who first scores four points, unless each player has won three points, when 'deuce' is called and play proceeds until one player is two points ahead of the other and so wins. If A's chance of winning any point is 2/3 and B's chance is 1/3 calculate the chance of:

(*i*) A winning the game without deuce being called;
(*ii*) a similar win by B;
(*iii*) 'deuce' being called.

If deuce is called, prove that A's subsequent chance of winning the game is 4/5.

Deduce that A's chance of winning the game is nearly six times that of B. (c)

9 An absent-minded mathematician writes three letters and addresses three envelopes. What is the chance that they are put:
(*i*) all in the right envelopes;
(*ii*) all in wrong envelopes?

How would these answers be changed if there were four letters and four envelopes; five letters and five envelopes? (see miscellaneous problems 6, no. 10).

10 Three events $\boldsymbol{A}$, $\boldsymbol{B}$, $\boldsymbol{C}$ are said to be *pairwise independent* if any two are independent of each other, and *completely independent* when

$$p(\boldsymbol{ABC}) = p(\boldsymbol{A})\,p(\boldsymbol{B})\,p(\boldsymbol{C})$$

Consider the spin of two coins, a 5 p piece and a 10 p piece and

let $\boldsymbol{A}$ be the event of 5 p showing head,
$\boldsymbol{B}$ be the event of 10 p showing head,
and $\boldsymbol{C}$ be the event of them being both heads or both tails.

Calculate:
(*i*) $p(\boldsymbol{A})$, $p(\boldsymbol{B})$, $p(\boldsymbol{C})$;
(*ii*) $p(\boldsymbol{BC})$, $p(\boldsymbol{CA})$, $p(\boldsymbol{AB})$;
(*iii*) $p(\boldsymbol{ABC})$.

Are three events which are pairwise independent necessarily completely independent?

11 (The St. Petersburg Paradox.)
A gambler spins a coin until it comes down heads. If this happens at the first spin the banker gives him 1 rouble, at the second spin 2 roubles, at the third spin 4 roubles, at the fourth spin 8 roubles, and so on. Show that his expectation is infinite.

If the total resources of the banker are 1 000 roubles, what is a fair price to pay for playing the game?

12 Public houses are scattered randomly along a road at *average* intervals of 1 km.
(*i*) What is the approximate chance that a stretch of length 100 m:
(*a*) has a pub, (*b*) doesn't have a pub?

Hence estimate the probability that a particular kilometre of road (made up of ten such stretches) doesn't have a pub.
(*ii*) Repeat with 100 stretches of length 10 m.
(*iii*) To what figure do these probabilities appear to be tending? Compare your answer with no. 9 and with Miscellaneous problems 2, no. 5.

[Take $\log_{10} 0.9 = \bar{1}.954\,243$
$\log_{10} 0.99 = \bar{1}.995\,635$
$\log_{10} 0.999 = \bar{1}.999\,565$]

8

Introduction to vectors and mechanics

8.1 Vectors and vector addition

Many physical quantities have no direction. If we enquire about the temperature of a bath, the pressure of the atmosphere, or the charge of an electron, we expect only to be told a magnitude, i.e., the number of whatever are the appropriate units for the quantity. It would be unusual to speak of one's bath having a temperature 42 °C vertically upwards, or of the atmospheric pressure being on a bearing of 241°, or of an electron having its charge of -1.6×10^{-19} coulomb in a direction North-North-West. Such quantities, which are completely described by their magnitudes, are called *scalars*.

But there are also others known as *vector* quantities, such as the pass of a football from one player to another, the velocity of a car and the intensity of a magnetic field, where information is quite inadequate unless it includes not only a magnitude (or *modulus*) but also a direction.

As an example, we shall begin by considering the possible operation of a helicopter service between five points: Aberdeen, Buckie, Crieff, Dundee, and Edinburgh.

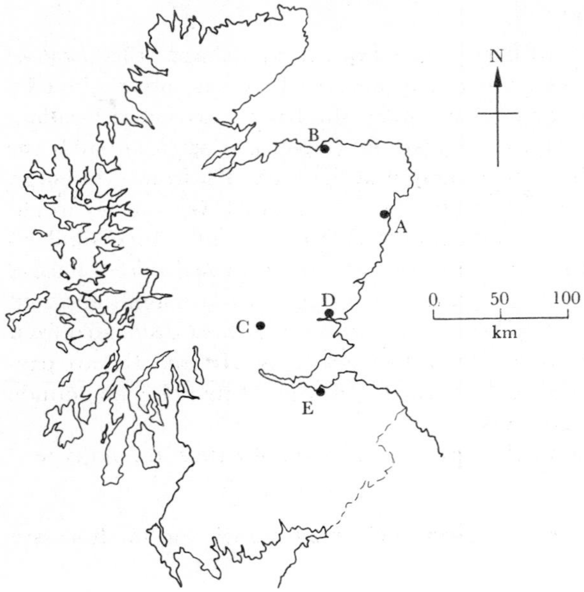

We shall represent a flight from Aberdeen to Buckie, called a *displacement vector*, by **AB** (or, in manuscript, by $\overrightarrow{AB}$), a flight from Buckie to Crieff by **BC** etc.

Then **AB** followed by **BC** achieves the same as **AC**. So we write **AB** + **BC** = **AC**.

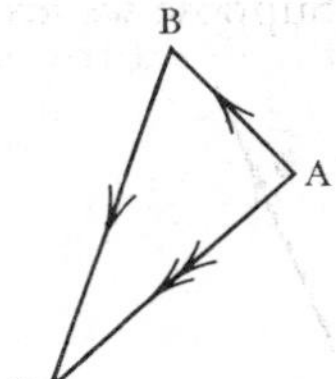

This is the *vector law of addition*, and **AC** is called the *vector sum* of **AB** and **BC**. The symbol + and the words 'addition' and 'sum' have, as yet, no connection with the symbol + and the same words in elementary arithmetic and algebra, but it will soon be clear that they are intimately related.

We are now able to *define* a vector quantity as one which (like displacement) has magnitude and direction and obeys the triangle law of addition; and we shall see later that among such quantities are velocity and acceleration, force and momentum.

Exercise 8.1a

1 Simplify:

(*i*) **AB** + **BE**; (*ii*) **AC** + (**CD** + **DE**);

(*iii*) (**AC** + **CD**) + **DE**; (*iv*) (**AB** + **BC**) + (**CD** + **DE**).

2 Use vectors to list ways of flying from Buckie to Crieff:

(*i*) with only one stop between;

(*ii*) with two stops.

Localised vectors and free vectors

Displacements like these which we have just considered are called *localised vectors*: *vectors* because they have magnitude and direction and combine by the triangle law, and *localised* because they also have a particular location. The flight $\boldsymbol{AB}$ from Aberdeen to Buckie, for instance, is approximately the same distance and in the same direction as the flight $\boldsymbol{EC}$ from Edinburgh to Crieff, but most passengers wishing to make a particular journey would regard the differences between them at least as important as their similarities, and almost certainly they would be expected to have different tickets.

Often, however, we wish to concentrate only on the magnitudes and directions of vectors. When we do so, by considering them dislocated from any particular position, we call them *free* vectors. So $\boldsymbol{AB}$ and $\boldsymbol{EC}$ are particular examples of the free vector which has (approximately) magnitude 70 km and direction North-West.

Such free vectors are usually represented by small letters in bold type

$\boldsymbol{a}, \boldsymbol{b}, \boldsymbol{c}, \ldots$

(or in manuscript $\underset{\sim}{a}, \underset{\sim}{b}, \underset{\sim}{c}, \ldots$), and their magnitudes (or moduli) are denoted by $a, b, c, \ldots$.

Addition of vectors

Suppose we are given two vectors $\boldsymbol{a}$ and $\boldsymbol{b}$

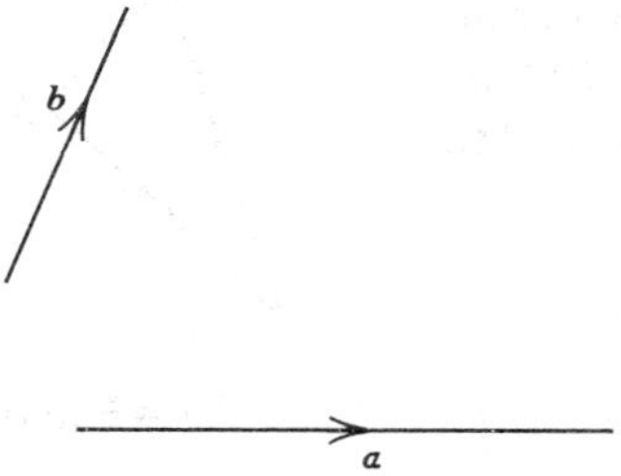

We can localise these in a variety of ways, and in particular both as

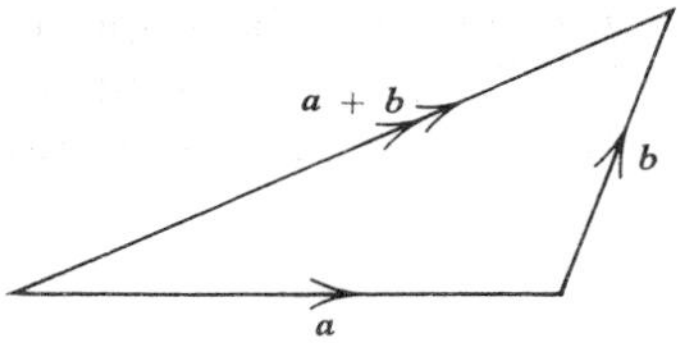

and

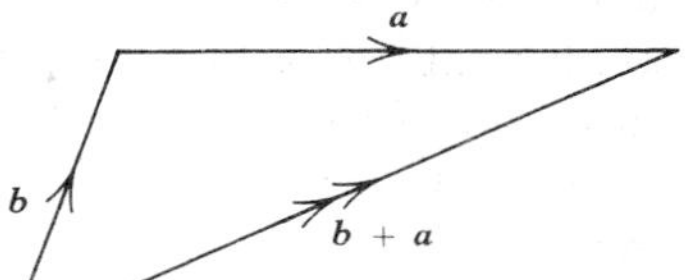

From what has been said about the combination of localised vectors, it is clear that the combined displacements may be called $\boldsymbol{a} + \boldsymbol{b}$ and $\boldsymbol{b} + \boldsymbol{a}$, respectively.

But these have the same magnitude and direction, so are the same free vector.

Hence $\quad \boldsymbol{a} + \boldsymbol{b} = \boldsymbol{b} + \boldsymbol{a}.$

This is the *commutative* law of vector addition.

By combining the two figures, we see that the sum of the two vectors $\boldsymbol{a}$ and $\boldsymbol{b}$ is represented by the diagonal of the parallelogram formed by $\boldsymbol{a}$ and $\boldsymbol{b}$:

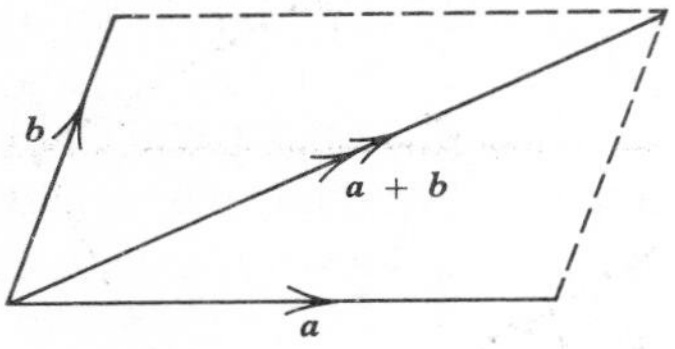

This is sometimes called the *parallelogram law of vector addition.*

Furthermore, given three vectors $\boldsymbol{a}$, $\boldsymbol{b}$, and $\boldsymbol{c}$, it is clear from the figures

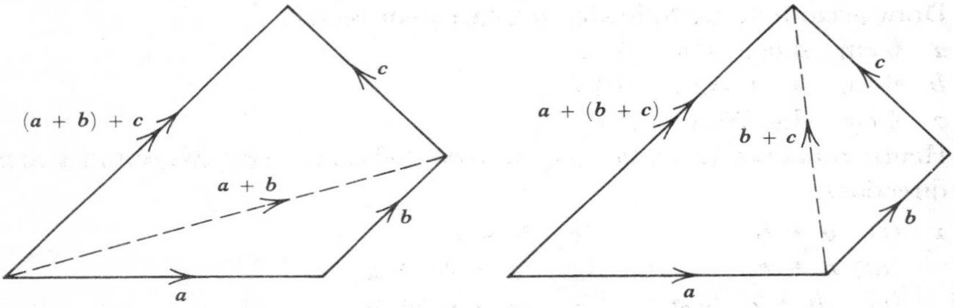

that $(\boldsymbol{a} + \boldsymbol{b}) + \boldsymbol{c} = \boldsymbol{a} + (\boldsymbol{b} + \boldsymbol{c})$.

This is the *associative* law of vector addition.

Negation

If we wish to consider the vector which has the same magnitude as $\boldsymbol{a}$ but the opposite direction, it is convenient to call it $-\boldsymbol{a}$:

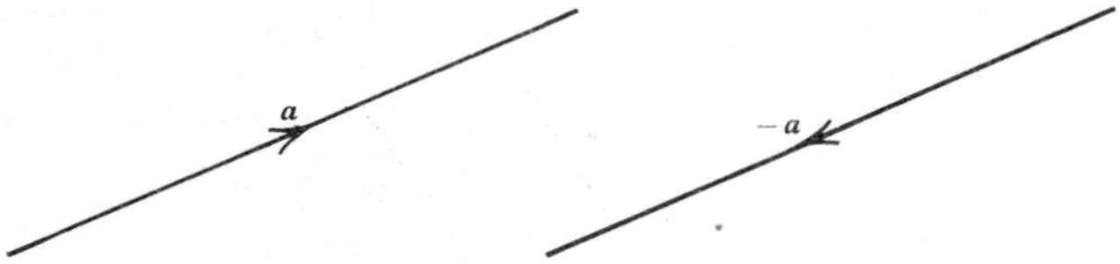

We can then write the addition of $\boldsymbol{a}$ and $-\boldsymbol{a}$ as

$$\boldsymbol{a} + (-\boldsymbol{a}) = \boldsymbol{0},$$

where $\boldsymbol{0}$ is called the *zero vector*.

Subtraction

We are now in a position to define $\boldsymbol{a} - \boldsymbol{b}$ as $\boldsymbol{a} + (-\boldsymbol{b})$. It can therefore be represented by either $\boldsymbol{OC}$ or $\boldsymbol{BA}$, both of these having the same magnitude and direction:

$\boldsymbol{b} + (\boldsymbol{a} - \boldsymbol{b}) = \boldsymbol{a}$

$\boldsymbol{a} + (-\boldsymbol{b}) = \boldsymbol{a} - \boldsymbol{b}$

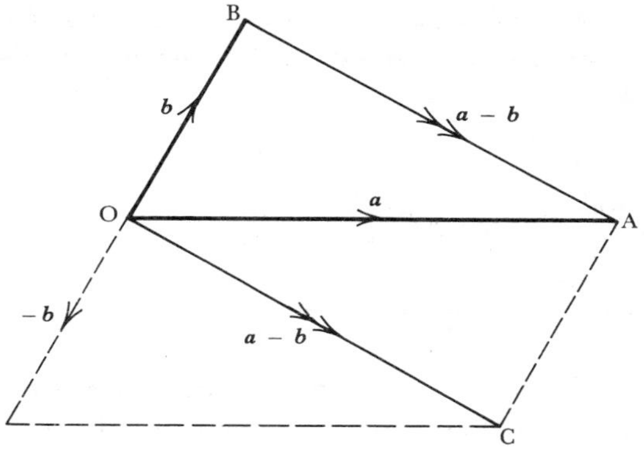

Exercise 8.1b

Draw accurately the following displacement vectors:

$\boldsymbol{a}$ 6 cm due North (0°);
$\boldsymbol{b}$ 8 cm North-East (045°);
$\boldsymbol{c}$ 4 cm due West (270°).

Hence construct the following vectors and state their magnitudes and directions:

1 (*i*) $\boldsymbol{a} + \boldsymbol{b}$; (*ii*) $\boldsymbol{b} + \boldsymbol{c}$;
(*iii*) $\boldsymbol{a} + \boldsymbol{c}$; (*iv*) $(\boldsymbol{a} + \boldsymbol{b}) + \boldsymbol{c}$;
(*v*) $(\boldsymbol{b} + \boldsymbol{c}) + \boldsymbol{a}$; (*vi*) $(\boldsymbol{a} + \boldsymbol{c}) + \boldsymbol{b}$.

2 (*i*) $-\boldsymbol{a}$; (*ii*) $-\boldsymbol{b}$; (*iii*) $-\boldsymbol{c}$;
(*iv*) $\boldsymbol{a} - \boldsymbol{b}$; (*v*) $\boldsymbol{a} - \boldsymbol{c}$; (*vi*) $\boldsymbol{b} - \boldsymbol{c}$;
(*vii*) $\boldsymbol{b} - \boldsymbol{a}$; (*viii*) $\boldsymbol{c} - \boldsymbol{a}$; (*ix*) $\boldsymbol{c} - \boldsymbol{b}$;
(*x*) $\boldsymbol{b} + \boldsymbol{c} - \boldsymbol{a}$.

Multiplication by a scalar

If λ is a scalar, then $\lambda\boldsymbol{a}$ is defined as the vector which has the same direction as $\boldsymbol{a}$ and λ times its magnitude:

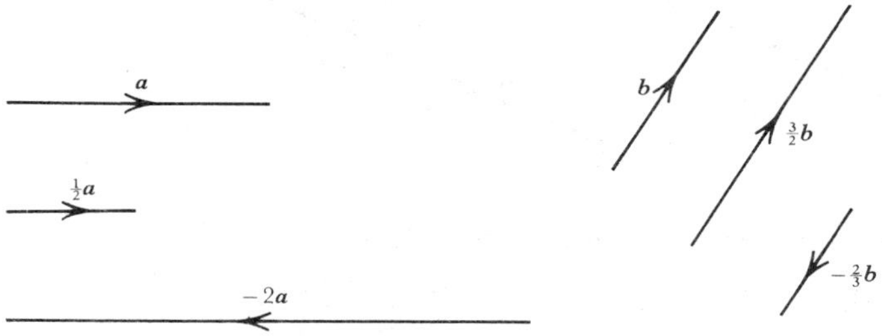

Now if we multiply by λ a vector which has already been multiplied by μ, the result is clearly the same as if we had multiplied the original vector by $\lambda\mu$:

$\lambda(\mu\boldsymbol{a}) = (\lambda\mu)\boldsymbol{a}$.

This is the *associative law* for multiplication by a scalar.

Fortunately, the operation of multiplying by a scalar also combines very conveniently with that of vector addition. For $2\boldsymbol{a} + 3\boldsymbol{a}$ is clearly the same as $(2 + 3)\boldsymbol{a}$; and more generally,

$$(\lambda + \mu)\boldsymbol{a} = \lambda\boldsymbol{a} + \mu\boldsymbol{a}$$

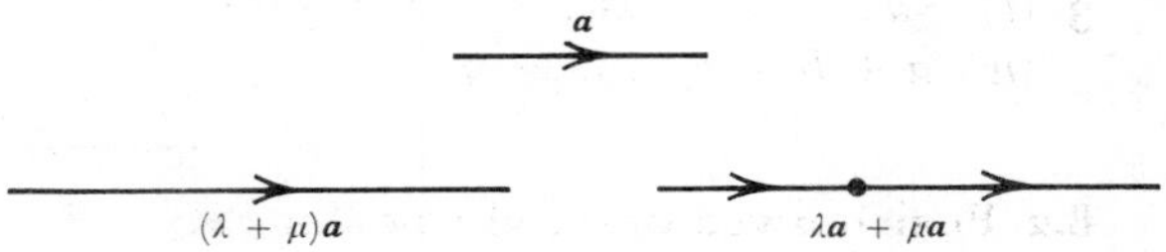

Secondly, multiplying the vector sum $\boldsymbol{a} + \boldsymbol{b}$ by λ produces the same result as adding $\lambda\boldsymbol{a}$ to $\lambda\boldsymbol{b}$:

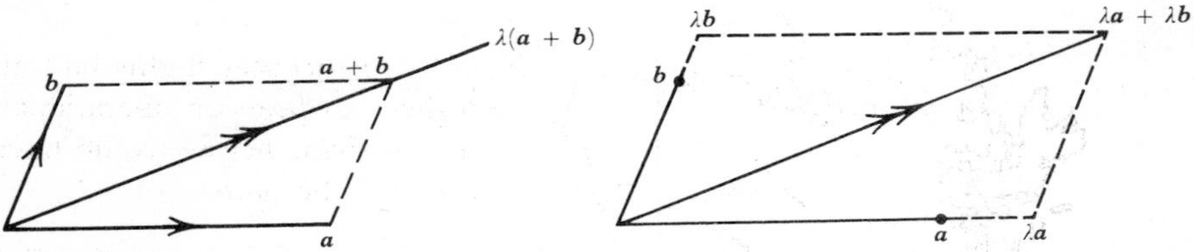

So $\lambda(\boldsymbol{a} + \boldsymbol{b}) = \lambda\boldsymbol{a} + \lambda\boldsymbol{b}$.

These two results,

$$(\lambda + \mu)\boldsymbol{a} = \lambda\boldsymbol{a} + \mu\boldsymbol{a}$$

and $$\lambda(\boldsymbol{a} + \boldsymbol{b}) = \lambda\boldsymbol{a} + \lambda\boldsymbol{b}$$

are the *distributive laws* which link multiplication by a number with vector addition.

We have already seen that the addition of vectors is different from the addition of numbers, and this could have been emphasised if we had used another symbol, say $\boldsymbol{a} \oplus \boldsymbol{b}$, to represent a vector 'sum'. This operation would then obey the *associative* and *commutative* laws

$$(\boldsymbol{a} \oplus \boldsymbol{b}) \oplus \boldsymbol{c} = \boldsymbol{a} \oplus (\boldsymbol{b} \oplus \boldsymbol{c})$$

and $$\boldsymbol{a} \oplus \boldsymbol{b} = \boldsymbol{b} \oplus \boldsymbol{a};$$

and furthermore, it is now clear that it would also obey the *distributive* laws

$$(\lambda + \mu)\boldsymbol{a} = \lambda\boldsymbol{a} \oplus \mu\boldsymbol{a}$$
$$\lambda(\boldsymbol{a} \oplus \boldsymbol{b}) = \lambda\boldsymbol{a} \oplus \lambda\boldsymbol{b}.$$

The first of this latter pair demonstrates the relationship between the symbols $\oplus$ and $+$, so that it is only natural to refer to the *addition* of $\boldsymbol{a}$ and $\boldsymbol{b}$, and to $\boldsymbol{a} \oplus \boldsymbol{b}$ as their *vector sum*, which for simplicity can be written $\boldsymbol{a} + \boldsymbol{b}$.

Exercise 8.1c

If $\boldsymbol{a}$ is the vector 4 cm due North and $\boldsymbol{b}$ is the vector 3 cm due East, construct the following vectors and state their magnitudes and directions.

1 (*i*) $2\boldsymbol{a}$; (*ii*) $3(2\boldsymbol{a})$; (*iii*) $6\boldsymbol{a}$.

2 (*i*) $3\boldsymbol{b}$; (*ii*) $4\boldsymbol{b}$; (*iii*) $7\boldsymbol{b}$.

3 (*i*) $\frac{1}{2}\boldsymbol{a}$; (*ii*) $\frac{1}{2}\boldsymbol{b}$; (*iii*) $\frac{1}{2}\boldsymbol{a} + \frac{1}{2}\boldsymbol{b}$;
(*iv*) $\boldsymbol{a} + \boldsymbol{b}$; (*v*) $\frac{1}{2}(\boldsymbol{a} + \boldsymbol{b})$.

8.2 Position vectors and components

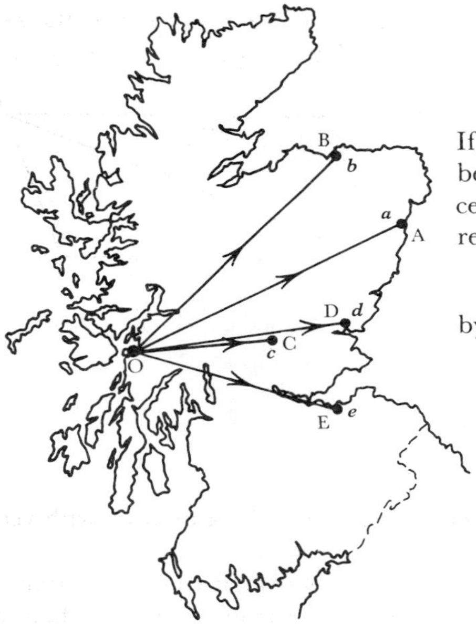

If, in 8.1, helicopter flights had all been directed from an operational centre at Oban, then we could have represented the *positions* of

	Aberdeen	Buckie	Crieff
by	***OA***	***OB***	***OC***

Dundee and Edinburgh
OD ***OE***

We call these the *position vectors*, referred to O, of A, B, C, D, E, and usually abbreviate them as $\boldsymbol{a}$, $\boldsymbol{b}$, $\boldsymbol{c}$, $\boldsymbol{d}$, $\boldsymbol{e}$.

Now if we are given the position vectors of two points A and B we can clearly use them to represent other points of their plane:

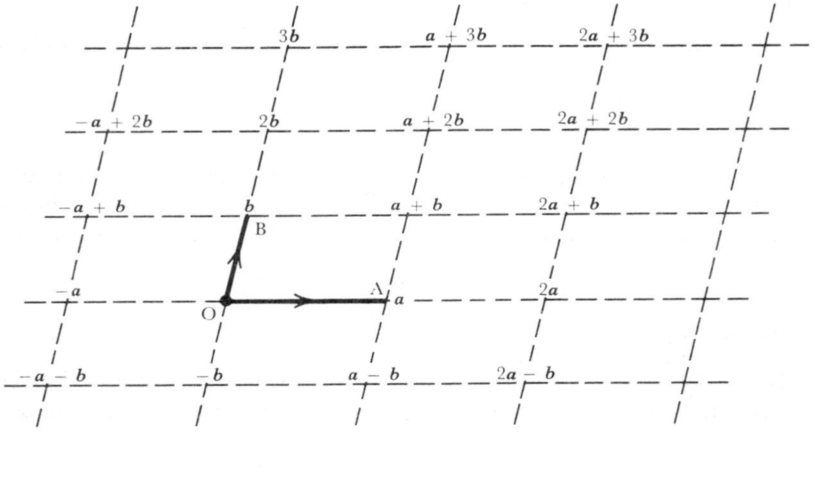

Indeed, providing that $\boldsymbol{a}$ and $\boldsymbol{b}$ are not parallel, any point P of the plane can be represented in terms of $\boldsymbol{a}$ and $\boldsymbol{b}$:

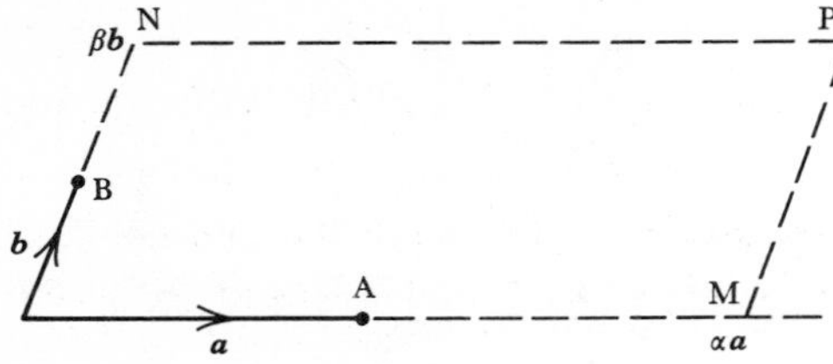

For we can always draw a parallelogram as shown.

Since M is on OA, $\boldsymbol{OM} = \alpha\boldsymbol{a}$
and since N is on OB, $\boldsymbol{ON} = \beta\boldsymbol{b}$ } where α, β are numbers.

So $\boldsymbol{OP} = \boldsymbol{OM} + \boldsymbol{ON}$

$\Rightarrow \quad \boldsymbol{p} = \alpha\boldsymbol{a} + \beta\boldsymbol{b}.$

Similarly, if we are given three non-coplanar vectors $\boldsymbol{a}$, $\boldsymbol{b}$, $\boldsymbol{c}$, then any point in space can be represented by its position vector referred to an origin O, and this vector is a linear combination of $\boldsymbol{a}$, $\boldsymbol{b}$, $\boldsymbol{c}$:

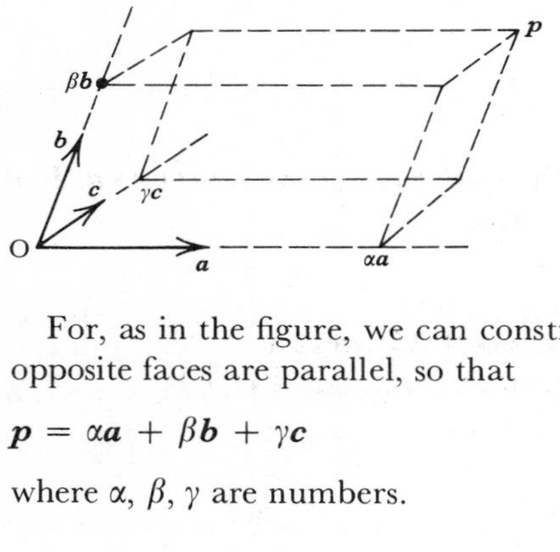

For, as in the figure, we can construct a *cuboid* (or *parallelepiped*), whose opposite faces are parallel, so that

$\boldsymbol{p} = \alpha\boldsymbol{a} + \beta\boldsymbol{b} + \gamma\boldsymbol{c}$

where α, β, γ are numbers.

Exercise 8.2a

1 O is an origin and points with position vectors $\boldsymbol{a}$, $\boldsymbol{b}$ are 6 km due East from O and 4 km North-East from O respectively. Sketch these points and mark the points which have position vectors:
(*i*) $2\boldsymbol{a}$, (*ii*) $\frac{1}{3}\boldsymbol{b}$; (*iii*) $-\frac{1}{2}\boldsymbol{a}$; (*iv*) $\boldsymbol{a} + 2\boldsymbol{b}$; (*v*) $2\boldsymbol{a} + \frac{1}{2}\boldsymbol{b}$; (*vi*) $\boldsymbol{a} - 2\boldsymbol{b}$.

2 $\boldsymbol{i}$, $\boldsymbol{j}$ are the position vectors (from O) of points whose Cartesian co-ordinates are (1, 0) and (0, 1). What are:
(*i*) the position vectors of (3, 2), (1, 3), (0, −2), $(\frac{1}{2}, \frac{1}{2})$?;
(*ii*) the Cartesian coordinates of points with position vectors:
$2\boldsymbol{i} + \boldsymbol{j}$, $\frac{1}{2}\boldsymbol{i} + \boldsymbol{j}$, $-3\boldsymbol{i}$, $-\boldsymbol{i} + \boldsymbol{j}$?

3 OAPB is a parallelogram. If, from O, A and B have position vectors $\boldsymbol{a}$ and $\boldsymbol{b}$, express in terms of $\boldsymbol{a}$ and $\boldsymbol{b}$:
(*i*) $\boldsymbol{AB}$ and $\boldsymbol{BA}$;
(*ii*) the position vector of P;
(*iii*) the position vectors of the mid-points of AB and OP. What do you deduçe?

4 ABCD is a skew quadrilateral and E, F, G, H are the mid-points of AB, BC, CD, and DA respectively. Denoting position vectors by corresponding letters, find:
(*i*) $\boldsymbol{e}, \boldsymbol{f}, \boldsymbol{g}, \boldsymbol{h}$ in terms of $\boldsymbol{a}, \boldsymbol{b}, \boldsymbol{c}, \boldsymbol{d}$;
(*ii*) $\boldsymbol{EF}$ and $\boldsymbol{HG}$ in terms of $\boldsymbol{a}, \boldsymbol{b}, \boldsymbol{c}, \boldsymbol{d}$;
(*iii*) the position vectors of the mid-points of EG and FH in terms of $\boldsymbol{a}, \boldsymbol{b}, \boldsymbol{c}, \boldsymbol{d}$.
What do you deduce?

5

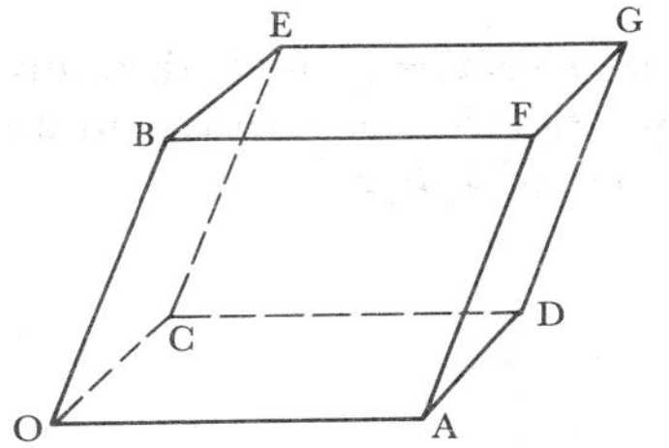

The corners A, B, C of this parallelepiped have position vectors $\boldsymbol{a}, \boldsymbol{b}, \boldsymbol{c}$ from the corner O. Find in terms of $\boldsymbol{a}, \boldsymbol{b}, \boldsymbol{c}$:
(*i*) the position vectors of D, E, F, G;
(*ii*) $\boldsymbol{AC}, \boldsymbol{FE}$;
(*iii*) the position vectors of the mid-points of AC and EF;
(*iv*) the position vectors of the mid-points of OG, AE, BD, CF.
What do you deduce?

6

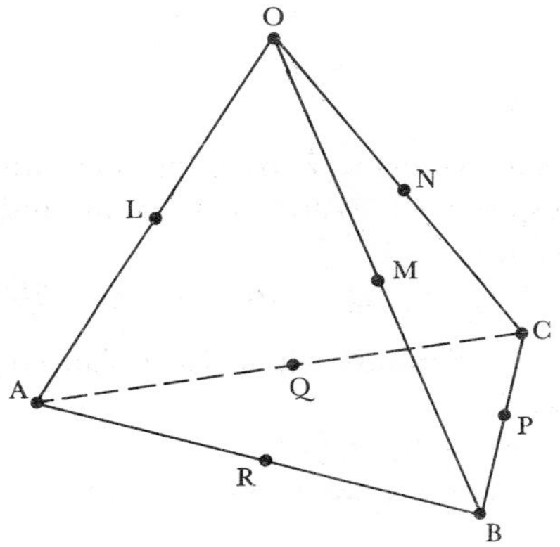

The corners A, B, C of a tetrahedron have position vectors $\boldsymbol{a}$, $\boldsymbol{b}$, $\boldsymbol{c}$ from the corner O. Find the position vectors of:

(*i*) the mid-points of OA, OB, OC (L, M, N);
(*ii*) the mid-points of BC, CA, AB (P, Q, R);
(*iii*) the mid-points of LP, MQ, NR.

What do you deduce?

Components

So far we have regarded a vector as a single entity, having both magnitude and direction and combining with similar vectors according to a parallelogram law of addition.

If, however, we take a point O as origin, we can refer to the position of another point P either by its position vector $\boldsymbol{r}$ or by its coordinates relative to a set of mutually perpendicular axes Ox, Oy, Oz. It is usual, for reasons which will be clear in 8.6, to take a right-hand set of axes, i.e., axes for which a rotation from Ox to Oy threads a right-handed screw in the direction Oz.

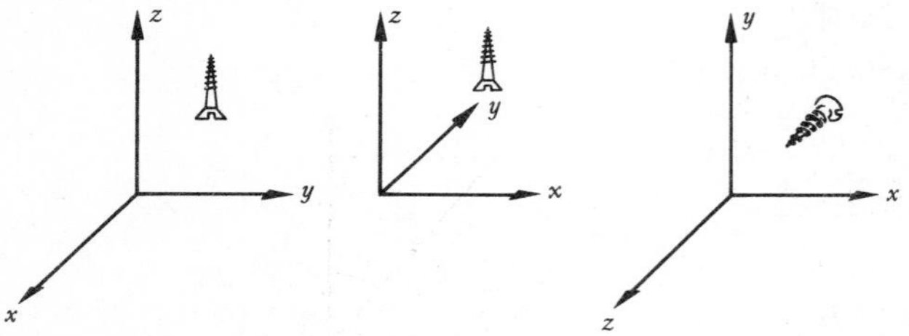

If we let $\boldsymbol{i}, \boldsymbol{j}$, and $\boldsymbol{k}$ be unit vectors along the axes Ox, Oy, Oz, we obtain the figure

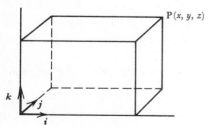

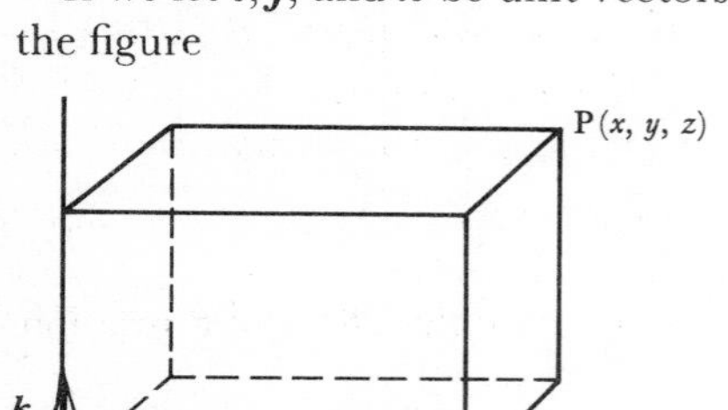

So the point $P(x, y, z)$ can also be described as

$$\boldsymbol{r} = x\boldsymbol{i} + y\boldsymbol{j} + z\boldsymbol{k}$$

where $x\boldsymbol{i}$, $y\boldsymbol{j}$, $z\boldsymbol{k}$ are said to be the *components* of $\boldsymbol{r}$.

Using Pythagoras' theorem, we see that the modulus of $\boldsymbol{r}$ is given by

$$r = \sqrt{(x^2 + y^2 + z^2)}.$$

In this way, every result which we have so far established can be expressed in terms of such components.

Example 1

If $\boldsymbol{i}, \boldsymbol{j}$ are unit vectors in the directions East and North respectively, find:
(*i*) the magnitude and direction of the vector $\boldsymbol{a} = 3\boldsymbol{i} - 4\boldsymbol{j}$;
(*ii*) the components along $\boldsymbol{i}$ and $\boldsymbol{j}$ of a displacement $\boldsymbol{b}$ which has magnitude 80 units on a bearing of 320°.

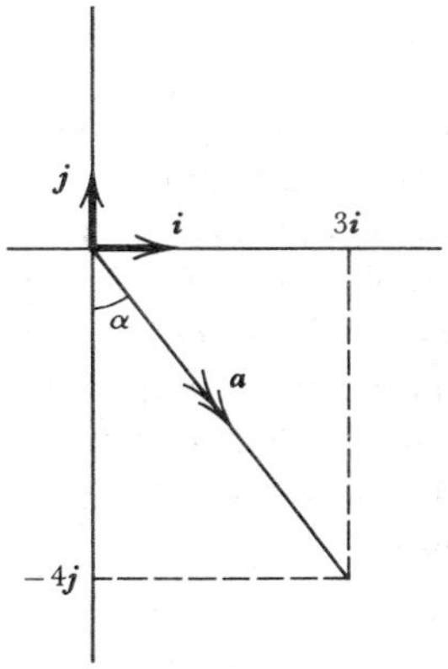

(*i*) From the figure,

$$a = \sqrt{(3^2 + 4^2)} = 5$$

and $\quad \tan\alpha = \frac{3}{4} \quad \Rightarrow \quad \alpha \approx 37°$

so the direction of $\boldsymbol{a}$ is on bearing 143°

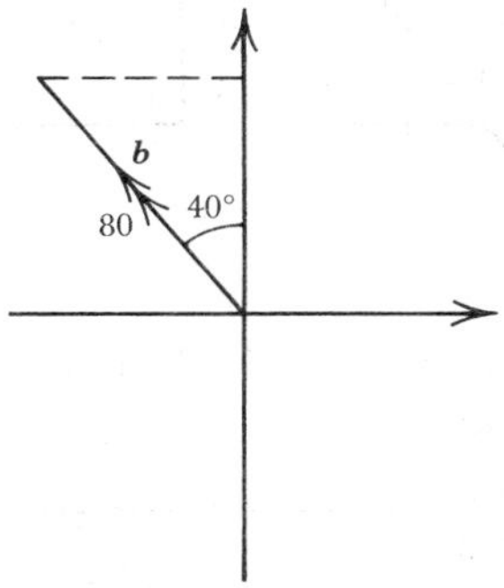

(*ii*) Again from the figure,
$\boldsymbol{b}$ is seen to have components

$$-80\sin 40° \approx -51.4 \quad \text{along } \boldsymbol{i}$$

and $\quad 80\cos 40° \approx +60.4 \quad \text{along } \boldsymbol{j}$

so $\quad \boldsymbol{b} = -51.4\boldsymbol{i} + 60.4\boldsymbol{j}.$

Example 2

We can also use the unit vectors $\boldsymbol{i}, \boldsymbol{j}$ to investigate the effect of a rotation about O through an angle ϕ.

It is clear from the figure that

$$\boldsymbol{i} \mapsto \cos\phi\, \boldsymbol{i} + \sin\phi\, \boldsymbol{j}$$

and $\quad \boldsymbol{j} \mapsto -\sin\phi\, \boldsymbol{i} + \cos\phi\, \boldsymbol{j}.$

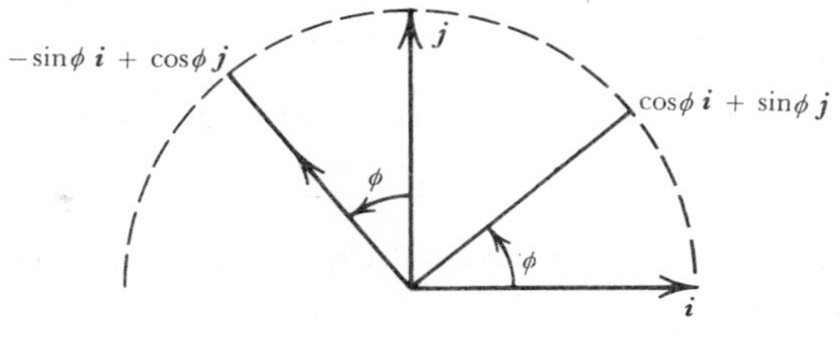

Furthermore, we can investigate the effect of this rotation on the unit vector $\boldsymbol{p}$ which makes an angle θ with $\boldsymbol{i}$:

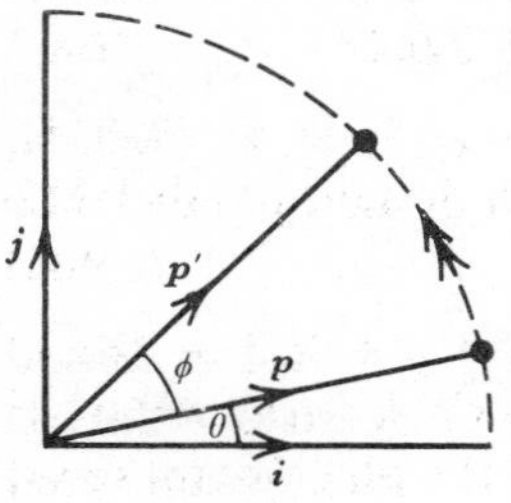

Now $\quad \boldsymbol{p} = \cos\theta\, \boldsymbol{i} + \sin\theta\, \boldsymbol{j}$,

So $\quad \boldsymbol{p} \mapsto \boldsymbol{p}' = \cos\theta\,(\cos\phi\, \boldsymbol{i} + \sin\phi\, \boldsymbol{j}) + \sin\theta\,(-\sin\phi\, \boldsymbol{i} + \cos\phi\, \boldsymbol{j})$

$$= (\cos\theta\cos\phi - \sin\theta\sin\phi)\boldsymbol{i} + (\sin\theta\cos\phi + \cos\theta\sin\phi)\boldsymbol{j}.$$

But $\boldsymbol{p}'$ is a unit vector in the direction $\theta + \phi$

So $\quad \boldsymbol{p}' = \cos(\theta + \phi)\boldsymbol{i} + \sin(\theta + \phi)\boldsymbol{j}$

and comparing the two expressions for $\boldsymbol{p}'$ we see that

$$\cos(\theta + \phi) = \cos\theta\cos\phi - \sin\theta\sin\phi$$
$$\sin(\theta + \phi) = \sin\theta\cos\phi + \cos\theta\sin\phi.$$

This is, of course, fundamentally the same proof of the addition formulae as that given in 5.5, though here expressed in the language of vectors rather than that of matrices.

Exercise 8.2b

1 If $\boldsymbol{i}, \boldsymbol{j}$ are unit vectors in the directions East and North respectively,
(*i*) find the magnitudes and directions of

$\boldsymbol{i} + 2\boldsymbol{j}, \quad -\boldsymbol{i} + 3\boldsymbol{j}, \quad 2\boldsymbol{i} - 5\boldsymbol{j}$;

(*ii*) express, in terms of $\boldsymbol{i}$ and $\boldsymbol{j}$, vectors
of magnitude 3 units in the direction North-East;
of magnitude 5 units in the direction South-East;
of magnitude 10 units on a bearing of 287°.

2 If $\quad \boldsymbol{a} = \boldsymbol{i} + 2\boldsymbol{j} + 3\boldsymbol{k}$

and $\quad \boldsymbol{b} = 2\boldsymbol{i} + 3\boldsymbol{j} + 4\boldsymbol{k}$

find the magnitudes of

(*i*) $\boldsymbol{a}$; (*ii*) $\boldsymbol{b}$; (*iii*) $\boldsymbol{a} + \boldsymbol{b}$; (*iv*) $\boldsymbol{a} - \boldsymbol{b}$.

Hence verify that the sum of the squares of the diagonals of a parallelogram is equal to the sum of the squares of its four sides.

3 The unit vectors $\boldsymbol{i}, \boldsymbol{j}$ are directed along the axes Ox, Oy respectively. Referred to these axes the points A, B, C have coordinates (1, 2), (4, 3), (3, −1) respectively. D is the mid-point of the line BC. Express the vectors $\boldsymbol{BA}$ and $\boldsymbol{CA}$ in terms of $\boldsymbol{i}, \boldsymbol{j}$. *Verify* that $\boldsymbol{BA} + \boldsymbol{CA} = 2\boldsymbol{DA}$. (O.C.)

4 Prove that the points with coordinates $(0, -2, -\frac{1}{2})$, (1, 1, 1), (5, 3, 4), and $(8, 2, 5\frac{1}{2})$ are the vertices of a trapezium in which the non-parallel sides are equal in length. (S.M.P.)

5 Three points A, B, and C have position vectors $\boldsymbol{i} + \boldsymbol{j} + \boldsymbol{k}$, $\boldsymbol{i} + 2\boldsymbol{k}$ and $3\boldsymbol{i} + 2\boldsymbol{j} + 3\boldsymbol{k}$ respectively, relative to a fixed origin O. A particle P starts from B at time $t = 0$, and moves along BC towards C with constant speed 1 unit per s. Find the position vector of P after t s (*a*) relative to O and (*b*) relative to A. (L.)

6 An aircraft takes off from the end of a runway in a southerly direction and climbs at an angle of $\tan^{-1}(\frac{1}{2})$ to the horizontal at a speed of $225\sqrt{5}$ km h^{-1}. Show that t s after take-off the position vector $\boldsymbol{r}$ of the aircraft with respect to the end of the runway is given by $\boldsymbol{r} = \frac{1}{16}t\,(2\boldsymbol{i} + \boldsymbol{k})$ where $\boldsymbol{i}, \boldsymbol{j}, \boldsymbol{k}$ represent vectors of length 1 km in directions South, East, and vertically upwards.

At time $t = 0$ a second aircraft flying horizontally South-West at $720\sqrt{2}$ km h^{-1} has a position vector $-1.2\boldsymbol{i} + 3.2\boldsymbol{j} + \boldsymbol{k}$. Find its position vector at time t in terms of $\boldsymbol{i}, \boldsymbol{j}, \boldsymbol{k}$ and t. Show that there will be a collision unless courses are changed and state at what time it will occur. (S.M.P.)

8.3 Differentiation of vectors

Suppose that a moving point goes from position P at time t to position P′ at time $t + \delta t$.

Its displacement in this interval is $\boldsymbol{PP'}$, so we call $(1/\delta t)\,\boldsymbol{PP'}$ its *average velocity* during the interval; and as this is a scalar multiple of $\boldsymbol{PP'}$, it also is a vector.

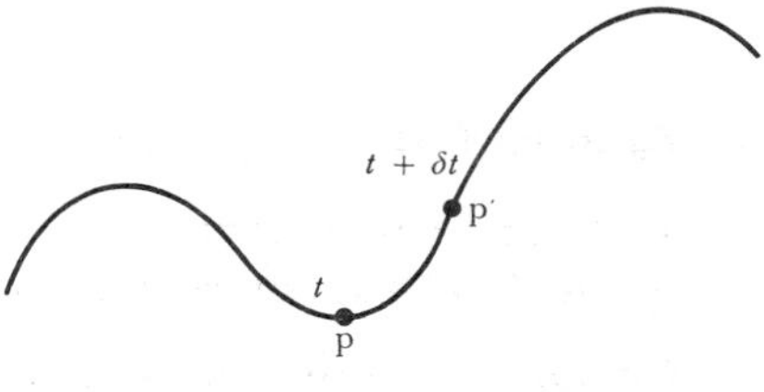

Now if, as $\delta t \to 0$ and P′ approaches P, the vector $(1/\delta t)\,\boldsymbol{PP'}$ (which we shall also denote by $\boldsymbol{PP'}/\delta t$) tends to a certain limiting vector $\boldsymbol{v}$, we call this limit the *velocity of P at the instant t*:

$$\boldsymbol{v} = \lim_{\delta t \to 0} \frac{\boldsymbol{PP'}}{\delta t}.$$

We now refer the positions P and P′ to an origin O, and suppose that

at time t, $\quad \boldsymbol{OP} = \boldsymbol{r}$;

at time $t + \delta t$, $\quad \boldsymbol{OP'} = \boldsymbol{r} + \delta\boldsymbol{r}$.

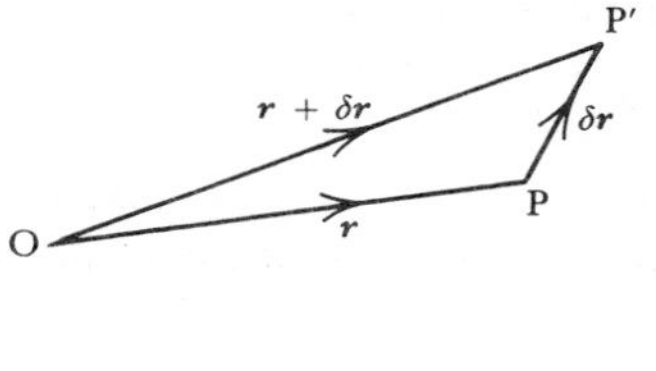

Then $\displaystyle \lim_{\delta t \to 0} \frac{\boldsymbol{PP'}}{\delta t} = \lim_{\delta t \to 0} \frac{\delta\boldsymbol{r}}{\delta t}.$

We naturally call this limit $\frac{\mathrm{d}\boldsymbol{r}}{\mathrm{d}t}$, so that $\boldsymbol{v} = \frac{\mathrm{d}\boldsymbol{r}}{\mathrm{d}t}$.

Similarly, if $\frac{\delta \boldsymbol{v}}{\delta t}$ tends to a limit as $\delta t \to 0$, we call this limit $\frac{\mathrm{d}\boldsymbol{v}}{\mathrm{d}t}$, the *acceleration* $\boldsymbol{a}$ of the point at time t.

So
$$\boldsymbol{v} = \frac{\mathrm{d}\boldsymbol{r}}{\mathrm{d}t}$$
and
$$\boldsymbol{a} = \frac{\mathrm{d}\boldsymbol{v}}{\mathrm{d}t} = \frac{\mathrm{d}^2\boldsymbol{r}}{\mathrm{d}t^2}.$$

When any new operation, like differentiation of vectors, is introduced, it is essential to discover how it combines with other operations.

Now $\delta(\boldsymbol{r} + \boldsymbol{s}) = (\boldsymbol{r} + \delta\boldsymbol{r} + \boldsymbol{s} + \delta\boldsymbol{s}) - (\boldsymbol{r} + \boldsymbol{s}) = \delta\boldsymbol{r} + \delta\boldsymbol{s}$.

So
$$\frac{\delta(\boldsymbol{r} + \boldsymbol{s})}{\delta t} = \frac{\delta\boldsymbol{r}}{\delta t} + \frac{\delta\boldsymbol{s}}{\delta t}$$
$$\Rightarrow \quad \lim_{\delta t \to 0} \frac{\delta(\boldsymbol{r} + \boldsymbol{s})}{\delta t} = \lim_{\delta t \to 0} \frac{\delta\boldsymbol{r}}{\delta t} + \lim_{\delta t \to 0} \frac{\delta\boldsymbol{s}}{\delta t}$$
$$\Rightarrow \quad \frac{\mathrm{d}}{\mathrm{d}t}(\boldsymbol{r} + \boldsymbol{s}) = \frac{\mathrm{d}\boldsymbol{r}}{\mathrm{d}t} + \frac{\mathrm{d}\boldsymbol{s}}{\mathrm{d}t}.$$

So the derivative of a vector sum is the vector sum of its derivatives; and similarly it can be proved that, if λ is a constant,

$$\frac{\mathrm{d}}{\mathrm{d}t}(\lambda\boldsymbol{r}) = \lambda\frac{\mathrm{d}\boldsymbol{r}}{\mathrm{d}t}.$$

When we are considering such derivatives with respect to time it is frequently convenient to denote them by means of dots placed above their symbols.

So $\boldsymbol{v} = \frac{\mathrm{d}\boldsymbol{r}}{\mathrm{d}t}$ is written $\dot{\boldsymbol{r}}$ and its components as $\dot{x}$, $\dot{y}$, $\dot{z}$.

Similarly $\boldsymbol{a} = \frac{\mathrm{d}\boldsymbol{v}}{\mathrm{d}t} = \frac{\mathrm{d}^2\boldsymbol{r}}{\mathrm{d}t^2}$ is written $\ddot{\boldsymbol{r}}$ and its components as $\ddot{x}$, $\ddot{y}$, $\ddot{z}$.

Summarising,

$$\boldsymbol{r} = x\boldsymbol{i} + y\boldsymbol{j} + z\boldsymbol{k}$$
$$\Rightarrow \quad \boldsymbol{v} = \dot{\boldsymbol{r}} = \dot{x}\boldsymbol{i} + \dot{y}\boldsymbol{j} + \dot{z}\boldsymbol{k}$$
$$\Rightarrow \quad \boldsymbol{a} = \ddot{\boldsymbol{r}} = \ddot{x}\boldsymbol{i} + \ddot{y}\boldsymbol{j} + \ddot{z}\boldsymbol{k}$$

Example 1

The position vector of a particle after time t is given by

$\boldsymbol{r} = 3t\boldsymbol{i} + 4t\boldsymbol{j} - 5t^2\boldsymbol{k}$.

Find its velocity and acceleration after time t and show that its acceleration is constant.

$$\boldsymbol{r} = 3t\boldsymbol{i} + 4t\boldsymbol{j} - 5t^2\boldsymbol{k}$$

$$\Rightarrow \quad \boldsymbol{v} = \dot{\boldsymbol{r}} = 3\boldsymbol{i} + 4\boldsymbol{j} - 10t\boldsymbol{k}$$

$$\Rightarrow \quad \boldsymbol{a} = \ddot{\boldsymbol{r}} = -10\boldsymbol{k}, \quad \text{which is constant.}$$

Exercise 8.3

1 The position vector $\boldsymbol{r}$ of a particle after time t is given by

$\boldsymbol{r} = t^2\boldsymbol{i} - 5t\boldsymbol{j} + (t^2 - 1)\boldsymbol{k}$.

Find expressions for its velocity $\boldsymbol{v}$ and acceleration $\boldsymbol{a}$ and calculate their magnitudes when $t = 2$.

2 The position vector $\boldsymbol{r}$ of a particle can be expressed in terms of the constant orthogonal unit vectors $\boldsymbol{i}, \boldsymbol{j}$ by means of the relation

$\boldsymbol{r} = \boldsymbol{i}\cos nt + \boldsymbol{j}\sin nt$,

Find the speed of the particle at time $t = 2$.
Show that its acceleration is the vector $-n^2\boldsymbol{r}$. (S.M.P.)

3 A particle moves in a plane so that its coordinates (x, y) at any time t are $(t + \cos t, 1 + \sin t)$. Find the values of t for which it is stationary.

Show that the acceleration of the particle has constant magnitude. (S.M.P.)

4 The position vector $\boldsymbol{p}$ of a passenger in a roundabout at a fairground is given by

$$\boldsymbol{p} = 9\boldsymbol{i}\cos\frac{2t}{3} + 9\boldsymbol{j}\sin\frac{2t}{3} + \frac{9}{2}\boldsymbol{k}\left(1 - \cos\frac{4t}{3}\right),$$

where $\boldsymbol{i}, \boldsymbol{j}, \boldsymbol{k}$ are mutually perpendicular vectors each of length 1 m, $\boldsymbol{k}$ being vertical, t is the time measured in seconds, and the angles are measured in radians.

How long does it take to perform a complete revolution, and what happens in the vertical direction while this is taking place?

Find the acceleration of the passenger at time t. What is the magnitude of the greatest acceleration he experiences? (S.M.P.)

5 If a particle moving on a smooth horizontal plane is connected to the origin O by means of an elastic spring, it can be shown that its position vector $\boldsymbol{r}$ obeys the equation

$\ddot{\boldsymbol{r}} + \omega^2\boldsymbol{r} = 0$

where ω is a constant.

Verify that a possible solution of this differential equation is

$\boldsymbol{r} = \boldsymbol{a} \sin \omega t + \boldsymbol{b} \cos \omega t$

where $\boldsymbol{a}$, $\boldsymbol{b}$ are constant vectors.

6 At a certain instant a body has a velocity of 3 m s^{-1} eastward and a constant acceleration of 2.5 m s^{-2} in a direction North 30° East. Find, graphically or otherwise, the magnitude and direction of its velocity two seconds later. When the body is moving in a direction due North-East find the magnitude of its velocity. (M.E.I.)

7 Two particles A and B start simultaneously from the origin. Particle A moves with constant velocity $2\boldsymbol{i} + 6\boldsymbol{j}$, while particle B has initial velocity $4\boldsymbol{i} - 8\boldsymbol{j}$ and a constant acceleration $\boldsymbol{i} + \boldsymbol{j}$.
(*i*) Find the position of each particle after t seconds;
(*ii*) find the time at which the velocities are perpendicular;
(*iii*) if C is the point with position vector $17\boldsymbol{i} + 25\boldsymbol{j}$, find the time at which the line joining the positions of A and B passes through C. (W.)

8 The acceleration of a particle moving in a plane is $n^2(x\boldsymbol{i} + y\boldsymbol{j})$ where $\boldsymbol{i}, \boldsymbol{j}$ are unit vectors parallel to the perpendicular fixed axes Ox, Oy; (x, y) are the coordinates of the particle at time t and $n^2 (>0)$ is a constant. When $t = 0$ the particle is at $(a, 0)$ and is projected parallel to the positive y-axis with speed $2na$. Find the equation of the path of the particle, and show that ultimately the particle will move off to infinity in a direction making an acute angle $\tan^{-1} 2$ with the x-axis.

Show also that there is one point only on the path at which the particle has a given speed $v(>2na)$ and find the coordinates of the point when $v = 3na$. (C.)

8.4 Motion with constant acceleration: projectiles

A stone is thrown horizontally out to sea from the top of a cliff. If its initial velocity is $\boldsymbol{u}$ and thereafter it has an acceleration $\boldsymbol{g}$ vertically downwards, find its position after time t.

Taking unit vectors $\boldsymbol{i}, \boldsymbol{j}$, horizontally and vertically downwards, we see that

$$\ddot{\boldsymbol{r}} = g\boldsymbol{j}$$

$$\Rightarrow \quad \dot{\boldsymbol{r}} = gt\boldsymbol{j} + \boldsymbol{c}$$

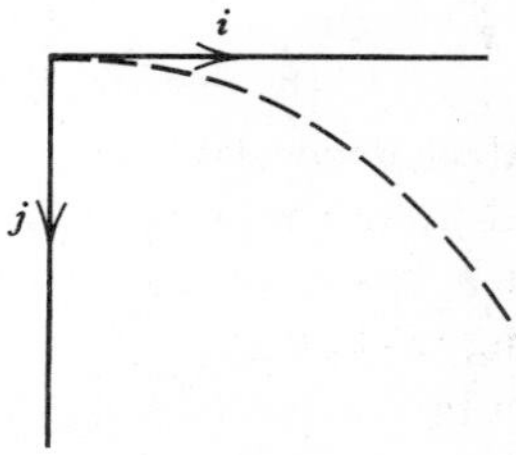

(where $\boldsymbol{c}$ is an arbitrary constant vector).

But when $t = 0$, $\dot{\boldsymbol{r}} = u\boldsymbol{i}$

So $\boldsymbol{c} = u\boldsymbol{i}$

Hence $\dot{\boldsymbol{r}} = u\boldsymbol{i} + gt\boldsymbol{j}$

$$\Rightarrow \quad \boldsymbol{r} = ut\boldsymbol{i} + \tfrac{1}{2}gt^2\boldsymbol{j} + \boldsymbol{d}$$

(where $\boldsymbol{d}$ is another arbitrary constant vector).

But when $t = 0$, $\boldsymbol{r} = \boldsymbol{0}$

So $\boldsymbol{d} = \boldsymbol{0}$

Hence $\boldsymbol{r} = ut\boldsymbol{i} + \frac{1}{2}gt^2\boldsymbol{j}$.

If, for instance, the stone was thrown with speed 20 m s^{-1}, then

$u = 20$ and $g \approx 10$.

So $\boldsymbol{r} \approx 20t\boldsymbol{i} + 5t^2\boldsymbol{j}$

$\dot{\boldsymbol{r}} \approx 20\boldsymbol{i} + 10t\boldsymbol{j}$

and $\ddot{\boldsymbol{r}} \approx 10\boldsymbol{j}$,

and the position and velocity of the stone can be found at successive instants:

t	$\boldsymbol{r}$	$\dot{\boldsymbol{r}}$
0	$\boldsymbol{0}$	$20\boldsymbol{i}$
1	$20\boldsymbol{i} + 5\boldsymbol{j}$	$20\boldsymbol{i} + 10\boldsymbol{j}$
2	$40\boldsymbol{i} + 20\boldsymbol{j}$	$20\boldsymbol{i} + 20\boldsymbol{j}$
3	$60\boldsymbol{i} + 45\boldsymbol{j}$	$20\boldsymbol{i} + 30\boldsymbol{j}$
4	$80\boldsymbol{i} + 80\boldsymbol{j}$	$20\boldsymbol{i} + 40\boldsymbol{j}$

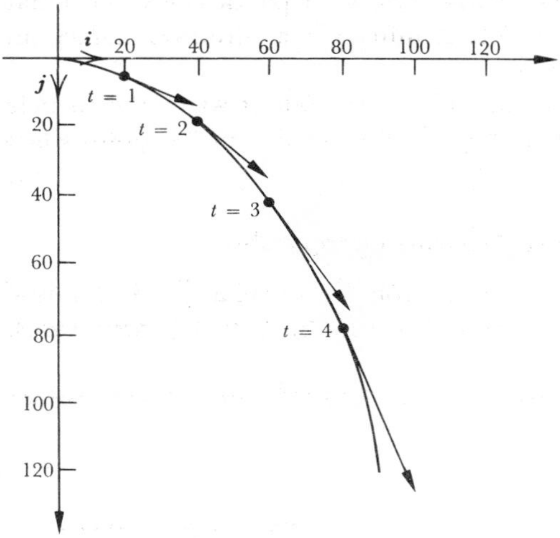

More generally, we now consider the motion of a bullet which has been fired from a point O with velocity $\boldsymbol{u}$ at an angle α to the horizontal and which has constant acceleration $\boldsymbol{g}$ vertically downwards, air resistance being negligible.

Suppose that after time t the bullet is at a point P, with position vector $\boldsymbol{r}$.

Then $\ddot{\boldsymbol{r}} = \boldsymbol{g}$

$\Rightarrow$ $\dot{\boldsymbol{r}} = t\boldsymbol{g} + \boldsymbol{c}$ (where $\boldsymbol{c}$ is a constant vector).

But when $t = 0, \quad \dot{\boldsymbol{r}} = \boldsymbol{u}$

So $\boldsymbol{c} = \boldsymbol{u}$

Hence $\dot{\boldsymbol{r}} = \boldsymbol{u} + t\boldsymbol{g}$

$\Rightarrow \quad \boldsymbol{r} = t\boldsymbol{u} + \frac{1}{2}t^2\boldsymbol{g} + \boldsymbol{d}$

(where $\boldsymbol{d}$ is another constant vector). But when $t = 0$, $\boldsymbol{r} = \boldsymbol{0}$. So $\boldsymbol{d} = \boldsymbol{0}$.

Hence $\boldsymbol{r} = t\boldsymbol{u} + \frac{1}{2}t^2\boldsymbol{g}$.

Summarising,

$$\begin{aligned} \boldsymbol{r} &= t\boldsymbol{u} + \tfrac{1}{2}t^2\boldsymbol{g} \\ \dot{\boldsymbol{r}} &= \boldsymbol{u} + t\boldsymbol{g} \\ \ddot{\boldsymbol{r}} &= \boldsymbol{g} \end{aligned}$$

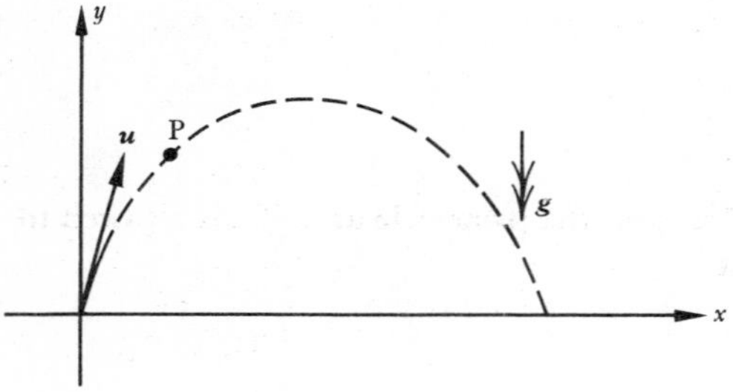

Alternatively, we could take axes Ox, Oy horizontally and vertically upwards, and carry out the above work in components:

$\ddot{\boldsymbol{r}} = \boldsymbol{g}$

$\Rightarrow \quad \ddot{x} = 0 \qquad \ddot{y} = -g$

Integrating with respect to t,

$\dot{x} = a \qquad \dot{y} = -gt + b$

where a and b are constants.

But when $t = 0, \quad \dot{x} = u\cos\alpha$ and $\dot{y} = u\sin\alpha$

So $u\cos\alpha = a, \quad u\sin\alpha = b$

$\Rightarrow \quad \dot{x} = u\cos\alpha, \quad \dot{y} = u\sin\alpha - gt.$

Integrating again, we obtain

$x = ut\cos\alpha + c \qquad y = ut\sin\alpha - \frac{1}{2}gt^2 + d$

But when $t = 0, \quad x = 0$ and $y = 0$

So $c = d = 0$

$$\Rightarrow \quad x = ut\cos\alpha, \qquad y = ut\sin\alpha - \tfrac{1}{2}gt^2.$$

Summarising, we see that the components of displacement, velocity, and acceleration are

$$\boxed{\begin{array}{ll} x = ut\cos\alpha & y = ut\sin\alpha - \tfrac{1}{2}gt^2 \\ \dot{x} = u\cos\alpha & \dot{y} = u\sin\alpha - gt \\ \ddot{x} = 0 & \ddot{y} = -g \end{array}}$$

Now the bullet hits the ground when $y = 0$

$$\Rightarrow \quad ut\sin\alpha - \tfrac{1}{2}gt^2 = 0$$

$$\Rightarrow \quad t = 0 \quad \text{or} \quad \frac{2u\sin\alpha}{g}.$$

So the time of flight is $(2u\sin\alpha)/g$, and when $t = (2u\sin\alpha)/g$,

$$x = u\frac{2u\sin\alpha}{g}\cos\alpha = \frac{u^2\sin 2\alpha}{g}.$$

This is therefore the *horizontal range* of the projectile and, if α is allowed to vary, this range is greatest when

$$\sin 2\alpha = 1 \quad \Rightarrow \quad \alpha = \tfrac{1}{4}\pi$$

So maximum range $= \dfrac{u^2}{g}$.

We also see that

$$x = ut\cos\alpha, \qquad y = ut\sin\alpha - \tfrac{1}{2}gt^2.$$

provide us with parametric equations for the trajectory, and that we can eliminate t to obtain its x, y equation.

For $\quad t = \dfrac{x}{u\cos\alpha}$

$$\Rightarrow \quad y = u\frac{x}{u\cos\alpha}\sin\alpha - \tfrac{1}{2}g\frac{x^2}{u^2\cos^2\alpha}$$

$$\Rightarrow \quad \boxed{y = x\tan\alpha - \frac{gx^2}{2u^2}\sec^2\alpha}$$

which is a parabola.

Exercise 8.4

(Take $g = 10 \text{ m s}^{-2}$.)

1 A golfer strikes a ball with velocity $25\boldsymbol{i} + 15\boldsymbol{j}$ (using SI units and unit vectors $\boldsymbol{i}, \boldsymbol{j}$ horizontally and vertically). Find:
(*i*) its position vector $\boldsymbol{r}$ after time t;
(*ii*) its velocity $\boldsymbol{v}$ at this instant;
(*iii*) its time of flight before pitching on to a horizontal fairway;
(*iv*) its horizontal range;
(*v*) its maximum height.

2 A ball is thrown with velocity 20 m s^{-1} at an angle of $60°$ to the horizontal. Find in terms of unit vectors $\boldsymbol{i}, \boldsymbol{j}$, horizontally and vertically:
(*i*) its displacement and velocity after t s;
(*ii*) its speed after 2 s and the direction in which it is travelling;
(*iii*) when it is travelling horizontally.

3 If a man could throw a ball at 30 m s^{-1} in any direction, find
(*i*) the greatest height it could reach;
(*ii*) its greatest horizontal range.

4 The record hit at cricket is reputedly (and incredibly) 175 yd, or approximately 160 m. What must have been the least initial speed of the ball for such a feat?

5 A man throws a ball with speed u at an angle α to the horizontal. Prove that:
(*i*) its maximum height is $(u^2 \sin^2 \alpha)/2g$;
(*ii*) its greatest horizontal range is four times the maximum height it would reach during such a trajectory.

6 A golfer strikes a golf ball so that its initial velocity makes an angle $\tan^{-1}(\frac{1}{2})$ with the horizontal. If the range of the ball on the horizontal plane through the point of projection is 156.8 m, calculate the greatest height of the ball above the plane and its time of flight. (Take g as 9.8 m s^{-2}.)
(O.C.)

7 A particle is projected from a point O with speed V at an angle of elevation α. Prove that the equation of its trajectory referred to horizontal and vertical axes through O is

$$y = x \tan \alpha - \frac{gx^2}{2V^2} \sec^2 \alpha.$$

Taking $g = 10$, find the two values of $\tan \alpha$ so that a particle projected with $V = 70$ will pass through the point (280, 40). Find the ranges, on the horizontal plane through O, corresponding to these two angles of projection.
(J.M.B.)

8.5 Circular motion

Angular velocity

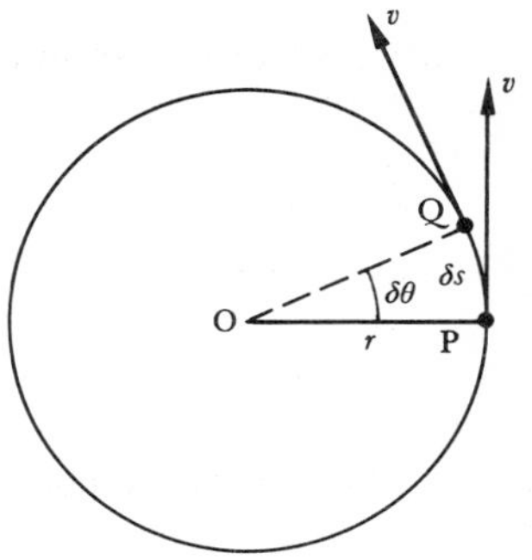

(i)

Suppose that a point is moving round a circle of radius r with constant speed v

Further, suppose that in a brief time δt, the point moves from P to Q through an angle $\delta\theta$ and distance δs.

Then $r\,\delta\theta \approx \delta s \approx v\,\delta t$

$$\Rightarrow \qquad \frac{d\theta}{dt} = \frac{v}{r}.$$

The quantity $\dfrac{d\theta}{dt}$ (or $\dot{\theta}$) is called the *angular velocity of* P *about* O, and is commonly denoted by ω.

So $\omega = \dot{\theta} = \dfrac{v}{r}$ and $v = r\dot{\theta} = r\omega$.

Acceleration of a particle moving in a circle with constant speed

Method 1 Although the speed of P is constant, its *velocity* (a vector) is always changing direction. So we can now proceed to find its rate of change, which is the acceleration of P.

Suppose that the velocity at P is $\boldsymbol{v}$ and at Q is $\boldsymbol{v} + \delta\boldsymbol{v}$.

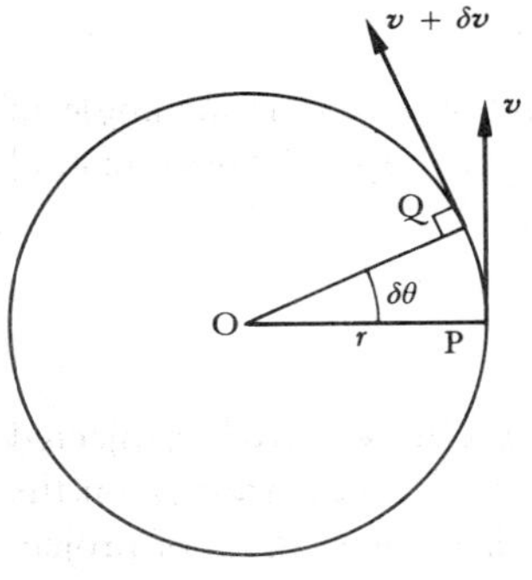

(ii)

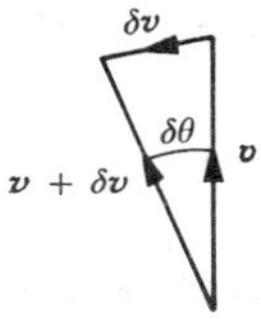

(iii)

Now the magnitudes of $\boldsymbol{v}$ and $\boldsymbol{v} + \delta\boldsymbol{v}$ are the same, and we see from fig. (*iii*) that $\delta\boldsymbol{v}$ is almost perpendicular to $\boldsymbol{v}$ and has magnitude $v\,\delta\theta$.

So $\dfrac{\delta\boldsymbol{v}}{\delta t}$ is almost perpendicular to $\boldsymbol{v}$ and has magnitude $v\dfrac{\delta\theta}{\delta t}$.

In the limit, $\boldsymbol{a} = \dfrac{\mathrm{d}\boldsymbol{v}}{\mathrm{d}t}$ is perpendicular to $\boldsymbol{v}$ and has magnitude

$$v\frac{\mathrm{d}\theta}{\mathrm{d}t} = \frac{v^2}{r}.$$

So a particle moving in a circle with constant speed v has acceleration $\dfrac{v^2}{r}$ directed towards its centre.

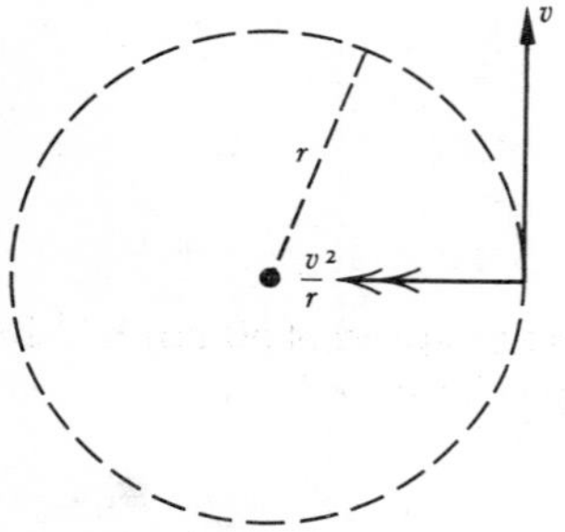

If the angular velocity of the point is ω, then $v = r\omega$, so that this acceleration can also be expressed as $r\omega^2$.

These are such important results that we shall now derive them by another method.

Method 2 Suppose that P is moving round a circle with constant angular velocity ω, and that initially $\boldsymbol{r}$ was along $\boldsymbol{i}$.

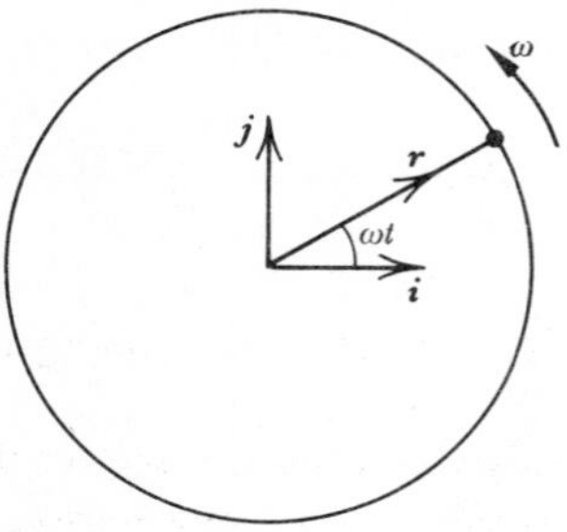

Then, after time t,

$$\boldsymbol{r} = r\cos\omega t\,\boldsymbol{i} + r\sin\omega t\,\boldsymbol{j}$$

$$\Rightarrow \quad \boldsymbol{v} = \dot{\boldsymbol{r}} = -r\omega\sin\omega t\,\boldsymbol{i} + r\omega\cos\omega t\,\boldsymbol{j}$$

and $\quad v = \sqrt{(r^2\omega^2\sin^2\omega t + r^2\omega^2\cos^2\omega t)} = r\omega.$

Furthermore,

$$\boldsymbol{a} = \ddot{\boldsymbol{r}} = -r\omega^2 \cos \omega t\, \boldsymbol{i} - r\omega^2 \sin \omega t\, \boldsymbol{j}$$
$$= -\omega^2 \boldsymbol{r}.$$

Hence the acceleration of P is directed towards O and has magnitude

$$r\omega^2 = \frac{v^2}{r}.$$

Summarising, for motion in a circle of radius r at constant angular velocity ω, the magnitudes of velocity and acceleration are

$$v = r\omega \quad \text{and} \quad a = r\omega^2 = \frac{v^2}{r}$$

Example 1

A motor-cyclist is going round a Wall of Death at a speed of 50 km h^{-1} in a horizontal circle of radius 20 m. What is his acceleration?

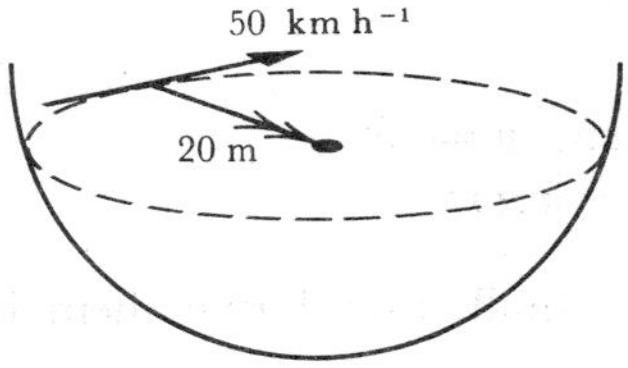

Using SI units,

$$\text{speed} = v = 50 \text{ km h}^{-1} = \frac{50 \times 10^3}{3\,600} \text{ m s}^{-1} \approx 14 \text{ m s}^{-1},$$

radius $= r = 20$ m.

$$\text{Hence} \quad \text{acceleration} = \frac{v^2}{r} \approx \frac{14^2}{20} = 9.8 \text{ m s}^{-2}$$

(and, by sheer coincidence, is the same in magnitude as he would experience if he were falling under gravity).

We shall soon see that any such acceleration necessitates a force on the motor-cycle in the same direction, which is provided by a combination of its weight and the thrust of the wall on its tyres.

Example 2

The Earth is moving round the Sun in an orbit which is approximately a circle of radius 1.50×10^8 km. What is its acceleration?

The Earth takes a year (of roughly 365 days) to complete the circumference which is approximately

$$2\pi \times 1.50 \times 10^{11} \text{ m} = 9.43 \times 10^{11} \text{ m}$$

$$\text{So the Earth's speed} = \frac{9.43 \times 10^{11}}{365 \times 24 \times 3\,600} \text{ m s}^{-1}$$

$$= 2.99 \times 10^{4} \text{ m s}^{-1}$$

$$\text{Hence its acceleration} = \frac{v^2}{r} = \frac{(2.99 \times 10^4)^2}{1.50 \times 10^{11}} \text{ m s}^{-2}$$

$$\approx 0.006 \text{ m s}^{-2}.$$

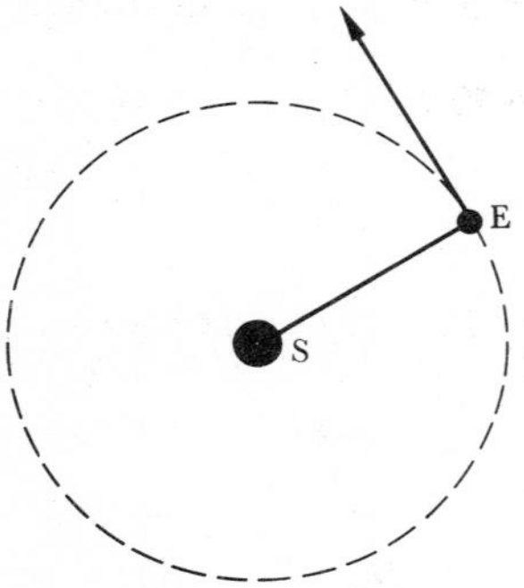

Again, this acceleration necessitates a force towards the Sun, called *gravitational attraction*.

Exercise 8.5

1 What is the acceleration of a particle which moves in a circle:
(*i*) with radius 2 m and constant speed 3 m s^{-1};
(*ii*) with radius 1 km and constant speed 20 m s^{-1};
(*iii*) with radius 4 m and constant angular velocity 2 rad s^{-1};
(*iv*) with radius 1 cm and constant angular velocity 20 rad s^{-1};
(*v*) with radius 1 m and constant angular velocity 80 rev min^{-1};
(*vi*) with radius 2 cm and constant angular velocity 100 rev min^{-1}?

2 A supersonic jet is travelling at a constant speed of 500 m s^{-1} and turning in a horizontal circle of radius 50 km. What is its acceleration? Find the turning radius of a light aircraft whose pilot experiences the same acceleration when travelling at 50 m s^{-1}.

3 A pilot can loop the loop without using a safety harness providing that his vertical acceleration at the top of the loop is greater than g (10 m s^{-2}). If the diameter of the loop is $\frac{1}{2}$ km, what is the minimum speed for safety?

4 A man swings a small bucket of water round and round in a vertical circle of radius 1 m at a constant speed of 1 rev. every 2 s. Show that the

magnitude of its acceleration is slightly greater than g, so that the water just stays in the bucket.

*8.6 Relative motion

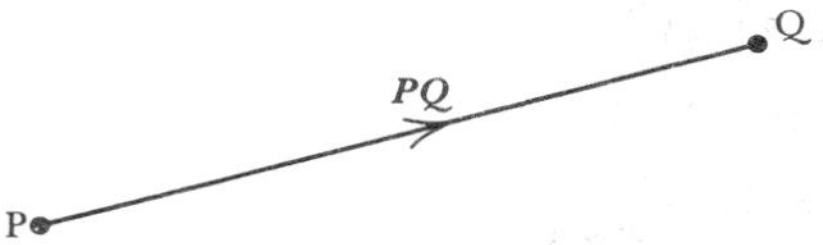

Suppose that P, Q are two moving points. Then $\boldsymbol{PQ}$ represents the position vector of Q relative to P, and $\frac{d}{dt}(\boldsymbol{PQ})$ represents its rate of change, which we call the *velocity of Q relative to P*, written $\boldsymbol{v}_P(Q)$.

So $\boldsymbol{v}_P(Q) = \frac{d}{dt}(\boldsymbol{PQ})$.

If we now consider three moving points, P, Q, R,

$\boldsymbol{PQ} + \boldsymbol{QR} = \boldsymbol{PR}$.

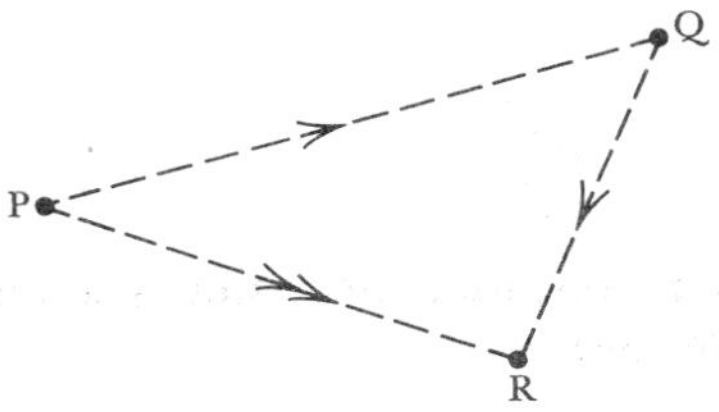

Hence $\frac{d}{dt}(\boldsymbol{PQ}) + \frac{d}{dt}(\boldsymbol{QR}) = \frac{d}{dt}(\boldsymbol{PQ} + \boldsymbol{QR}) = \frac{d}{dt}(\boldsymbol{PR})$

$\Rightarrow$ $\boxed{\boldsymbol{v}_P(Q) + \boldsymbol{v}_Q(R) = \boldsymbol{v}_P(R)}$

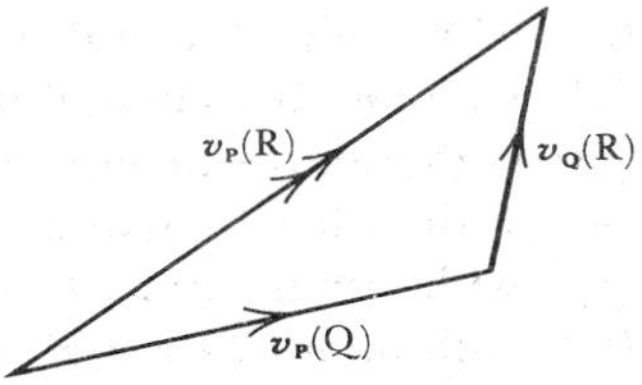

Furthermore, the rate of change of $\boldsymbol{v}_P(Q)$ can be called the *acceleration of Q relative to* P, written $\boldsymbol{a}_P(Q)$.

So $\boldsymbol{a}_P(Q) = \frac{d}{dt}\boldsymbol{v}_P(Q)$

and it immediately follows that

$\boxed{\boldsymbol{a}_P(Q) + \boldsymbol{a}_Q(R) = \boldsymbol{a}_P(R)}$

Example 1

A train is travelling at 40 m s^{-1} and rain is falling vertically at a speed of 10 m s^{-1}. Find the velocity of the rain relative to the train and hence the direction of its streaks on the carriage windows.

Letting T, R, G denote the train, rain, and ground respectively, and taking unit vectors $\boldsymbol{i}, \boldsymbol{j}$ horizontally and vertically,

$$v_G(T) = 40\boldsymbol{i}$$

$$\Rightarrow \quad v_T(G) = -40\boldsymbol{i}$$

and $v_G(R) = -10\boldsymbol{j}$.

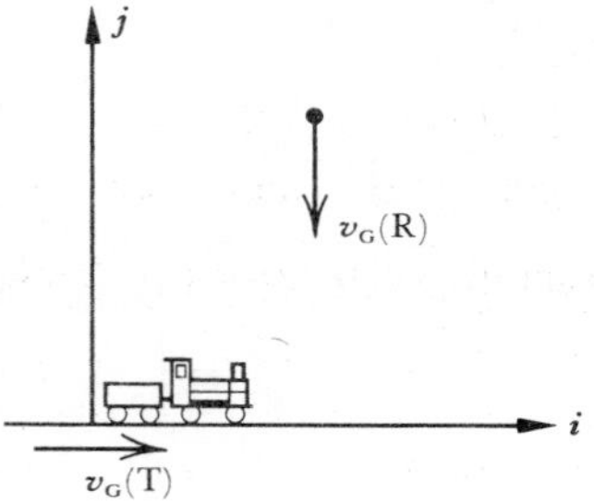

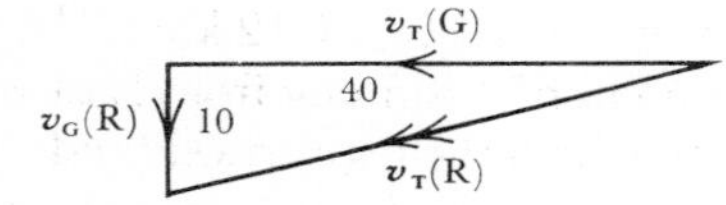

So $v_T(R) = v_T(G) + v_G(R)$ and we see that the streaks of rain make an angle θ with the horizontal,

where $\tan\theta = \frac{10}{40} = \frac{1}{4} \quad \Rightarrow \quad \theta = 14° \, 2'$.

Example 2

A ship, which steams at 18 km h^{-1}, has to travel North-East through water in which there is a current flowing from the West at 4 km h^{-1}. Find, by drawing or calculation, the direction in which the ship should be headed and the actual speed of the ship towards the North-East. (M.E.I.)

Letting S, W, L represent the ship, water, and land respectively, we know that

$v_W(S)$ has magnitude 18 km h^{-1}

$v_L(W)$ has magnitude 4 km h^{-1} due East

$v_L(S)$ has to be in the direction North-East.

But $v_W(S) = v_W(L) + v_L(S)$.

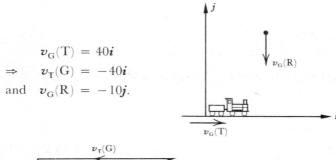

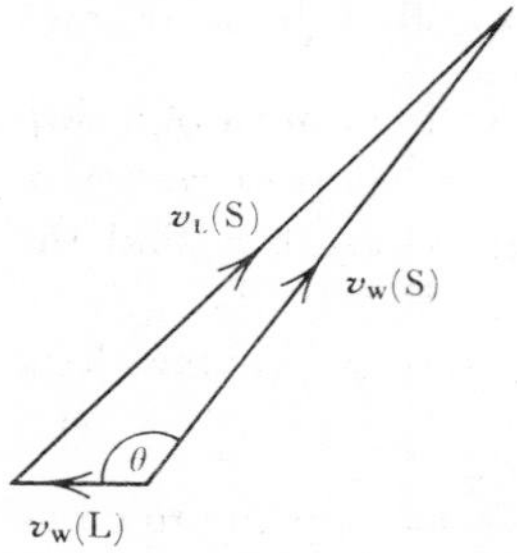

By drawing this to scale, we see that $\theta = 126°$, so that the ship should be set on course 036° and $\boldsymbol{v}_L(S) = 20.6$, which is its actual speed towards North-East.

Alternatively, if we let $\boldsymbol{v}_L(S) = \boldsymbol{v}$, we see from the cosine rule that

$$18^2 = v^2 + 4^2 - 2 \times v \times 4 \cos 45°$$

$$\Rightarrow \quad v^2 - 4\sqrt{2}v - 308 = 0$$

$$\Rightarrow \quad v = 2\sqrt{2} \pm \sqrt{(8 + 308)}$$

$$= 2\sqrt{2} \pm \sqrt{316}$$

$$= 2.83 \pm 17.77 = 20.6 \quad (\text{or } -14.94).$$

Hence the ship should be headed on course 036° and her actual speed will be 20.6 km h^{-1}.

Exercise 8.6

1 A motor-boat is steered due North and has a speed of 12 km h^{-1} through the water, in which there is a tide of 5 km h^{-1} running from East to West. Obtain by calculation the direction in which the boat travels and the time the boat will take to travel $6\frac{1}{2}$ km in this direction. (S.M.P.)

2 A canoeist, who can paddle at 2 m s^{-1} in still water, is on a river which is flowing at 1.5 m s^{-1}. He wishes to make a 'ferry glide', that is to go straight across the river on a track perpendicular to the banks and current. By calculation or scale drawing find the direction in which he must paddle.

How long will he take to cross if the river is 60 m wide? (S.M.P.)

3 On a day when there is a 20 km h^{-1} breeze blowing from the South-West, a cyclist is travelling in a direction 110° at 24 km h^{-1}. By scale drawing, find the direction from which the wind appears to the cyclist to be coming. (S.M.P.)

4 A destroyer with speed 60 km h^{-1} sights an enemy ship 10 km due East proceeding on a course 120° at 20 km h^{-1}. Find, by drawing, the course the destroyer should take to intercept the enemy. (S.M.P.)

5 A man travelling due East finds that the wind appears to blow from the North. On doubling his velocity he finds that the wind appears to blow from the North-East. Find the true direction from which the wind is blowing. (S.M.P.)

6 A ship X, whose maximum speed is 30 km h^{-1}, is due South of a ship Y which is sailing at a constant speed of 36 km h^{-1} on a fixed course of 150°. If X sails immediately to intercept Y as soon as possible, find the required course.

If initially the ships are 50 km apart, find to the nearest minute how long it takes X to reach Y. (L.)

7 When a boat travels due West at v km h^{-1}, the wind appears to come

from the North-North-West. When the boat travels North-West at the same speed, the wind appears to come from the North. Find the speed and direction of the wind.

When the boat travels due South at v km h^{-1}, from what direction does the wind appear to come? (O.C.)

8 A hawk flying slightly above level ground sees a mouse 100 m due West. The mouse is running Northwards at a speed of 5 m s^{-1}, and there is a 10 m s^{-1} wind blowing *from* the South-West. The hawk flies in a straight line and catches the mouse after 5 s. Find the magnitude and direction of its velocity relative to the air. (O.C.)

9 The equation of the path of a particle P is $\boldsymbol{r} = \boldsymbol{i}t + \boldsymbol{k}t^2$,

where t is the time. Show that the acceleration of P is constant.

The velocity of another particle Q relative to P is $(\boldsymbol{i} - \boldsymbol{j})$ and when $t = 0$, $\boldsymbol{PQ} = \boldsymbol{j}$. Find the equation of the path of Q and the time at which Q is nearest to P. (L.)

10 The flight controller at an airport is about to 'talk-down' an aircraft A whose position vector relative to the control tower is $(10\boldsymbol{i} + 20\boldsymbol{j} + 5\boldsymbol{k})$ km and whose constant velocity is $(-210\boldsymbol{i} - 50\boldsymbol{j})$ km h^{-1}.

At this instant a second aircraft B appears on his radar screen with position vector $(-20\boldsymbol{i} - 10\boldsymbol{j} + 3\boldsymbol{k})$ km and constant velocity $(150\boldsymbol{i} + 250\boldsymbol{j} + 60\boldsymbol{k})$ km h^{-1}. Find:

(*i*) the velocity of A relative to B,
(*ii*) the position vector of A relative to B t min after B first appeared on the radar screen,
(*iii*) the time that elapses, to the nearest 10 s, until the two aircraft are nearest to one another. (J.M.B.)

11 A particle P moves in a horizontal circle with fixed centre O and radius 3 m with angular velocity $\frac{1}{4}\pi$ rad s^{-1} in a clockwise sense. A second particle Q moves in the same plane in a concentric circle of radius 2 m with angular velocity $\frac{1}{2}\pi$ rad s^{-1} in a clockwise sense. Initially O, P, and Q are collinear with P and Q due North of O. Find the magnitude and direction of the velocity of P relative to Q (*i*) after 2 s and (*ii*) after 3 s.

(A graphical method may be used.) (C.)

8.7 Introduction to mechanics: Newton's first law

Nature, and Nature's laws, lay hid by night.
God said 'let Newton be', and all was light

It is a commonplace to speak of the Scientific Revolution of the seventeenth century, and in many ways this is an oversimplification. Much was known before 1600, much else remained unknown in 1700, and such movements rarely fall neatly into centuries. Nevertheless, and particularly in the science of mechanics, the century witnessed a revolution whose importance can hardly be exaggerated.

In 1600, men's understanding of motion depended on a tradition of nearly two thousand years, since the writings of Aristotle in the fourth century BC. They believed that earthly objects have a natural desire to fall to the ground, but can be diverted from this tendency if an *impetus* is imparted to them. When a stone is thrown it is given an impetus which, until exhausted, causes it to defy its natural destiny. As for the movements of the planets, they are subject to quite different laws. They are celestial objects and so move in orbits composed of systems of perfect circles, and are propelled along their heavenly spheres by supernatural forces.

In 1700, however, man's understanding of motion was utterly different. Thirteen years earlier, Isaac Newton had published his great work, *Philosophiae naturalis principia mathematica*: 'The mathematical foundations of science', or 'Theoretical mechanics', as it might now be called. This was the culmination of twenty five years' work, ranging from the discovery of the Calculus to optical investigations, the fundamental laws of motion and the law of universal gravitation. The first volume of *Principia* was devoted to the laws of motion, universal gravitation, and the orbits of the planets. In the second volume Newton discussed the motion of a body in a resisting medium, the motion of fluids and of waves. His third volume began:

> In the preceding books I have laid down the principles of philosophy; principles not philosophical but mathematical. But lest they should appear dry and barren, I have illustrated them here and there, giving an account of such things as the density and the resistance of bodies and the motion of light and sounds. It remains that, from the same principles, I now demonstrate the frame of the system of the World.

Newton then set out to investigate the movement of satellites round their planets, to calculate the mass of the sun and of the planets which have satellites, to explain and calculate the flattened shape of the Earth and the wobble of its axis, to discuss the irregularities of the Moon's motion due to the attraction of the Sun, to inaugurate the study of tides, to explain the movement of comets and to calculate their return. The inscription beneath his statue in Cambridge, 'Genus humanum ingenio superavit' was no exaggeration.

What then were Newton's laws of motion, why were they so revolutionary, and how were they discovered? The laws themselves, like the rest of *Principia*, were in Latin, and can be translated as:

1 Every object remains in a state of rest or of uniform motion in a straight line, unless a force is acting on it.

2 When an object is accelerating there is a force acting upon it which is proportional to the acceleration and in the same direction.

3 If two objects act upon each other, the forces between them are equal and opposite.

Before considering the significance of these laws, a preliminary remark should be made about the word 'object'. Newton meant by this a quantity of matter which can be regarded as concentrated at a point, and which we shall usually call a *particle*. A particle, therefore, is any object whose dimensions are small compared with all other distances under consideration; in one instance this might be a bullet moving through the air or, in another, the Earth relative to the rest of the solar system.

What, then, was so revolutionary about these laws, and why are they of such immense importance?

Perhaps the two most striking features occur in the first law:

'Every object'. The very first two words marked the end of the ancient distinction between the mechanics of earthly objects and the mechanics of heavenly objects: the motion of every object, terrestrial and celestial, was to be subject to the same laws.

'In a state of rest or of uniform motion in a straight line.' No longer, when confronted with an object moving with uniform velocity, were men to look for a force or 'impetus' sustaining its motion. Forces are to be sought not when there is motion, but only when there is *change* of motion.

This was the major revolution, and its principal agents were the concepts of force and acceleration. So far as the latter is concerned, it is assumed that we always know the acceleration of an object. But a little reflection shows that this is so only relative to a particular frame of reference. Are we to measure the acceleration of a car relative to the road, to axes fixed within the spinning Earth or fixed within the solar system? This was one of the questions which ultimately, for the consideration of very high velocities, led to the theory of relativity; but mercifully it will not be necessary for us to be detained by such doubts, providing we decide upon a particular frame of reference (usually a set of axes fixed in the Earth) and then use it consistently.

The other concept, force, has not been defined, though we all have intuitive ideas of push and pull, thrust and tension and pressure. Newton's second law can be regarded as a refinement of these ideas, relating them to the acceleration of the object on which they are acting; and Newton's third law states that such forces always occur in pairs, as inter-actions between two objects.

Applying Newton's second law to the case of a falling stone, we see that its acceleration downwards must necessitate a downward force acting upon it, which we call its *weight*. Now it will also be recalled that when a particle is moving with constant speed round a circle, its acceleration is directed towards the centre of the circle. So if we think of the Moon revolving round the Earth, we should expect there to be a force drawing it towards the Earth. It was Newton's insight to recognise that this could be regarded as the same kind of force as the weight of an object on the Earth. So the

Earth's pull on both the apple and the Moon was called *gravitation*, which he later understood as a force of attraction between every pair of particles in the universe.

This is a highly condensed account of Newton's revolutionary view of force and motion. How was it achieved?

Firstly, it must be regarded as the climax of nearly a century and a half of patient observation, calculation, and reflection. In 1543 was published, while its author was dying, Copernicus' book *De revolutionibus orbium caelestium*, 'On the revolutions of the heavenly spheres', explaining the simplifications that can be achieved if the Sun, rather than the Earth, is regarded as the centre round which the planets turn. Sixty years later, Kepler was to devote himself to years of patient calculation based on observations of the Danish astronomer Tycho Brahé. In 1618 Kepler published *Harmonice mundi*, 'On the harmony of the world', in which he stated his three laws of planetary motion:

1 Each planet moves in an ellipse with one focus at the Sun.
2 The line joining a planet to the Sun sweeps out equal areas in equal times.
3 If the mean distance of a planet from the Sun is r and its period (or time of revolution) is T, then T^2 is proportional to r^3: $T^2 = kr^3$.

Meanwhile, in Italy, Galileo was combining astronomical discoveries with investigation of the accelerations, as well as the velocities, of moving objects; and only a few months after he died, Newton was born. The final words are his, written when he was an old man about the work he accomplished by the age of twenty three:

> In the same year I began to think of gravity extending to the orb of the moon, and from Kepler's third law I deduced that the forces which keep the planets in their orbs must be reciprocally as the squares of their distances from the centres about which they revolve; and thereby compared the force requisite to keep the moon in her orb with the force of gravity at the surface of the earth, and found them answer pretty nearly. All this was in the two plague years of 1665 and 1666, for in those days I was in the prime of my age for invention, and minded mathematics and philosophy more than at any time since.

8.8 Force, mass, and weight: Newton's second law

The force acting on a particle is proportional to its acceleration, and in the same direction.

Newton's second law implies that force, like acceleration, must be a vector quantity. So we call these $\boldsymbol{F}$ and $\boldsymbol{a}$ respectively, and write

$$\boldsymbol{F} = m\boldsymbol{a}$$

where the constant of proportionality m is a scalar quantity, called the *mass* of the particle. This equation is also commonly called *the equation of motion*.

The SI unit of mass is the kilogram (kg), being the mass of a standard cylinder kept in Paris; and the unit of force is the newton (N), the force required to give a mass of 1 kg an acceleration of 1 m s^{-2}. (So 1 N = 1 kg m s^{-2}.)

Exercise 8.8a

1 What force must act on a mass of 3 kg to give it an acceleration of:
(*i*) 2 m s^{-2}; (*ii*) 40 m s^{-2}; (*iii*) 0.05 m s^{-2}?

2 What would be the acceleration of a mass of 40 kg when acted upon by a force of:
(*i*) 120 N; (*ii*) 24 N; (*iii*) 0.1 N?

3 What would be the mass of an object that is given an acceleration 2 m s^{-2} by a force of:
(*i*) 40 N; (*ii*) 5×10^3 N; (*iii*) 0.03 N?

4 A 3 kg mass is moving at 4 m s^{-1} round a horizontal circle of radius 2 m. Find:
(*i*) its acceleration;
(*ii*) the force which must be exerted upon it to produce this acceleration.

5 Repeat no. **4** if:
(*i*) its mass were doubled;
(*ii*) its speed were doubled;
(*iii*) its radius were doubled.

6 An electron of mass 9×10^{-31} kg is moving at 4×10^6 m s^{-1} in a field which produces a force of 2×10^{-15} N in the same direction. Find its velocity after an interval of 2×10^{-9} s.

7 A 4 kg mass has the vector position $\boldsymbol{r} = 10t\boldsymbol{i} - 5t^2\boldsymbol{j}$.

Find its position, velocity, and acceleration, and also the force acting on it when:
(*i*) $t = 0$; (*ii*) $t = 2$; (*iii*) $t = 10$.

8 As a car of mass 1 000 kg travels along a level road its motion is opposed by a constant resistance of 100 N. The values of the pull of the engine for given values of the time t s are as follows:

Time t (s)	0	1	2	3	4	5	6
Pull (N)	1 100	1 200	1 350	1 550	1 850	2 300	3 100

Find the acceleration of the car at each of the given values of t. When $t = 0$ the speed of the car is 50 km h^{-1}. Find graphically the speed of the car in km h^{-1} when $t = 6$.

State how you would find graphically the distance travelled by the car in the period from $t = 0$ to $t = 6$. (A.E.B.)

Resultants

As forces are vector quantities, it follows immediately that they are combined by the vector (or parallelogram) law of addition, and their vector sum is usually called their *resultant*.

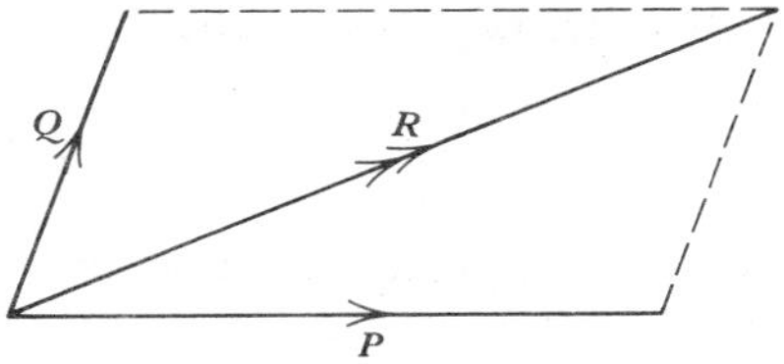

With a number of concurrent forces, this addition is most convenient by means of a *polygon of forces*:

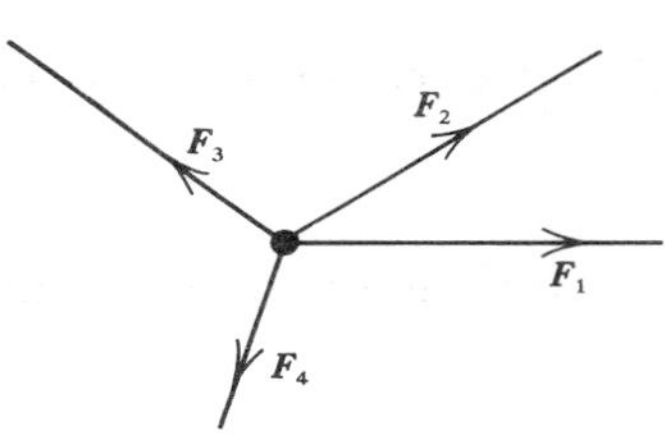

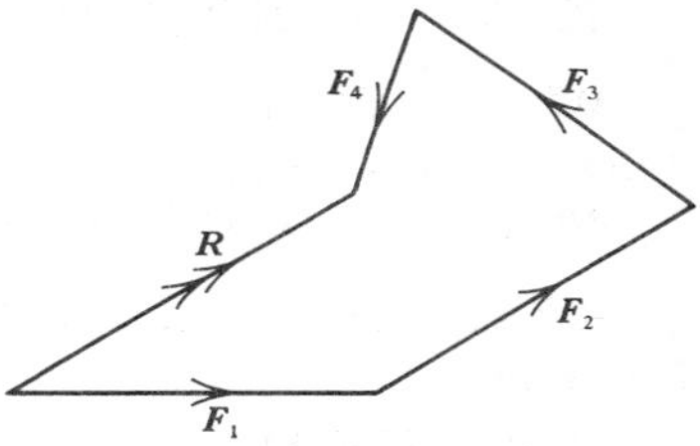

Equilibrium

If a particle is at rest or moving with constant velocity (i.e., it has no acceleration), it is said to be *in equilibrium*. This is clearly so when the resultant force acting upon it is zero and the polygon of forces is *closed*.

In the particular case of two forces in equilibrium, they must clearly be equal and opposite:

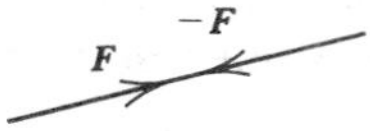

Furthermore, three forces acting on a particle are in equilibrium provided that they can be represented in magnitude and direction by a closed triangle, called a *triangle of forces*:

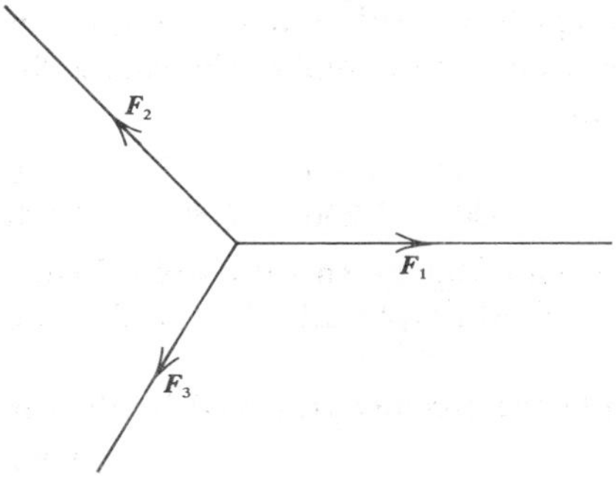

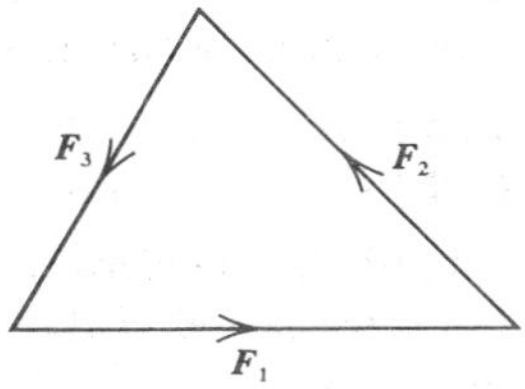

Example 2

A mass of 10 kg is acted upon by two forces, 20 N and 10 N, which are inclined to each other at 60°. Find:

(*i*) the magnitude of the resultant force;
(*ii*) the resulting acceleration.

As the two forces are vectors, they must be combined by the vector law of addition.

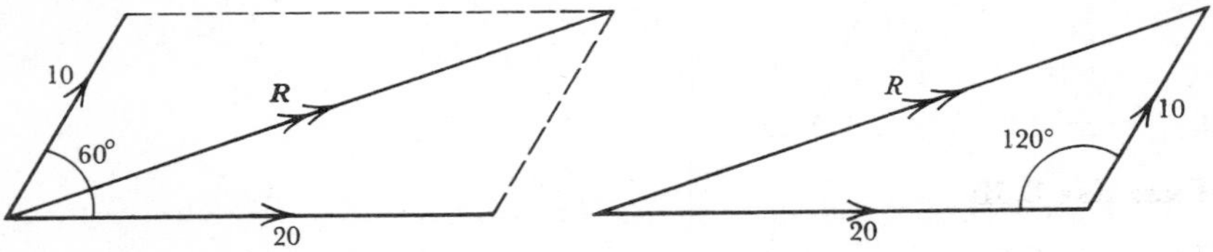

The resultant force $\boldsymbol{R}$ can be found by any of three methods:

(*a*) scale drawing;
(*b*) use of the cosine rule:

$$\begin{aligned} R^2 &= 20^2 + 10^2 - 2 \times 20 \times 10 \times \cos 120° \\ &= 400 + 100 - 2 \times 20 \times 10 \times -\tfrac{1}{2} \\ &= 500 + 200 = 700 \end{aligned}$$

$\Rightarrow \quad R = 26.5;$

(*c*) use of components:

The two forces can be represented as $20\boldsymbol{i}$
and $10 \cos 60\boldsymbol{i} + 10 \sin 60\boldsymbol{j} = 5\boldsymbol{i} + 5\sqrt{3}\boldsymbol{j}$

Hence $\boldsymbol{R} = 20\boldsymbol{i} + [5\boldsymbol{i} + 5\sqrt{3}\boldsymbol{j}]$

$\quad = 25\boldsymbol{i} + 5\sqrt{3}\boldsymbol{j}$

$\Rightarrow \quad R = \sqrt{(625 + 75)} = \sqrt{700} = 26.5.$

Hence the resultant force is 26.5 N and so the resultant acceleration has magnitude a, where

$26.5 = 10\,a$

$\Rightarrow \quad a = 2.65 \text{ m s}^{-2}.$

Example 3 (Lami's theorem)

If three concurrent forces $\boldsymbol{P}$, $\boldsymbol{Q}$, $\boldsymbol{R}$ are in equilibrium, show that their magnitudes and the angles α, β, γ between them are such that

$$\frac{P}{\sin \alpha} = \frac{Q}{\sin \beta} = \frac{R}{\sin \gamma}.$$

Since the forces are in equilibrium, they can be represented in magnitude and direction by the sides of a triangle as shown. Hence, using the sine rule:

$$\frac{P}{\sin(\pi - \alpha)} = \frac{Q}{\sin(\pi - \beta)} = \frac{R}{\sin(\pi - \gamma)}$$

$$\Rightarrow \quad \frac{P}{\sin \alpha} = \frac{Q}{\sin \beta} = \frac{R}{\sin \gamma}.$$

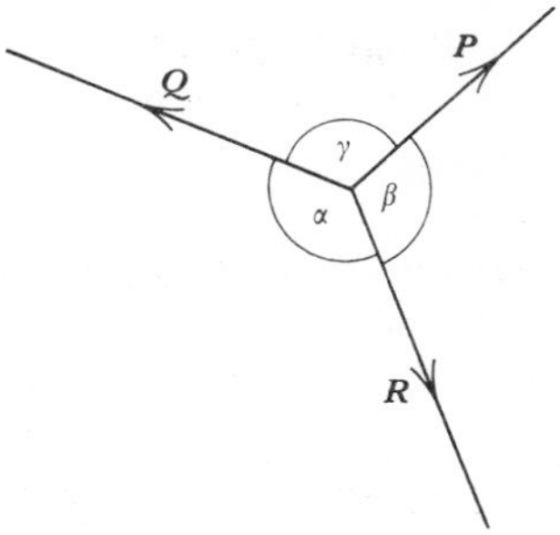

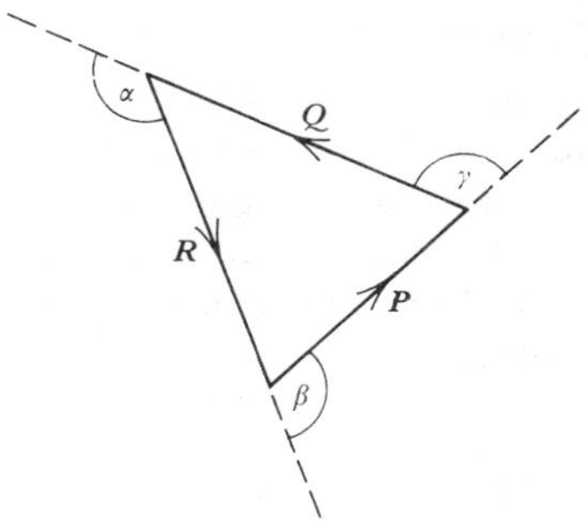

Exercise 8.8b

In nos. 1–6 find:
(*i*) the resultant force;
(*ii*) the resulting acceleration of the particle;
(*iii*) the force necessary to maintain the particle in equilibrium.
(SI units are used throughout.)

1 Forces $8\boldsymbol{i} - 2\boldsymbol{j}$, $3\boldsymbol{i} + 3\boldsymbol{j}$, $-\boldsymbol{i} + 4\boldsymbol{j}$, acting on a particle of mass 5 kg;

2 Forces $3\boldsymbol{i} + 2\boldsymbol{j} + \boldsymbol{k}$, $2\boldsymbol{i} + 3\boldsymbol{j} - \boldsymbol{k}$, $\boldsymbol{i} + \boldsymbol{j} + 3\boldsymbol{k}$, acting on a particle of mass 2 kg;

3 Forces 20 N due East and 30 N due South, acting on a particle of mass 10 kg;

4 Forces 10 N due West and 5 N in direction South-West, acting on a particle of mass 4 kg;

5 Forces 10 N due West and 5 N in direction North-East, acting on a particle of mass 20 kg;

6 Forces 10 N on bearing 080°
5 N on bearing 165°
12 N on bearing 284°
acting on a particle of mass 0.2 kg.

7 A mass of 3 kg is acted on by a constant force of $\begin{pmatrix} 6 \\ 3 \end{pmatrix}$ newtons. Find the acceleration.

At time $t = 0$ it is travelling with velocity $\begin{pmatrix} -3 \\ 2 \end{pmatrix}$ m s^{-1}. Find the displacement from $t = 0$ to $t = 4$ (s), and the velocity at the end of that time. (S.M.P.)

8 An electron of mass 9×10^{-31} kg is moving at 8×10^{6} m s^{-1} when it enters a field, which produces a force of 1.8×10^{-15} N at right angles to its initial direction of motion for a period of 3×10^{-9} s. Find its final velocity. (M.E.I.)

Acceleration due to gravity: weight

At the Earth's surface, a freely falling body accelerates at roughly 9.8 m s^{-2}. So, using Newton's second law, a mass of 1 kg must be acted upon by a force of approximately 9.8 N vertically downwards. This is called the *weight* of the body, and its magnitude is sometimes called 1 kg wt.

The acceleration due to gravity is usually denoted by $\mathbf{g}$ (which has magnitude g), and its magnitude varies from place to place. So the weight of a mass of 1 kg will also vary in magnitude:

	Acceleration due to gravity (m s^{-2})	*Weight of 1 kg* (N)
At Equator	9.781	9.781
London	9.812	9.812
North Pole	9.832	9.832
On Moon	1.6	1.6

On the Earth, therefore, 1 N is the weight of a mass of roughly 0.1 kg; or, not inappropriately, of a medium-sized apple.

Example 4

A mass of 10 kg is supported in equilibrium by two strings which are inclined to the vertical at 30° and 20°. Find their tensions.

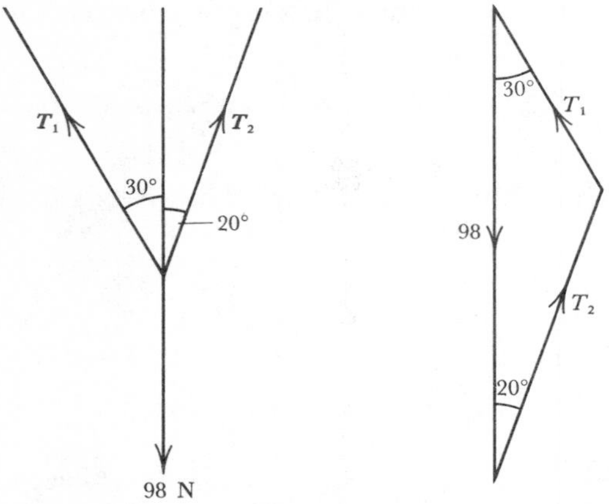

Taking $g = 9.8$ m s^{-2}, the weight of the 10 kg mass is 98 N. Hence the tensions T_1, T_2 are such that

$$\frac{T_1}{\sin 20^\circ} = \frac{T_2}{\sin 30^\circ} = \frac{98}{\sin 130^\circ}$$

$$\Rightarrow \qquad T_1 = \frac{98 \sin 20^\circ}{\sin 50^\circ} \approx 44 \text{ N}$$

$$\text{and} \qquad T_2 = \frac{98 \sin 30^\circ}{\sin 50^\circ} \approx 64 \text{ N}.$$

Example 5

A man of mass 100 kg is standing in a lift. What is the force of the floor on the man when the lift has:

(*i*) acceleration 2 m s^{-2} upwards;
(*ii*) constant velocity;
(*iii*) acceleration 2 m s^{-2} downwards?
(Take $g = 9.81$ m s^{-2}.)

In each case there are two external forces acting on the man:
(*a*) his weight 981 N vertically down;
(*b*) the force of the floor R N vertically up.

Applying Newton's second law in all three cases:

(*i*) Acceleration 2 m s^{-2} upwards:

$$R - 981 = 100 \times 2$$
$$\Rightarrow \quad R = 1181.$$

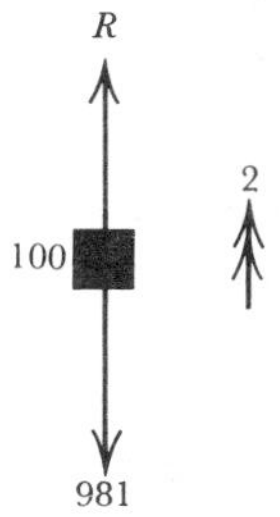

(*ii*) Constant velocity:

$$R - 981 = 100 \times 0$$
$$\Rightarrow \quad R = 981.$$

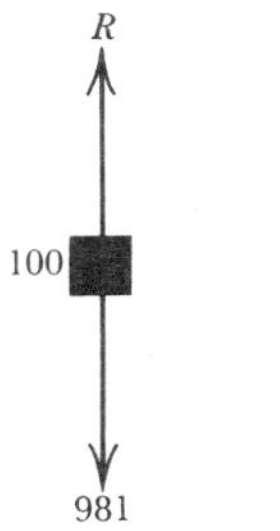

(*iii*) Acceleration 2 m s^{-2} downwards:

$$R - 981 = 100 \times -2 = -200$$
$$\Rightarrow \quad R = 781.$$

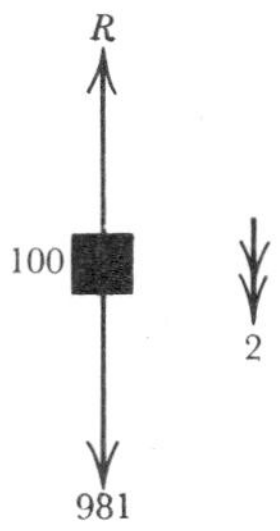

Hence the force of the floor is:

1 181 N whilst the lift is accelerating upwards;
981 N whilst it is rising at constant velocity;
and 781 N whilst it is decelerating.

Example 6

A man jumps from an aeroplane and falls freely under gravity. Until he opens his parachute the air resistance per unit mass is $-\frac{1}{5}\boldsymbol{v}$ (measured in newton), where $\boldsymbol{v}$ is his velocity (in m s^{-1}). What is his terminal velocity (when he ceases to accelerate)?

Suppose that his mass is m (kg) and that at a certain point on his path his acceleration (indicated by a double arrow) is $\boldsymbol{a}$.

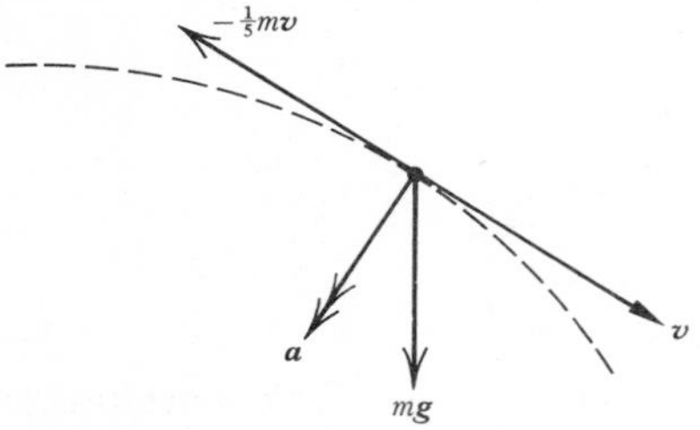

The two forces acting on the man are (*i*) his weight $m\boldsymbol{g}$ vertically down; and (*ii*) the resistance $-\frac{1}{5}m\boldsymbol{v}$.

By Newton's second law,

$$m\boldsymbol{g} - \tfrac{1}{5}m\boldsymbol{v} = m\boldsymbol{a}$$

$$\Rightarrow \qquad \boldsymbol{a} = \boldsymbol{g} - \tfrac{1}{5}\boldsymbol{v}$$

So as $\boldsymbol{a} \to \mathbf{0}$, $\quad \boldsymbol{g} - \frac{1}{5}\boldsymbol{v} \to \mathbf{0}$

$$\Rightarrow \qquad \boldsymbol{v} \to 5\boldsymbol{g}.$$

Hence, taking $g \approx 10$, we see that the terminal velocity is 50 m s^{-1} vertically downwards.

Example 7 (The 'conical pendulum')

A mass m is attached to a fixed point by means of an inextensible string and rotates in a horizontal circle with constant speed v. If the string makes an angle θ with the vertical, show that $v^2 = gr \tan \theta$.

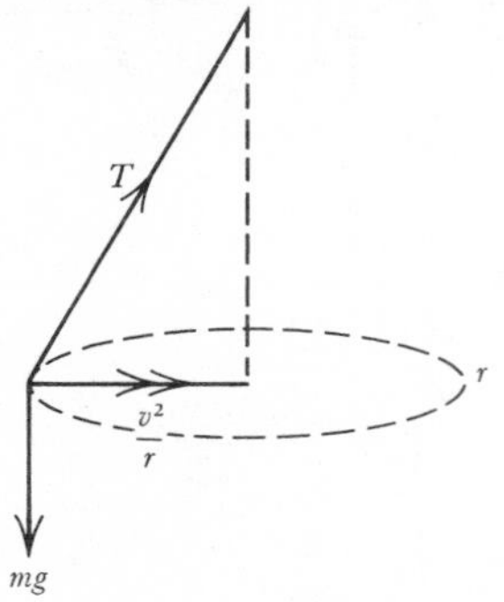

Let the tension in the string have magnitude T and suppose that the mass moves in a circle of radius r. Then the acceleration of m is directed towards the centre and has magnitude v^2/r.

Applying Newton's second law, and resolving horizontally and vertically:

$$T \sin \theta = m \frac{v^2}{r} \tag{1}$$

and $\quad -mg + T \cos \theta = 0$

$$\Rightarrow \qquad T \cos \theta = mg. \tag{2}$$

From (1) and (2),

$$\tan \theta = \frac{v^2}{gr} \quad \Rightarrow \quad v^2 = gr \tan \theta.$$

Example 8

Find the period of revolution of an artificial satellite which is orbiting the Earth just outside its atmosphere. (Assume the orbit to be approximately circular, the acceleration due to gravity as approximately 10 m s^{-2} and the radius of the Earth as 6 400 km.)

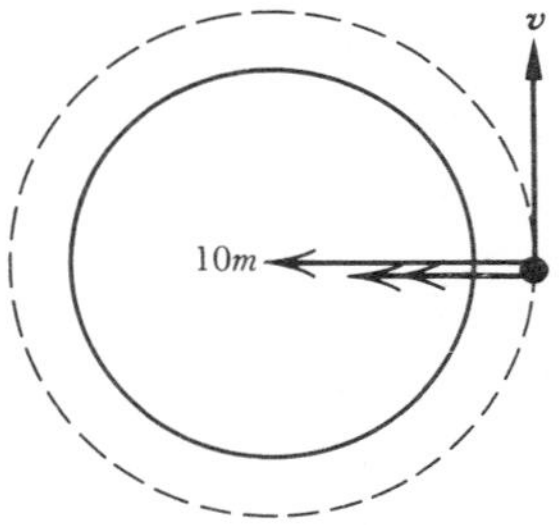

Let the mass of the satellite be m kg, so that its weight will be approximately $10m$ N.

Furthermore, suppose that its speed is v m s^{-1}. Now its distance from the centre of the Earth is approximately 6.4×10^6 m, so its acceleration, which is directed towards the Earth's centre, has magnitude $v^2/(6.4 \times 10^6)$.

Hence, using Newton's second law,

$$10m = m \frac{v^2}{6.4 \times 10^6}$$

$$\Rightarrow \quad v^2 = 64 \times 10^6$$

$$\Rightarrow \quad v = 8 \times 10^3$$

Hence $\quad$ the satellite's speed $\approx$ 8 km s^{-1}

But $\quad$ the circumference of its orbit $\approx 2\pi \times 6\,400$ km

$\approx 40\,000$ km

$$\Rightarrow \qquad \text{Period of orbit} \approx \frac{40\,000}{8} \approx 5\,000 \text{ s} \approx 1.4 \text{ h}.$$

Exercise 8.8*c*

(Take $g = 9.8 \text{ m s}^{-2}$ unless otherwise stated.)

1 A miner's cage has mass 1 tonne. What is the tension in its cable if the cage is:
(*i*) accelerating upwards at 2 m s^{-2};
(*ii*) accelerating downwards at 2 m s^{-2};
(*iii*) going up at constant velocity;
(*iv*) going down at constant velocity.

2 If the maximum allowable tension of the cable in no. **1** is 1.2×10^4 N, find the acceleration permitted
(*i*) when the cage is empty;
(*ii*) when it is carrying 5 boys of average mass 40 kg.

3 A ball of mass 2 kg is falling vertically through the air with an acceleration of only 8 m s^{-2}. Find the air resistance.

When the ball is moving faster, this resistance increases to 10 N. What is then the acceleration of the ball?

4 A mass of 10 kg is supported by two strings inclined to the vertical at 40° and 50°. Find their tensions.

5 A mass of 20 kg is suspended by two wires of lengths 3 m and 4 m attached to two hooks at the same level and a distance 5 m apart. Find the tensions in the wires.

6 $\boldsymbol{i}$, $\boldsymbol{j}$, and $\boldsymbol{k}$ are three mutually perpendicular unit vectors; $\boldsymbol{i}$ and $\boldsymbol{j}$ are horizontal and $\boldsymbol{k}$ is vertically upwards. A particle of weight 6 N is in equilibrium, supported by three strings. The tension in one of the strings, in newtons, is represented by $-\boldsymbol{i} - 2\boldsymbol{j} + 3\boldsymbol{k}$, and the other strings are each inclined at an angle θ to $\boldsymbol{k}$, one in the $\boldsymbol{i}$–$\boldsymbol{k}$ plane and the other in the $\boldsymbol{j}$–$\boldsymbol{k}$ plane. Show that $\theta = 45°$, and find the magnitudes of all three tensions. (C.)

7 Show how the acceleration of a train can be estimated by means of attaching a string with a heavy mass on its end to the ceiling of a coach.
(*i*) What is the acceleration when the string is inclined at a constant angle of 10° to the vertical?
(*ii*) At what constant angle would the string be inclined if the acceleration is 1 m s^{-2}?

8 A man of mass 100 kg is hanging by a rope attached to a winch in a helicopter which is hovering. The helicopter begins to ascend vertically with an acceleration of 0.75 m s^{-2}. Determine the tension in the rope in newtons.

If the rope were to be wound in at a constant rate of 1 m s^{-1}, what change would there be in the value of the tension in the rope? Comment briefly.

After a short time the helicopter ceases to rise vertically but subsequently moves horizontally with an acceleration of 0.75 m s^{-2}. It is observed that with the winch not operating, the rope hangs at a constant angle θ to the vertical where $\tan\theta = \frac{3}{4}$. What is the tension in the rope now, and what is the horizontal force on the man due to the resistance of the air? (M.E.I.)

9 An aircraft of mass 2×10^4 kg is taking off at 20° to the horizontal with an acceleration of 1.5 m s^{-2}. Find the forces
(*i*) perpendicular to the plane of its wings (the *lift*);
(*ii*) along its direction of motion.

10 A man is swinging a metal ball of mass 0.1 kg in a vertical circle by means of a light wire of length $\frac{1}{2}$ m. Find the tension in the wire if the ball
(*i*) is at the lowest point and has speed 4 m s^{-1};
(*ii*) is at the highest point and has speed 3 m s^{-1}.

11 With what period would the Earth need to rotate in order that gravity would appear to vanish at the equator? (Take g as 10 m s^{-2} and the circumference of the Earth as 4×10^7 m.)

Under these conditions what would be the angle at latitude 60° between a string suspending a weight in relative equilibrium and the 'true' vertical?

Show that all apparent verticals would be in the same direction in space. (M.E.I.)

12 In a 'golf driving practice kit' a ball of mass 0.05 kg is attached by a length of light elastic to the tee, so that when the ball is r metre from the tee the tension in the elastic is $0.2r$ newton. The ball is struck with velocity $\boldsymbol{u} \text{ m s}^{-1}$ at time $t = 0$. If $\boldsymbol{r}$ metre denotes the position vector at a subsequent time t second before it strikes the ground again, and if $\boldsymbol{j}$ denotes a unit vector vertically upwards and the acceleration of gravity is taken as 10 m s^{-2}, obtain the differential equation

$$\ddot{\boldsymbol{r}} + 4\boldsymbol{r} + 10\boldsymbol{j} = \boldsymbol{0}.$$

Verify that all the conditions of the problem are satisfied by the solution

$$\boldsymbol{r} = \tfrac{1}{2}\boldsymbol{u}\sin 2t + \tfrac{5}{2}(\cos 2t - 1)\boldsymbol{j}.$$ (S.M.P.)

8.9 Reactions: Newton's third law

Some of the objects with which we have so far been concerned, such as the planets, have been free to move anywhere in space. But it frequently happens that we wish to study an object which is under some form of constraint, perhaps merely by having to remain on a particular surface. It is clear that its behaviour would usually be very different if the surface were removed, so the surface must be exerting a force on the object, which we call a *reaction*; and, by Newton's third law, the object must be exerting an equal and opposite force on the surface.

When this force is perpendicular, or normal, to the surface, it is called a *normal reaction* and the surface is said to be *smooth*. Otherwise—to a greater or lesser extent—it is *rough*, and in our next section we shall investigate the consequent frictional forces.

Example 1

A toboggan of mass 50 kg is hauled along horizontal smooth ice by a force of 100 N at 20° to the horizontal. Taking $g \approx 9.8 \text{ m s}^{-2}$, find the acceleration of the toboggan and the force between ice and toboggan.

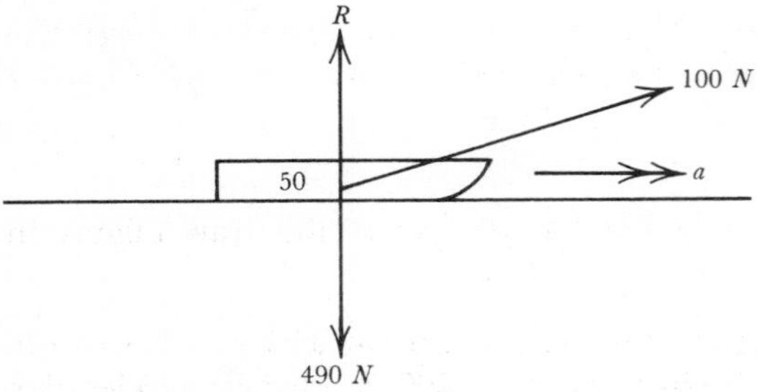

There are three forces acting on the toboggan

(*a*) its weight 490 N;

(*b*) the normal reaction R N;

(*c*) the tractive force 100 N.

Their resultant has horizontal and vertical components

$$(100 \cos 20°, \quad R + 100 \sin 20° - 490)$$
$$\equiv (94, \quad R - 456)$$

and the acceleration has components $(a, 0)$. So, by Newton's second law,

$$94 = 50a \qquad R - 456 = 50 \times 0$$
$$\Rightarrow \quad a = \tfrac{94}{50} = 1.88 \qquad R = 456.$$

Hence the acceleration is 1.88 m s^{-2} and the normal reaction is 456 N.

Example 2

If a particle is sliding down a smooth plane which is inclined at α to the horizontal, find

(*i*) its acceleration;

(*ii*) the reaction of the plane on the particle.

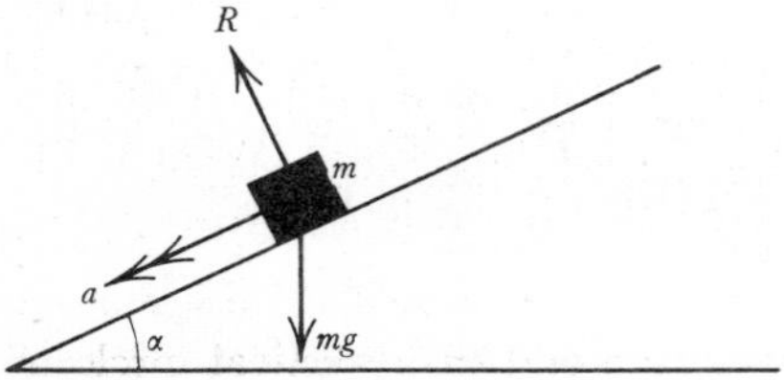

The two forces acting on a particle of mass m are:

(*a*) its weight mg vertically down;

and (*b*) a reaction R perpendicular to the plane.

Letting the acceleration of the particle be a down the plane, we see from Newton's second law

down the plane: $mg \sin \alpha = ma$

perpendicular to the plane: $R - mg \cos \alpha = 0.$

Hence $a = g \sin \alpha$

and $R = mg \cos \alpha.$

Exercise 8.9a

In each question state all the forces and accelerations, and draw a figure on which they are clearly indicated.

1 A car of mass 1 000 kg is acted upon by a tractive force (i.e., from the road and tangential to the driving wheels) of 500 N. Find its acceleration on a horizontal stretch of road

(*i*) if road and air resistances are negligible;

(*ii*) if there is a total resistance of 200 N.

2 Repeat no. 1 if the car is going up an incline of 1 in 200 (i.e., at θ to the horizontal, where $\sin \theta = \frac{1}{200}$).

3 Repeat no. 1 if the car is going down an incline of 1 in 200.

4 A mass of 50 kg on a smooth surface is acted upon by a force of 20 N. Find its acceleration and the reaction of the surface when

(*i*) both the surface and the force are horizontal;

(*ii*) the surface is horizontal, but the force is inclined at 20° to the horizontal;

(*iii*) the force is horizontal, but the surface is inclined at 20° to the horizontal;

(*iv*) both the force and the surface are inclined at 20° to the horizontal.

5 A water-skier can only perform effectively if the thrust of the water acts along the line of her body, assumed straight. If her mass is 50 kg, find the tension in the horizontal tow-rope if she leans back at 10° to the vertical in steady motion.

What will be her acceleration (assumed horizontal) if the tension is doubled, but she leans back at 15°? (S.M.P.)

6 A car of mass 800 kg is travelling along a horizontal road, curved in a circle of radius 200 m and banked at 5° to the horizontal. What is the speed of the car if it has no tendency to slip either inwards or outwards (i.e., with the road as smooth as can be)?

7 A man swings a small bucket of water round in a vertical circle of

radius 1 m. If the bucket travels at constant speed without spilling, show that the least speed possible is $\sqrt{g}$ m s^{-1}. (M.E.I.)

8 At what angle should a race track corner be banked if its radius is 200 m and vehicles are to have no tendency to side-slip when travelling at 50 m s^{-1}?

9 An aircraft travelling at 400 m s^{-1} banks at 10° to turn in a horizontal circle. Assuming that the force of the air on the plane is perpendicular to its wings, find the radius of turn if there is no side-slip.

10 A particle is moving with constant speed v in a horizontal circle on the inside of a smooth hemispherical bowl of radius a. If the line from the particle to the centre of the sphere is inclined to the vertical at an angle θ, show that

$$v^2 = \frac{ag \sin^2 \theta}{\cos \theta}.$$

11 A charged particle, mass 2×10^{-3} kg, initially at the origin, is attracted to charges at (4, 0) and (0, 3) by forces of constant magnitude 6×10^{-3} and 3×10^{-3} N respectively, and there are no other forces acting on it.
(*i*) Calculate its initial acceleration, giving the magnitude and direction. Will the particle travel in a straight line?
(*ii*) If it were constrained by a smooth tube so that it could only move freely along the line $y = 2x$, what would its initial acceleration then be?
When the particle is at the point (2, 4) show that the attractive force on it is $\begin{pmatrix} 0 \\ -3\sqrt{5} \end{pmatrix} \times 10^{-3}$ N and hence find the force it exerts on the side of the tube. (S.M.P.)

Connected particles

Newton's third law states that a body cannot exert a force upon another without itself experiencing an equal and opposite force. If a railway engine is pulling a coach forwards, then the coach is inevitably pulling the engine backwards with an equal and opposite force; and just as the Earth is exerting a gravitational attraction on the Moon, so the Moon is exerting an equal and opposite attraction on the Earth. Such pairs are usually called *action and reaction.*

Example 3

An engine of mass 10 tonne is hauling wagons of total mass 50 tonne along a horizontal track, and together they are accelerating at $\frac{1}{10}$ m s^{-2}. Ignoring all resistances, find the tractive force of the engine and the tension in the engine coupling.

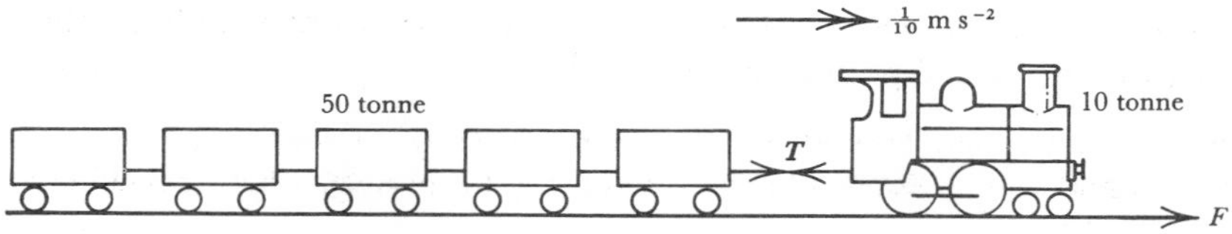

The horizontal forces with which we are concerned are:

(*i*) the tractive force F, which is the horizontal reaction of the rails on the driving wheels (the wheels themselves, thanks to friction, exert a backwards force on the rails);

(*ii*) the tension T in the coupling which connects the engine to the rest of the train, so that T is a forwards force on the wagons but a backwards force on the engine.

By Newton's second law,

For the wagons: $T = 50\,000 \times \frac{1}{10} = 5\,000$

For the engine: $F - T = 10\,000 \times \frac{1}{10} = 1\,000$

$\Rightarrow$ $F = 6\,000.$

Hence the tractive force is 6 000 N and the tension in the coupling is 5 000 N.

If we had merely required the tractive force, this could have been obtained by considering the motion of the train as a whole:

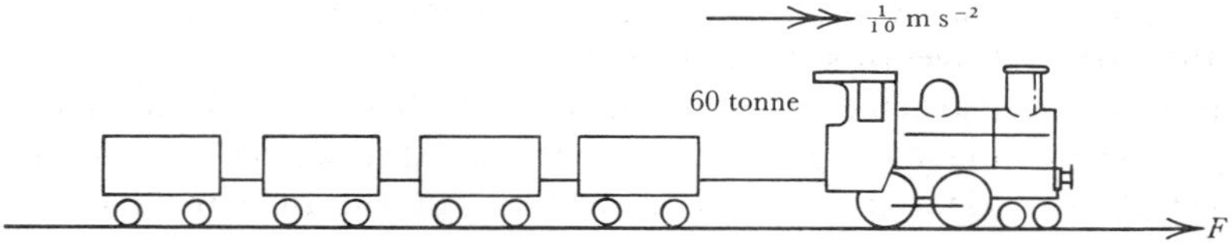

$\mathrm{F} = 60\,000 \times \frac{1}{10} = 6\,000$ N.

Example 4

A mass m lies on a plane inclined at α to the horizontal and is connected to a mass M which hangs freely by means of a light inextensible string passing over a pulley at the top of the plane. Supposing that frictional forces are negligible, find the resulting accelerations and the tension in the string.

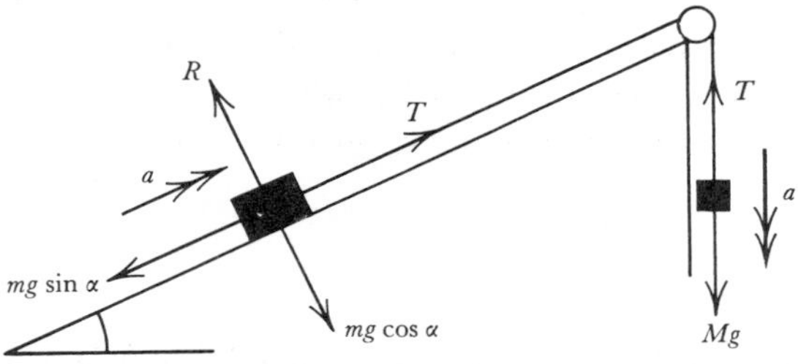

Since the pulley is smooth and the string is light, its tension T is constant along its length; and since the string is inextensible, the magnitudes of the accelerations of the two masses are equal.

Now the forces acting are as follows:

On m: (*a*) its weight mg, which can be resolved into components $mg \sin \alpha$ and $mg \cos \alpha$, down and perpendicular to the plane;
(*b*) a normal reaction R;
(*c*) the tension T of the string.

On M: (*a*) its weight Mg;
(*b*) the tension T of the string.

Furthermore, we suppose that the two masses have accelerations of magnitude a. Then, applying Newton's second law,

For m: Up plane $T - mg \sin \alpha = ma$ (1)
Perpendicular to plane $R - mg \cos \alpha = 0$ (2)
For M: $Mg - T = Ma.$ (3)

From (1) and (3), $(M - m \sin \alpha)g = (M + m)a$

$$\Rightarrow \qquad a = \frac{M - m \sin \alpha}{M + m} g$$

and $T = ma + mg \sin \alpha$

$$= \frac{m(M - m \sin \alpha)g}{M + m} + mg \sin \alpha$$

$$\Rightarrow \qquad T = \frac{Mm(1 + \sin \alpha)g}{M + m}.$$

Exercise 8.9b

In each question state all the forces and accelerations which act on every body and draw a figure in which these are clearly indicated.

1 A car of mass 1 tonne is pulling a trailer of mass $\frac{1}{4}$ tonne along a horizontal road. Find their acceleration and the tension in the tow-bar in the following cases:

(*i*) when the tractive force on the car is 2 000 N and the road and air resistances are negligible;

(*ii*) when the tractive force on the car is 2 000 N but it suffers a total resistance of 500 N, and the trailer's resistance is negligible;

(*iii*) when the tractive force on the car is 2 000 N but it suffers a total resistance of 500 N and the trailer suffers a resistance of 200 N.

In nos. **2** to **7** all surfaces and pulleys are supposed to be smooth and the masses indicated are in kilograms. In each case specify all the forces which are acting on the separate bodies, draw a figure to indicate these and the accelerations and proceed to calculate the tensions and accelerations. (Take $g \approx 10 \text{ m s}^{-2}$.)

2

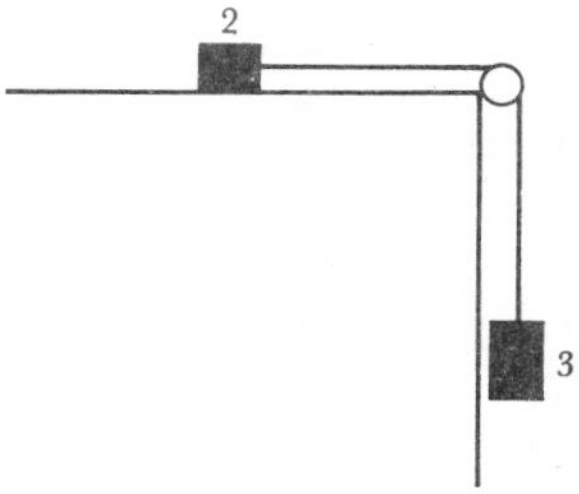

3

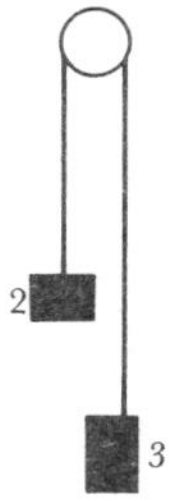

4

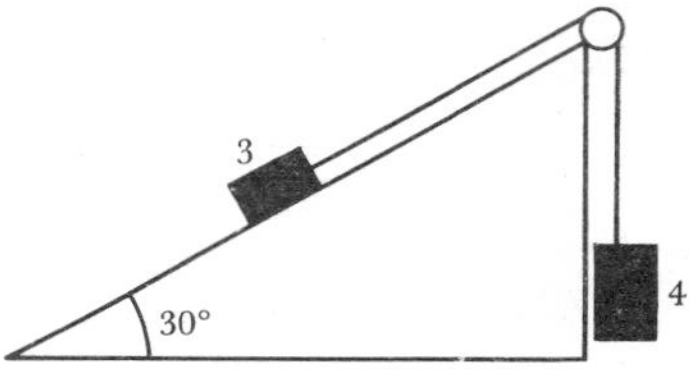

5

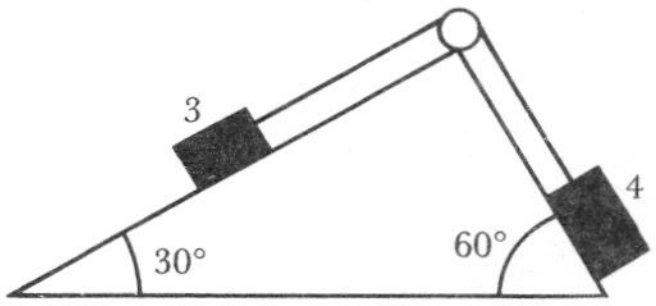

6

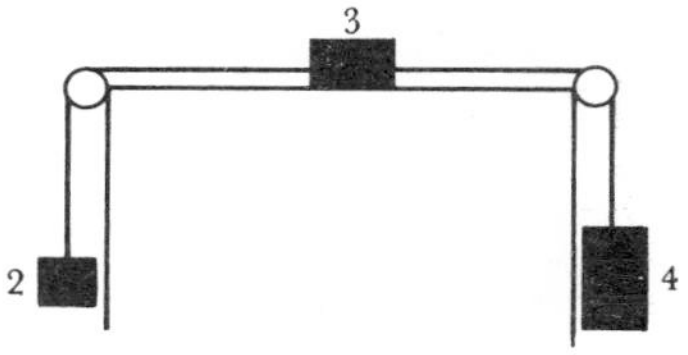

7

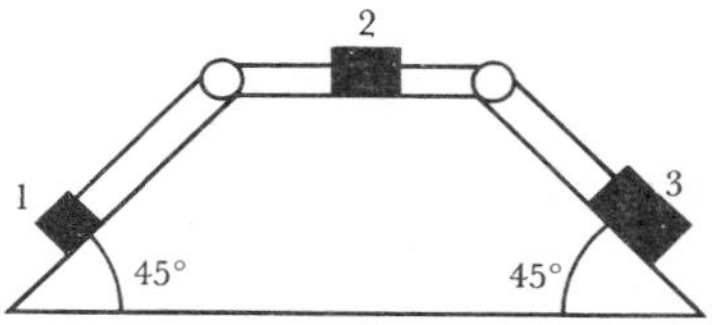

8 Repeat no. **1** when the car is towing the trailer up a slope of 1 in 100 (i.e., at θ to the horizontal, where $\sin\theta = \frac{1}{100}$).

8.10 Friction

We have seen, in the last section, that the reaction of a smooth surface is always perpendicular, or *normal*, to the surface. More usually, however, a surface is rough and capable of exerting a reaction which has a tangential component, called *friction*. In such a case the resultant reaction $\boldsymbol{R}$ has two components:

a frictional component $\boldsymbol{F}$

and a normal component $\boldsymbol{N}$.

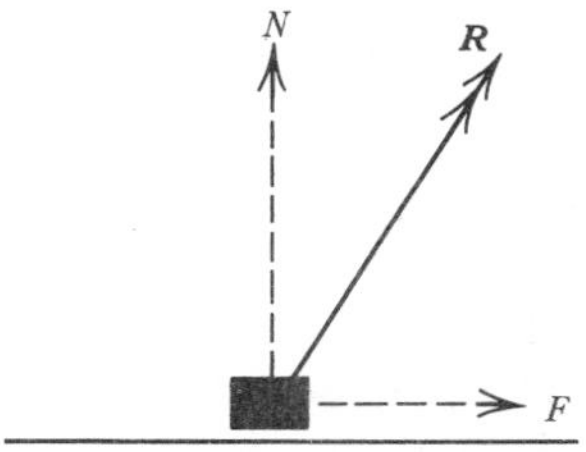

The relations between these quantities depend on the degree of roughness of the surfaces and on whether the particle is moving or at rest. They were first investigated by G. Coulomb who in 1821 summarised his experimental results in the so-called *Laws of statical and dynamical friction*. Though only very approximate, these still form the basis for the study of rough surfaces.

Dynamical friction

When sliding takes place, the component ***F*** *is in a direction that opposes relative motion and has magnitude* μN, *where* μ *is a constant.* This constant is called the *coefficient of friction* and depends only on the nature of the surfaces and *not* on the area of contact or the normal reaction.

Example 1

A block of wood of mass 20 kg is being pulled along a rough horizontal surface by means of a rope inclined to the horizontal at 30° and under tension 100 N. If $\mu = 0.3$, find the acceleration of the block. (Take $g \approx 9.8 \text{ m s}^{-2}$.)

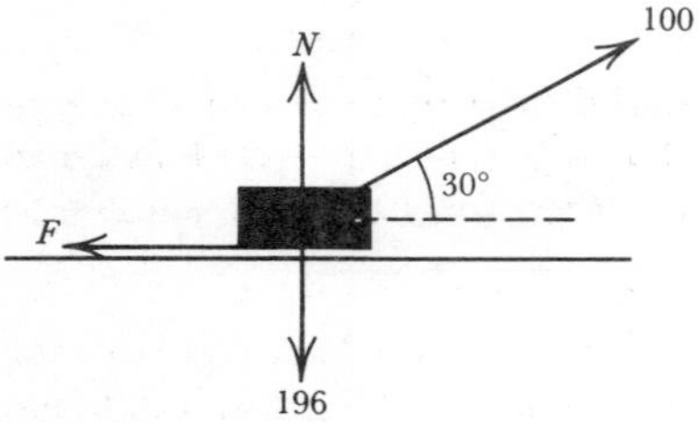

The forces acting on the block are:
(*i*) weight 196 N;
(*ii*) tension 100 N;
(*iii*) normal reaction N;
(*iv*) friction F.
As the block is moving,

$F = 0.3$ N.

If the acceleration of the block is $\boldsymbol{a}$, then (by Newton's second law)

Vertically

$$\begin{aligned} N + 100 \sin 30^\circ &= 196 \\ \Rightarrow \quad N + 50 &= 196 \\ \Rightarrow \quad N &= 146 \end{aligned}$$

Hence $F = 0.3 \times 146 = 43.8$.

Horizontally

$$\begin{aligned} 100 \cos 30 - F &= 20a \\ \Rightarrow \quad 86.6 - 43.8 &= 20a \\ \Rightarrow \quad 42.8 &= 20a \\ \Rightarrow \quad a &= 2.14. \end{aligned}$$

Hence the block has an acceleration of 2.14 m s^{-2}.

Exercise 8.10a

In each question state all forces and accelerations and draw a diagram on which these are clearly indicated. Take $g \approx 10 \text{ m s}^{-2}$.

1 A boy of mass 40 kg is sliding across a horizontal sheet of ice ($\mu = 0.1$) at a speed of 2 m s^{-1}. Find his deceleration and the distance travelled before he comes to rest.

2 A car travelling at 10 m s^{-1} skids to a halt in a distance of 5 m. Find:
(*i*) its deceleration (supposed constant);
(*ii*) the coefficient of friction;
(*iii*) its stopping distance from a speed of 20 m s^{-1}.

3 A mass of 10 kg is resting on a wooden floor and a horizontal force of 40 N is just sufficient to move it. What is the coefficient of friction? What force is required to give it an acceleration of 2 m s^{-2}?

4 An ice-hockey puck is hit with a speed of 20 m s^{-1} and travels 100 m before coming to rest. Calculate μ.

5 A mass of 2 kg is pulled along a horizontal floor by means of a force of 40 N acting at 20° above the horizontal. If $\mu = 0.2$, find the acceleration of the mass. What would be the acceleration if the force acted *downwards* at 20° to the horizontal?

6 A girl at a swimming pool is sliding down a chute which is 10 m long and inclined at 30° to the horizontal. If her coefficient of friction is 0.2, find her acceleration down the chute and her speed on leaving it.

7 A mass m is sliding down a plane inclined at α to the horizontal whose coefficient of friction is μ. Find its acceleration.

8 A mass m rests on a horizontal surface whose coefficient of friction is μ and is connected by a taut string perpendicular to the edge of the table to a mass M which hangs vertically over the edge of the table and which is falling downwards. Find the acceleration of the system and the tension in the string.

9 The triangle ABC, in which the lengths of AB, BC, CA are 4 m, 3 m and 5 m, is the vertical cross-section of a wedge fixed with AB in contact with a horizontal surface. BC is vertical with C above B, and AC is a line of greatest slope. A light inextensible string passes over a smooth pulley at C and connects a mass m on AC with a mass $3m$ hanging freely. The coefficient of friction between the mass m and the wedge is $\frac{1}{2}$. Initially the system is at rest with m at A and $3m$ just below the pulley at C. Show that each mass moves with acceleration $g/2$ until the mass $3m$ hits the horizontal surface. Show that the velocity of the mass m is then $\sqrt{96} \text{ m s}^{-1}$ up the slope. Find the further distance which m moves up AC and its velocity just before the string again becomes taut, assuming that the mass $3m$ is then in contact with the horizontal surface. (M.E.I.)

10 A parcel rests on the horizontal floor at the back of a van which is travelling along a level road at a steady speed of 63 km h^{-1}. The van is brought to rest in a stopping distance of 20 m by a uniform application of the brakes. Show that the parcel will slide forward on the floor if the coefficient of friction between the parcel and the floor is $< \frac{25}{32}$.

If this coefficient is $\frac{3}{4}$, show that the parcel will slide forward $\frac{5}{6}$ m before coming to rest. (W.)

Statical friction

When there is equilibrium, the amount of friction is just sufficient to prevent relative motion and cannot exceed μN (which is therefore called the limiting friction):

$F \leqslant \mu N.$

Example 2

A block of mass m is at rest on a plane inclined to the horizontal at an angle α. If the coefficient of friction is μ, show that $\mu \geqslant \tan \alpha$.

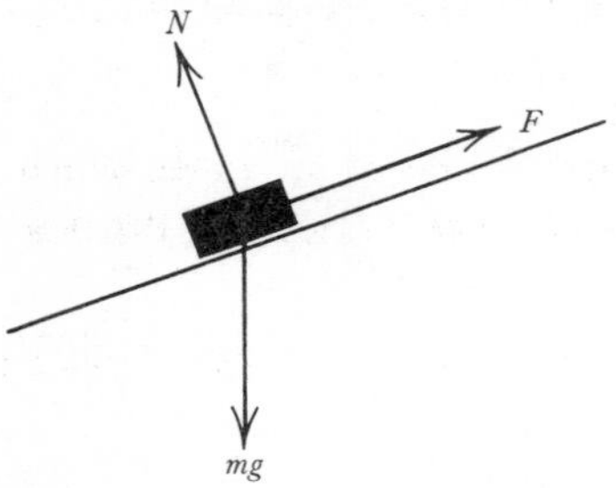

The forces acting on the block are
(*i*) its weight mg;
(*ii*) the normal reaction N;
(*iii*) the frictional force F.

Since the body is in equilibrium, we obtain by resolving up the plane,

$F = mg \sin \alpha;$

resolving perpendicular to the plane,

$N = mg \cos \alpha.$

Hence $\dfrac{F}{N} = \tan \alpha$

But $\dfrac{F}{N} \leqslant \mu$, so $\mu \geqslant \tan \alpha$.

Angle of friction

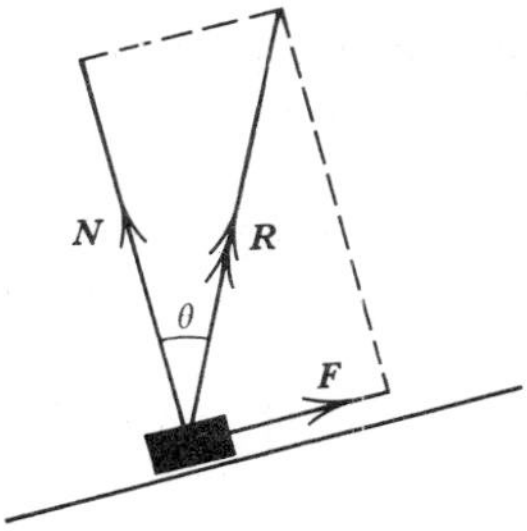

Suppose that a block lies on a rough surface and that its resultant reaction $\boldsymbol{R}$ makes an angle θ with the normal. Then if the normal and frictional components of $\boldsymbol{R}$ are $\boldsymbol{N}$, $\boldsymbol{F}$ respectively, it follows that

$$\frac{F}{N} = \tan\theta$$

But $\dfrac{F}{N} \leqslant \mu \quad \Rightarrow \quad \tan\theta \leqslant \mu.$

If we now introduce an angle λ such that $\mu = \tan\lambda$, it follows that

$$\tan\theta \leqslant \tan\lambda \quad \Rightarrow \quad \theta \leqslant \lambda.$$

Hence the reaction must be inclined to the normal at an angle which does not exceed λ. This is therefore called the *angle of friction*, and the possible limiting positions of $\boldsymbol{R}$ are said to lie on the *cone of friction*.

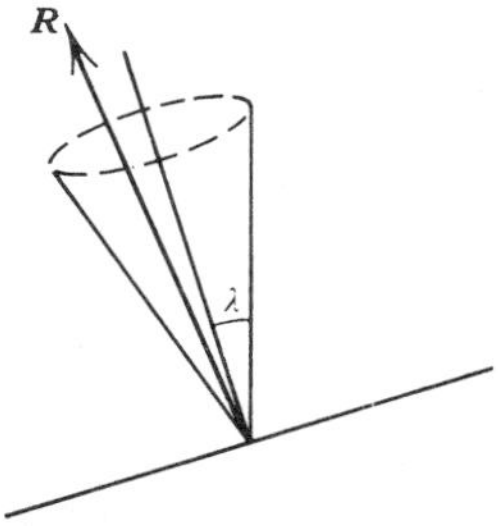

Example 3

A block of weight $\boldsymbol{W}$ is placed on a rough plane (coefficient of friction μ) which is inclined to the horizontal at an angle α, where $\tan\alpha > \mu$. Find the magnitude and direction of the *least* force which will prevent the block from slipping down the plane.
Since $\tan\alpha > \mu$, then (as we saw in Example 2) the block cannot remain in equilibrium unless another force $\boldsymbol{P}$ is applied to prevent it from sliding down the plane.

Hence the forces on the block are:

(*i*) its weight $\boldsymbol{W}$ vertically down;

(*ii*) the reaction of the plane, $\boldsymbol{R}$;

(*iii*) the additional force $\boldsymbol{P}$.

Furthermore, the resultant reaction $\boldsymbol{R}$ (if it is just to prevent sliding downwards) must lie on the cone of friction, and so be inclined to the normal at an angle λ, where $\tan \lambda = \mu$.

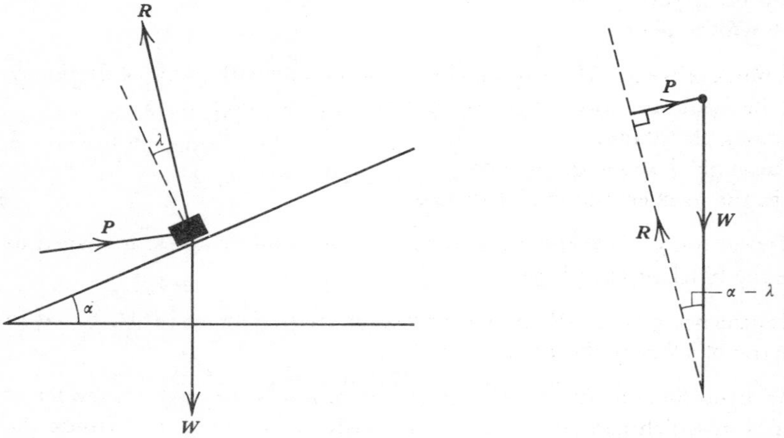

Hence $\boldsymbol{R}$ makes an angle $\alpha - \lambda$ with the vertical and we see from the triangle of forces that $\boldsymbol{P}$, if it is to be as small as possible, must be perpendicular to $\boldsymbol{R}$.

So, from the triangle of forces, $P = W \sin (\alpha - \lambda)$ is the smallest force that will maintain equilibrium and is inclined to the horizontal at an angle $\alpha - \lambda$.

Exercise 8.10b

1 A block of mass 20 kg rests on a horizontal plane whose coefficient of friction is 0.4. Find the least force required to move the block if it acts:

(*i*) horizontally;

(*ii*) at 30° above the horizontal;

(*iii*) at 30° below the horizontal;

(*iv*) in the most favourable direction.

2 A block of mass 50 kg lies on a horizontal plane and can just be moved by a force of 100 N acting horizontally. Find:

(*i*) the coefficient of friction;

(*ii*) the least force required to move the block if its direction of action can be chosen at will.

3 A block of mass 10 kg rests on a plane which is inclined at $10°$ to the horizontal and has $\mu = 0.3$. Find the least force required to move the body if the force acts:
(*i*) up the plane;
(*ii*) down the plane.

4 In no. 3 what are the least forces (if there is complete freedom of choice of their direction) required to move the block:
(*i*) up the plane;
(*ii*) down the plane?

5 A block of weight W rests on a horizontal plane with angle of friction λ. Find the minimum force required to move it if the force acts:
(*i*) along the plane;
(*ii*) at angle θ above the plane;
(*iii*) in the most economical direction.

6 Repeat no. 5 when the plane is tilted at an angle α and it is wished to move the block *up* the plane.

7 Repeat no. 5 when the plane is tilted at an angle α and it is wished to move the block *down* the plane.

8 At a fun-fair one of the amusements consists of a circular cylinder of radius 4 m which can be made to rotate with its axis vertical. Inside the cylinder is a horizontal floor which rotates with the cylinder and which can also be lowered and raised a distance of 2 m. When the cylinder is stationary a door in its wall allows access to the floor; the door is closed and forms part of the cylinder before it begins to rotate.

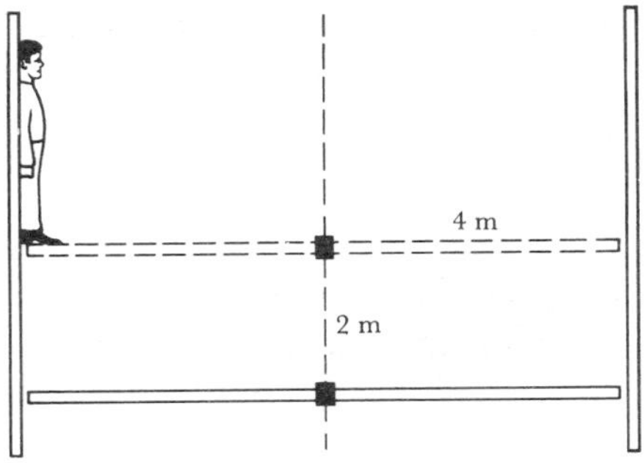

A man enters and stands with his back to the wall of the cylinder. The cylinder is then made to rotate more and more quickly until the friction between the man's back and the wall is sufficient to support his weight. The floor is then lowered 2 m, as in the diagram.

If the coefficient of friction between the man's body and the wall of the cylinder is 0.5, find the least angular velocity of the cylinder for which the man will not slip down. (Take g as 9.8 m s^{-2}.)

If the angular velocity of the cylinder is then quickly reduced to four-fifths of its previous value, find how long the man will take to slide down to the floor. (M.E.I.)

*8.11 Universal gravitation

We have seen that in 1665 Newton deduced from Kepler's laws that the gravitational attraction of the Sun diminishes with distance according to an inverse square law. This calculation was based upon the assumption that the planets move in ellipses round the Sun as a focus, but we can repeat Newton's work in the simplified case of orbits which are exactly circular.

Example 9

Suppose that a planet moves with speed v in a circular orbit of radius r which it completes in time T.

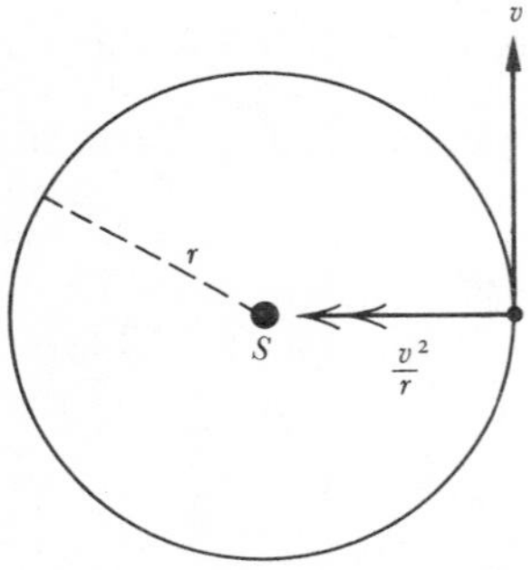

Now the acceleration of the planet towards the Sun is v^2/r.

But $v = \dfrac{2\pi r}{T}$.

Also, by Kepler's third law (see 8.7),

$T^2 = kr^3$ (where k is a constant for all planets).

So acceleration $= \dfrac{v^2}{r} = \dfrac{4\pi^2 r}{T^2} = \dfrac{4\pi^2 r}{kr^3} = \dfrac{4\pi^2}{k} \times \dfrac{1}{r^2}$

and we see that the force of attraction by the Sun must be inversely proportional to the square of the planet's distance.

We also saw that Newton checked this discovery by comparing 'the force requisite to keep the Moon in her orbit with the force of gravity at the surface of the Earth, and found them to answer pretty nearly'. We can now repeat Newton's calculations.

Example 10

Find the acceleration due to gravity at the Earth's surface in terms of the Earth's radius, r; the Moon's distance, R; and the period of the Moon's orbit round the Earth, T.

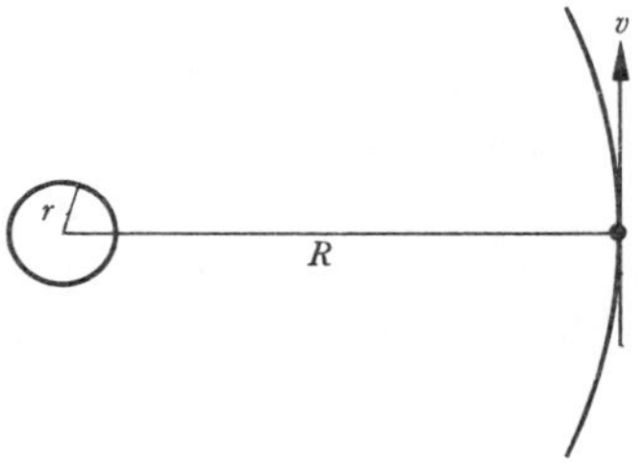

Speed of Moon $= v = \dfrac{2\pi R}{T}$.

So acceleration due to Earth's gravity $= \dfrac{v^2}{R} = \dfrac{4\pi^2 R}{T^2}$.

Hence, using the inverse square law, the acceleration at the Earth's surface due to gravity is

$$\frac{4\pi^2 R}{T^2} \times \frac{R^2}{r^2} = \frac{4\pi^2 R^3}{r^2 T^2}.$$

But $r = 6.37 \times 10^6$, $R = 3.84 \times 10^8$

and $T = 27.3 \text{ days} = 27.3 \times 24 \times 3\,600 \text{ s}$

$= 2.36 \times 10^6 \text{ s}$.

So the acceleration due to gravity

$$= \frac{4\pi^2 \times (3.84 \times 10^8)^3}{(6.37 \times 10^6)^2 \times (2.36 \times 10^6)^2}$$

$\approx 9.89 \text{ m s}^{-2}$.

Like Newton, we too find this 'to answer pretty nearly'.

We have also seen that this encouraged Newton to state his law of universal gravitation, that if two masses m_1 and m_2 are distance r apart, there is a force of attraction, F, between them which is proportional to $m_1 m_2/r^2$.

So $F = \dfrac{G m_1 m_2}{r^2}$,

where G is a constant, known as the *constant of gravitation.* (Putting $m_1 = m_2 = 1$ and $r = 1$, we see that G is the force of attraction, in N, between two masses of 1 kg placed one metre apart.)

As Gm_1m_2/r^2 is a force, it must have the same units as mass × acceleration, so in SI units is measured in kg m s^{-2}.

But m_1m_2/r^2 has units (kg)^2/m^2.

So G has units $\dfrac{\text{kg m s}^{-2}}{(\text{kg})^2/\text{m}^2} = (\text{kg})^{-1}\ \text{m}^3\ \text{s}^{-2}$.

The magnitude of G was first determined by Henry Cavendish in 1798 and is now generally taken, in SI, to be approximately 6.67×10^{-11}. Using this value of G, we can now proceed to:

Example 11

Calculate the mass of the Earth.

The force of attraction on 1 kg at the Earth's surface is GM/r^2 (where r is the Earth's radius).

So $\dfrac{GM}{r^2} = g$

$\Rightarrow \quad M = \dfrac{gr^2}{G}$.

But $g = 9.81, \quad r = 6.37 \times 10^6, \quad G = 6.67 \times 10^{-11}$.

$$\text{So} \quad M = \frac{9.81 \times (6.37 \times 10^6)^2}{6.67 \times 10^{-11}}$$
$$= 5.97 \times 10^{24} \text{ kg}.$$

Example 12

Assuming the inverse square law of gravitation and ignoring air resistance, how fast must a stone be projected vertically upwards in order to escape from the Earth's gravitational field?

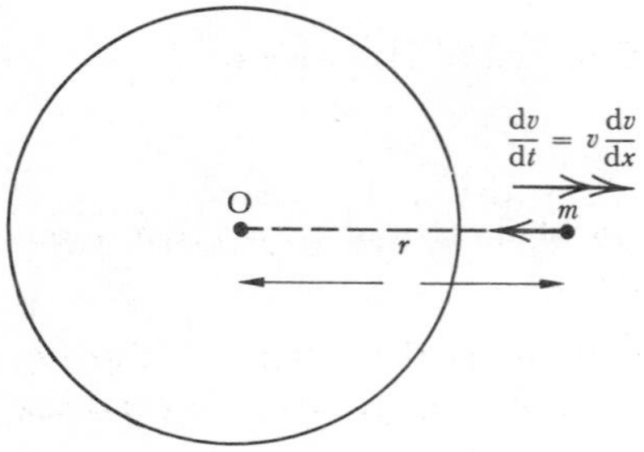

Suppose that the

Earth's radius $= r \approx 6.4 \times 10^6$ m; and that at its surface, $g \approx 10$ m s^{-2}.

By the inverse square law, the gravitational force acting on a mass m when it is at a distance x from the centre of the Earth will be

$$mg \times \left(\frac{r}{x}\right)^2 = \frac{mgr^2}{x^2}.$$

Also, if the stone has velocity v, then its acceleration can be expressed as

$$\frac{dv}{dt} = \frac{dx}{dt} \times \frac{dv}{dx} = v\frac{dv}{dx}, \quad \text{away from O.}$$

Using Newton's second law,

$$-\frac{mgr^2}{x^2} = mv\frac{dv}{dx}$$

$$\Rightarrow \quad 2v\,dv = -\frac{2gr^2}{x^2}\,dx$$

$$\Rightarrow \quad v^2 = \frac{2gr^2}{x} + c.$$

Now suppose that the initial speed was u, i.e., that when $x = r$, $v = u$.

Then $u^2 = 2gr + c \quad \Rightarrow \quad c = u^2 - 2gr.$

So $$v^2 = \frac{2gr^2}{x} + u^2 - 2gr.$$

As $x \to +\infty, \quad v^2 \to u^2 - 2gr.$

Now if the stone is to escape from the Earth this must be positive.

Hence $u^2 - 2gr > 0$

$$\Rightarrow \quad u > \sqrt{2gr} \approx \sqrt{(2 \times 9.8 \times 6.4 \times 10^6)}$$
$$\approx 11 \times 10^3 \text{ m s}^{-1}.$$

So the minimum escape velocity is about 11 km s^{-1}.

Miscellaneous problems 8

1 If P, Q are the mid-points of AB, CD respectively, prove that
(*i*.) $\boldsymbol{AC} + \boldsymbol{BD} = 2\boldsymbol{PQ}$;
(*ii*) $\boldsymbol{AC} + \boldsymbol{BC} + \boldsymbol{AD} + \boldsymbol{BD} = 4\boldsymbol{PQ}$.

2 If the opposite sides of a hexagon are parallel, what can you prove about its diagonals? Need the hexagon be plane?

3 A particle of constant mass m is moving under the action of a force $\boldsymbol{F}$. The position vector of the particle at time t is $\boldsymbol{r}$. State Newton's second law of motion in terms of $\boldsymbol{F}$, $\boldsymbol{r}$, m, t.

A particle is moving in a plane in such a way that

$$\boldsymbol{r} = \boldsymbol{a} \cos nt + \boldsymbol{b} \sin nt,$$

where n is constant and $\boldsymbol{a}$ and $\boldsymbol{b}$ are fixed non-parallel vectors. Show that the particle describes an ellipse, and that the force acting on it is directed towards the centre O of the ellipse and is proportional to the distance of the particle from O. (O.C.)

4 The coordinates of a moving point P are

$(a \cos \omega t, a \sin \omega t)$

where a and ω are constants. Find the magnitude and direction of the velocity and acceleration of P at time t.

There is a speed limit of 80 km h^{-1} round an unbanked curve on a railway track. The curve is in the shape of an arc of a circle and the angle turned through is 30° in 500 m.

Find:

(*i*) the radius of the circle;

(*ii*) the lateral force on the rails at maximum speed as a fraction of the weight;

(*iii*) the least distance before the curve that a warning sign must be placed if the trains cruise at 150 km h^{-1} and a comfortable braking retardation is $\frac{1}{2}$ m s^{-2}. (M.E.I.)

5 An object of unit mass is moving in a plane and its position vector from O is $\boldsymbol{r}$ and its coordinates (r, θ). If

$$\boldsymbol{r} = \frac{\cos \theta}{2 + \cos \theta}\boldsymbol{i} + \frac{\sin \theta}{2 + \cos \theta}\boldsymbol{j},$$

obtain an expression for r in terms of θ.

By differentiating find $\dot{\boldsymbol{r}}$ in terms of $\dot{\theta}$, θ, $\boldsymbol{i}$ and $\boldsymbol{j}$, and if $r^2\dot{\theta} = a$, where a is a constant, prove that

$$\dot{\boldsymbol{r}} = -2a \sin \theta\boldsymbol{i} + a(2 \cos \theta + 1)\boldsymbol{j}.$$

Hence, by differentiating again, prove that the resultant force acting on the object is of magnitude $2a^2/r^2$ and directed towards O. (S.M.P.)

6 A shell is projected from O with fixed speed v and the initial gradient of its trajectory is m. Show that its path is

$$y = mx - \frac{g(1 + m^2)}{2v^2}x^2.$$

Hence show that possible initial directions of a trajectory that goes through (x, y) are given by the roots of the quadratic equation

$$gx^2m^2 - 2v^2xm + (gx^2 + 2v^2y) = 0;$$

and that there are two such trajectories provided that

$$y < \frac{v^2}{2g} - \frac{gx^2}{2v^2}.$$

What happens if (x, y) is such that this inequality is not satisfied?

Sketch the *parabola of safety*

$$y = \frac{v^2}{2g} - \frac{gx^2}{2v^2},$$

showing its relationship to the various trajectories for different values of m. Why is it important?

7 A, B are fixed points in a vertical line with A above B and AB $= l$. A string ACB of length $2l$ is attached to A and B and carries a heavy bead C, free to slide on the string, which describes a horizontal circle about AB in a plane at depth y below A with angular velocity ω. Prove that

$$y = \tfrac{1}{2}l + \frac{4g}{3\omega^2}.$$

8 A particle starts from a point A whose position vector is $\boldsymbol{a}$ with velocity $\boldsymbol{u}$ under a constant acceleration $\boldsymbol{g}$. Obtain an expression for the position vector of the particle after time t.

Two particles move freely under gravity; show that their relative velocity remains constant.

Two boys stand on level ground, b m apart. One throws a stone vertically upwards with velocity u m s^{-1}, the other waits until this stone is at its highest point then throws a stone directly at it with speed v m s^{-1}. Show that the stones will collide provided v is big enough and find the lower limit for v. Find u to make this lower limit least.

List the assumptions you are making. (M.E.I.)

9 A river flows between parallel banks which are at a distance $2a$ apart. The speed of the current at any point is proportional to the distance, x, of the point from the nearer bank and has the value U at midstream. A motor boat whose speed is V in still water ($V > U$) crosses the river.

(*i*) If the boat is steered so as always to point in a direction perpendicular to the current, find how far it is carried downstream in making the complete crossing.

(*ii*) If the boat is steered so as to cross in a straight line from a point on one bank to a point directly opposite to it on the other bank, find an expression for the retardation of the boat in terms of U, x and a during the first half of the crossing and calculate the time taken to cross the river completely. (J.M.B.)

10 An aircraft whose speed in still air is V km h^{-1} describes a circuit which may be taken as a horizontal square of side a km. A wind of magnitude v km h^{-1} (where $v < V$) blows parallel to one of the sides of the square. Assuming that no time is lost on corners, find the time taken by the aircraft to complete the circuit. (O.C.)

11 The apparent diameters of the Sun from the Earth and of the Earth from the Moon are 32′ and 1° 54′ respectively. Given, further, that each year there are 13.4 lunar revolutions round the Earth, find the ratio of the mean densities of the Sun and the Earth.

*Appendix: parabolas, ellipses, and hyperbolas

The present book has no pretence to being even an introduction to geometry. Nevertheless it is convenient to use its methods (and particularly those of chapters 1 and 3) to obtain a few of the geometrical properties of some simple curves.

For a variety of reasons, those which have been most fully investigated are the so-called *conic sections*, obtained by taking plane slices of the surface of a circular double-cone (composed of straight lines extending on both sides of its vertex). Clearly one possible section, when the plane is perpendicular to the cone's axis, is a circle; and another possibility, if the plane

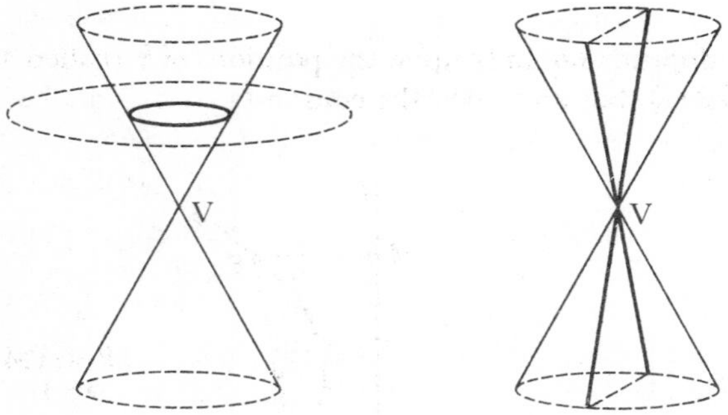

passes through the vertex V, is a pair of straight lines. But we shall concentrate upon oblique sections which do not pass through the vertex and lead to three types of curve, depending on the inclination of the plane to the cone:

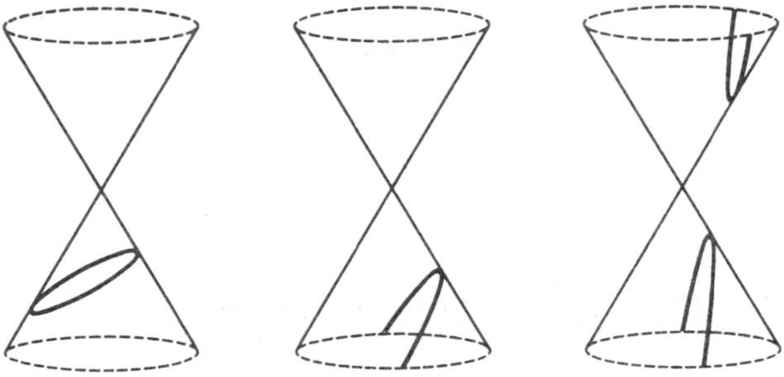

These curves, which were first described by Euclid (ca. 300 BC), gained great importance nearly two thousand years later when John Kepler investigated the orbits of planets around the Sun, and more recently when the Danish mathematician and physicist Niels Bohr used them to describe the motion of electrons round a nucleus.

There is a wide variety of approaches to the study of these curves and we shall start by using the so-called focus-directrix property as our definition, finally (in section 4) returning to the original sections of a cone.

If we are given a fixed point S and a fixed line l (not through S), we can consider the locus of a point P which moves so that its distance from S is proportional to its perpendicular distance from l,

that is $SP = ePM$, where e is a constant.

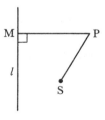

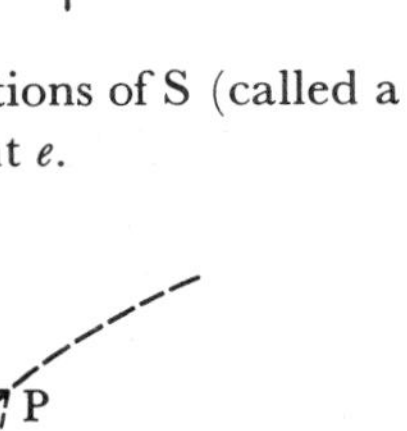

Clearly the locus of P depends not only upon the positions of S (called a *focus*) and l (called a *directrix*), but also upon the constant e.

If $e = 1$,

then $SP = PM$

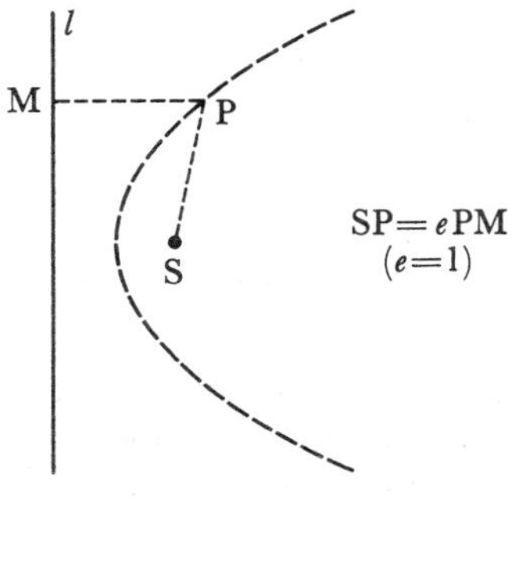

and we can easily see that the locus of P is a curve, as shown, which we call a *parabola*.

If $e < 1$,

then $\mathrm{SP} = e\mathrm{PM}$

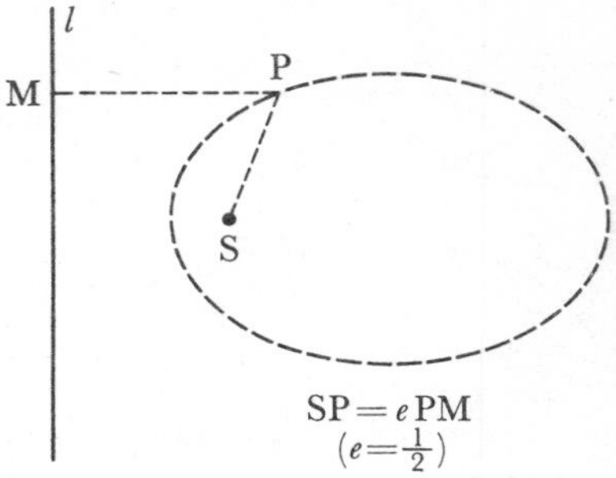

can similarly be shown to define a closed curve, called an *ellipse*, whose shape depends on the value of e.

If $e > 1$,

then $\mathrm{SP} = e\mathrm{PM}$

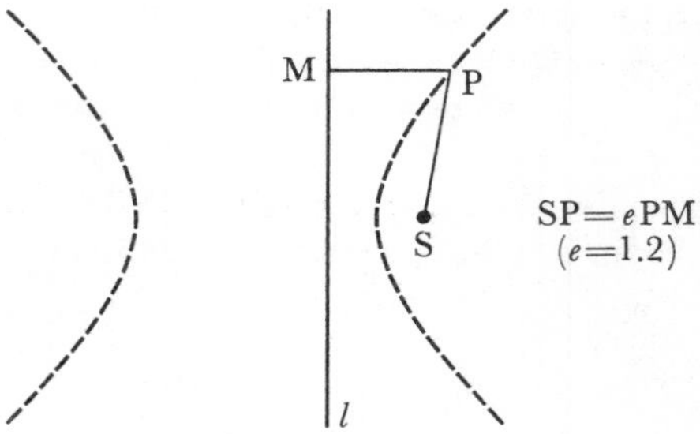

defines a curve with two separate branches, known as a *hyperbola*, and again the precise shape of the curve depends upon the value of e.

Polar equations

Using polar coordinates r, θ referred to S as pole, with angle θ measured from the axis of symmetry, $\mathrm{SP} = r$.

So $\mathrm{SP} = e\mathrm{PM} \Rightarrow \mathrm{PM} = \frac{r}{e}$.

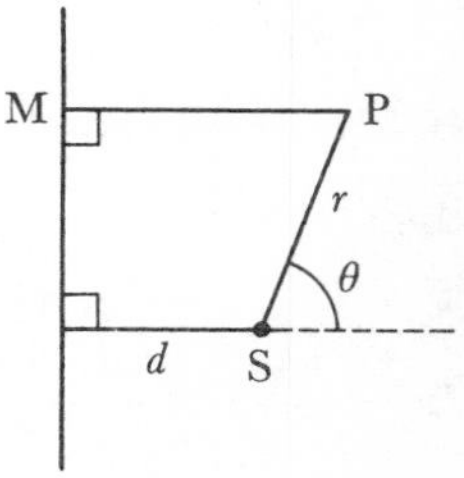

If the distance from S to l is d,

then $$\frac{r}{e} = d + r\cos\theta$$

$$\Rightarrow \quad r(1 - e\cos\theta) = ed$$

$$\Rightarrow \quad r = \frac{l}{1 - e\cos\theta} \quad (\text{where } l = ed)$$

and this can very conveniently be used for plotting the ellipse $(e < 1)$, parabola $(e = 1)$ and hyperbola $(e > 1)$.

We now proceed to investigate some of the properties of these curves.

A.1 The parabola

If SP = PM,

it is immediately seen that the point mid-way between S and l is a possible position of P.

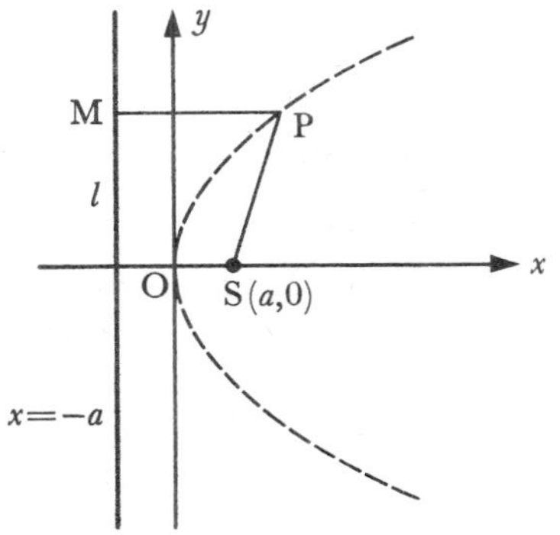

We now take this as the origin O and the axis of symmetry as Ox. Letting OS $= a$, we see that S is $(a, 0)$ and l is the line $x = -a$.

Then $$SP = PM$$

$$\Rightarrow \quad SP^2 = PM^2$$

$$\Rightarrow \quad (x - a)^2 + y^2 = (x + a)^2$$

$$\Rightarrow \quad \boxed{y^2 = 4ax}$$

This is taken as the standard form of the parabola, so that its focus is $(a, 0)$ and its directrix $x = -a$.

Furthermore,

the point O, (0, 0) is called its *vertex*

and the line of symmetry ($y = 0$) its *axis*.

Parametric representation

If $x = at^2$ and $y = 2at$,

we notice that $y^2 = 4a^2t^2$

and $4ax = 4a^2t^2$

So, for all values of t, the point $(at^2, 2at)$ lies on the parabola $y^2 = 4ax$.

Hence $\boxed{x = at^2, \quad y = 2at}$

provides a representation of the parabola in terms of a parameter t:

t	-2	-1	0	$+1$	$+2$	$+3$
$x = at^2$	$4a$	$-a$	0	a	$4a$	$9a$
$y = 2at$	$-4a$	$-2a$	0	$2a$	$4a$	$6a$

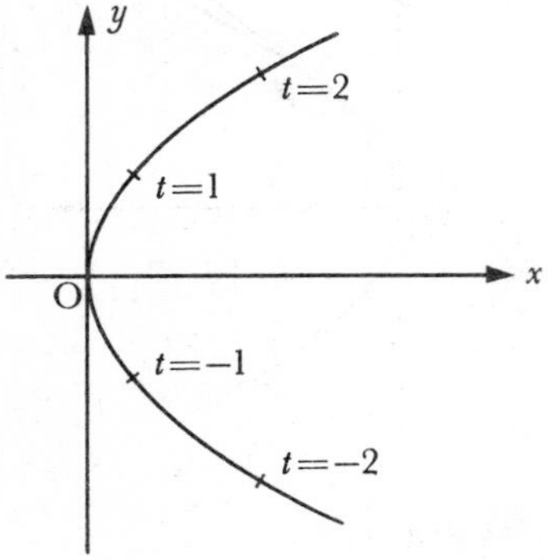

We now proceed to use this parametric form to investigate some of the properties of the parabola.

Chords, tangents, and normals

Suppose that P_1, P_2 are two points on the parabola and that their parameters are t_1, t_2.

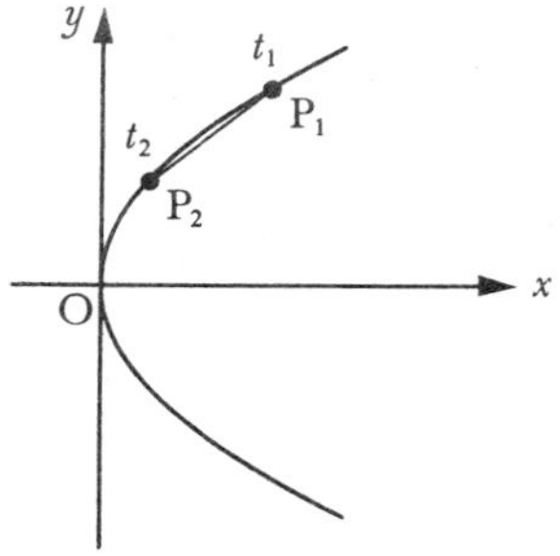

Then the line (or *chord*) P_1P_2 has gradient

$$\frac{2at_1 - 2at_2}{at_1^2 - at_2^2} = \frac{2a(t_1 - t_2)}{a(t_1^2 - t_2^2)} = \frac{2}{t_1 + t_2}$$

Now the line through P_1 $(at_1^2, 2at_1)$ with this gradient is

$$\frac{y - 2at_1}{x - at_1^2} = \frac{2}{t_1 + t_2}$$

$$\Rightarrow \quad 2x - 2at_1^2 = (t_1 + t_2)y - 2at_1(t_1 + t_2)$$

$$\Rightarrow \quad \boxed{2x - (t_1 + t_2)y + 2at_1t_2 = 0}$$

which is the equation of P_1P_2.

In the particular case when P_1 and P_2 are coincident at a point P and have parameter t, we see that this chord becomes the tangent at P and has the equation

$$2x - 2ty + 2at^2 = 0$$

$$\Rightarrow \quad \boxed{x - ty + at^2 = 0}$$

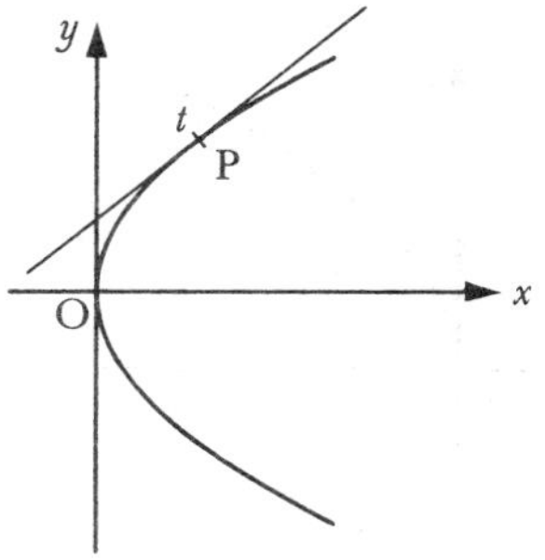

Alternatively, we see that the gradient of the tangent at t is

$$\frac{dy}{dx} = \frac{dy/dt}{dx/dt} = \frac{2a}{2at} = \frac{1}{t}$$

So the equation of the tangent is

$$\frac{y - 2at}{x - at^2} = \frac{1}{t}$$

$$\Rightarrow \quad x - ty + at^2 = 0$$

Similarly the normal at P has gradient $-t$, so that its equation is

$$\frac{y - 2at}{x - at^2} = -t$$

$$\Rightarrow \quad \boxed{tx + y = at^3 + 2at}$$

Example 1

Show that if a perpendicular is drawn from the focus S to any tangent to a parabola, the foot N of this perpendicular always lies on the tangent at the vertex.

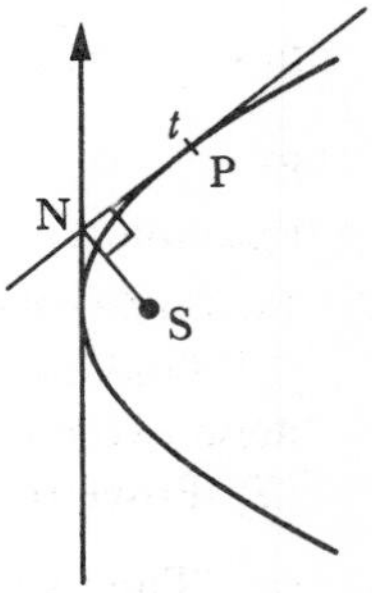

If P has parameter t,

PN is $x - ty + at^2 = 0$

and SN has gradient $-t$, so that its equation is

$$\frac{y - 0}{x - a} = -t$$

$$\Rightarrow \quad tx + y - at = 0$$

Now the point N lies on both PN: $x - ty + at^2 = 0$

and SN: $tx + y - at = 0$

Multiplying the second equation by t and adding,

$$(1 + t^2)x + 0y + 0 = 0 \quad \Rightarrow \quad x = 0$$

Hence N always lies on the tangent at the vertex and, as P varies, the locus of N is the tangent at the vertex.

Example 2

If P is any point on a parabolic mirror, prove that two rays of light through P, one from the focus and the other parallel to the axis, are equally inclined to the tangent at P. (The so-called 'reflector' property of the parabola.) If P has parameter t, and the angles α, β, γ are as shown in the figure,

$$\tan\beta = \text{gradient of tangent} = \frac{1}{t}$$

and $$\tan\gamma = \frac{2at - 0}{at^2 - a}$$

$$= \frac{2t}{t^2 - 1}$$

$$= \frac{2\frac{1}{t}}{1 - \left(\frac{1}{t}\right)^2}$$

$$= \frac{2\tan\beta}{1 - \tan^2\beta}$$

$$= \tan 2\beta$$

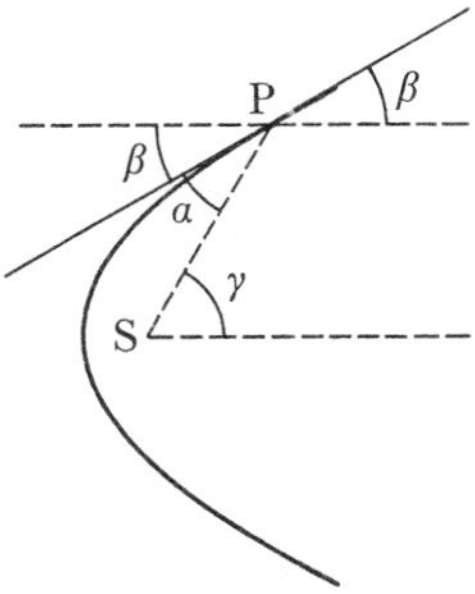

Hence $\gamma = 2\beta$.

But $\gamma = \alpha + \beta$,

so $\alpha = \beta$.

Example 3

Two points on the parabola $y^2 = 4ax$ have parameters t_1, t_2.
(*i*) Find the coordinates of the point of intersection of the tangents at these two points;
(*ii*) Prove that perpendicular tangents meet on the directrix.

(*i*) The two tangents are

$$x - t_1 y + at_1^2 = 0$$

and $$x - t_2 y + at_2^2 = 0$$

Subtracting these, we find that at their point of intersection

$$(t_1 - t_2)\,y = a(t_1^2 - t_2^2)$$

$$\Rightarrow \quad y = a(t_1 + t_2)$$

and $$x = t_1 a(t_1 + t_2) - at_1^2 = at_1t_2$$

So their point of intersection is

$$x = at_1t_2, \quad y = a(t_1 + t_2)$$

(*ii*) If the two tangents are perpendicular, then

$$\frac{1}{t_1}\frac{1}{t_2} = -1$$

$$\Rightarrow \quad t_1t_2 = -1$$

Hence at their point of intersection, $x = -a$, so that this point lies on the directrix.

Alternatively, for part (*i*), we can look more closely at the equation of the tangent to the parabola. Normally we think of this as a linear relation

$$x - ty + at^2 = 0$$

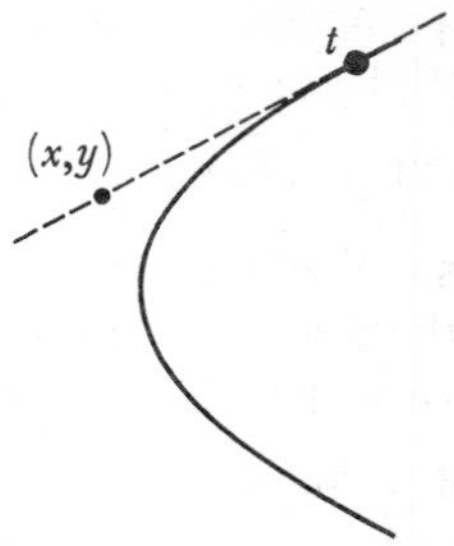

between the two coordinates of a point which lies on a given tangent.

But it can also be written as

$$at^2 - yt + x = 0$$

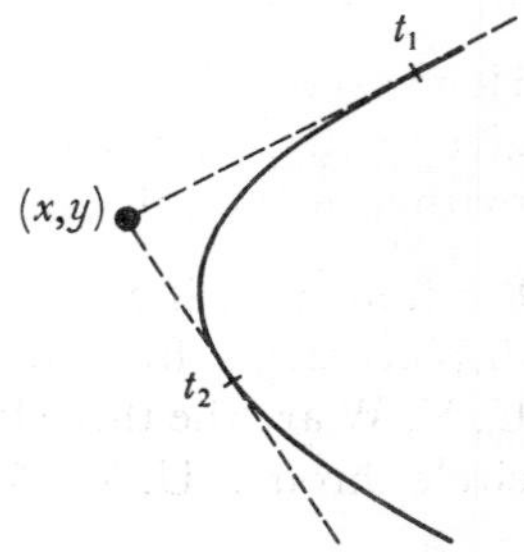

and we see that this quadratic equation has roots t_1, t_2 which are the parameters of the two points (distinct, coincident or imaginary) at which the tangents pass through a particular point (x, y).

Now we know from the theory of quadratic equations that

$$t_1 + t_2 = \frac{y}{a} \quad \text{and} \quad t_1 t_2 = \frac{x}{a}$$

So, again, $x = at_1t_2$ and $y = a(t_1 + t_2)$

This second proof, though less obvious than its predecessor, begins to show something of the power of the alliance between algebra and geometry.

Exercise A.1

1 Find the equation of the parabola whose focus is (0, 1) and directrix $y = 0$.

2 The tangents to a parabola at points P, Q meet at the point T and a line is drawn through T parallel to the axis of the parabola meeting the parabola in another point U. Prove that the tangent at U is parallel to PQ.

3 Two tangents to a parabola are perpendicular. Show that the line joining their points of contact passes through the focus.

4 A chord P_1P_2 of a parabola meets the axis at K, and the tangents at P_1, P_2 meet it at T_1, T_2. Prove that if O is the vertex, then

$$OT_1 \times OT_2 = OK^2.$$

5 The normal at a variable point P of a parabola meets its axis at G, and the foot of the perpendicular from P to the axis is N. Prove that NG (the *subnormal*) is of constant length, independent of the position of P.

6 The tangents to a parabola at points P, Q are perpendicular and intersect in a point T. If their corresponding normals meet at a point K, prove that TK is parallel to the axis of the parabola.

7 Show that if the normal at the point t of a parabola passes through a given point (h, k), then t must be a root of the equation

$$at^3 + (2a - h)t - k = 0$$

Hence show that not more than three normals can be drawn through a given point; and that if there are three such normals, the triangle formed by their feet on the parabola has its centroid on the axis.

8 Show that if a circle meets a parabola in four points the sum of their displacements from its axis is zero. Hence (using no. 7) prove that if U, V, W are the three feet of normals to a parabola from a given point, the circle through U, V, W must pass through the vertex of the parabola.

A.2 The ellipse

The ellipse has been defined as the locus of a point P which moves so that its distance from a fixed point S is proportional to its perpendicular distance from a fixed line l, with a constant of proportionality which is less than 1:

$$SP = ePM \ (e < 1)$$

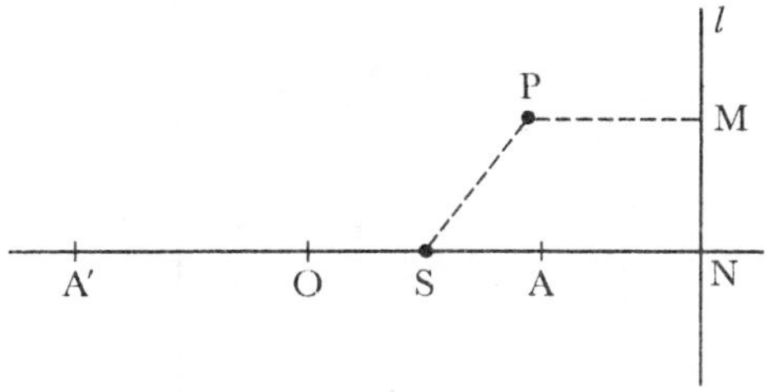

Clearly the locus of P will be symmetrical about the line through S which is perpendicular to l. We let this axis of symmetry meet l in the point N and notice that there will be two possible positions of P on this axis, one on each side of S. We call these A and A′ and take origin O at their mid-point, letting OA $= a$.

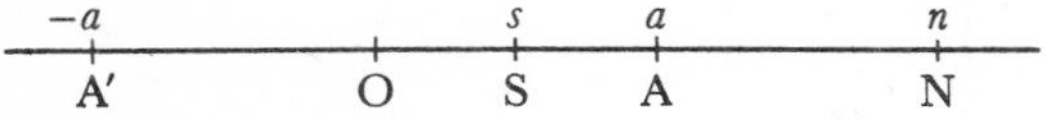

We also let OS $= s$ and ON $= n$.

So SA $= a - s$, AN $= n - a$

and SA′ $= a + s$, A′N $= n + a$

Now A and A′ are on the locus.

So $a - s = e(n - a)$

and $a + s = e(n + a)$

Hence $s = ae$

and $a = en \Rightarrow n = a/e$

We can now express the equation of the ellipse in terms of a and e.

For $\text{SP} = e\text{PM}$

$$\Rightarrow \text{SP}^2 = e^2\text{PM}^2$$

$$\Rightarrow (x - ae)^2 + y^2 = e^2\left(\frac{a}{e} - x\right)^2$$

$$\Rightarrow (x - ae)^2 + y^2 = (a - ex)^2$$

$$\Rightarrow (1 - e^2)x^2 + y^2 = a^2(1 - e^2)$$

$$\Rightarrow \frac{x^2}{a^2} + \frac{y^2}{a^2(1 - e^2)} = 1$$

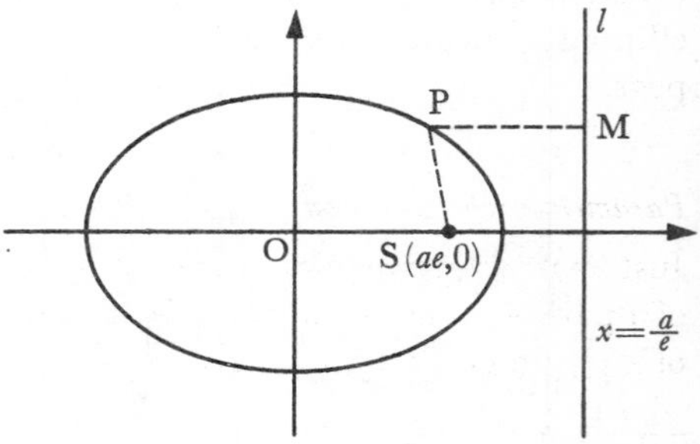

If we now write $b^2 = a^2(1 - e^2)$, it follows that

$$\boxed{\frac{x^2}{a^2} + \frac{y^2}{b^2} = 1}$$

which we take as the standard form of the ellipse.

We immediately see that the ellipse is symmetrical both about Ox (its major axis) and Oy (its minor axis), the lengths of these two axes being $2a$, $2b$ respectively; and that it also has point-symmetry about its centre O.

One consequence of these symmetries is that the ellipse must have a focus and directrix not only at S $(ae, 0)$ and l $(x = a/e)$, but also at S′ $(-ae, 0)$ and l' $(x = -a/e)$.

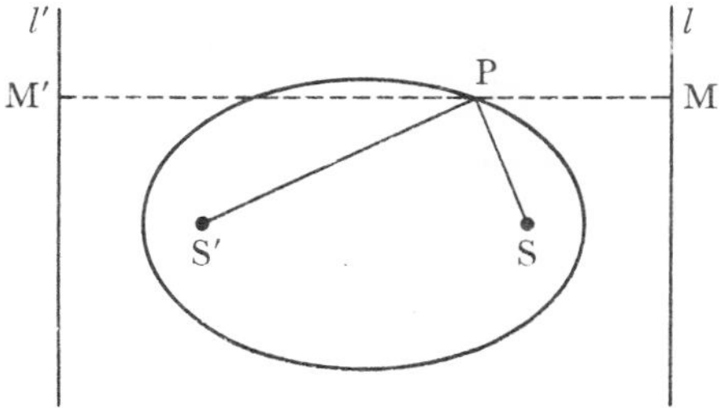

Furthermore, if M and M' are the feet of the perpendiculars from a point P on to l and l', it follows that

$$SP = ePM$$

and $$S'P = ePM'$$

$$\Rightarrow \quad SP + S'P = e(PM + PM')$$

$$= eMM' = e\frac{2a}{e} = 2a$$

So as P moves around an ellipse, the sum of its focal distances remains constant. This is the basis of the well-known method of constructing an ellipse by means of a pencil moving within a loop of string tied to two fixed pegs.

Parametric representation

Just as it was convenient to use a parametric representation for the points of a parabola, so the points of the standard ellipse can be expressed in terms of a parameter θ by the equations

$$x = a\cos\theta, \quad y = b\sin\theta$$

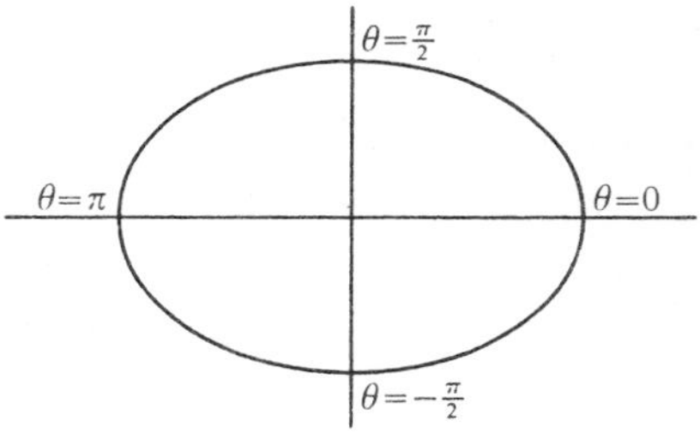

We immediately see that

$$\frac{x^2}{a^2} + \frac{y^2}{b^2}$$

$$\Rightarrow \frac{a^2\cos^2\theta}{a^2} + \frac{b^2\sin^2\theta}{b^2} = 1$$

so that such a point always lies on the ellipse.

Moreover, we can easily find the equation of the tangent at the point whose parameter is θ.

For $$\frac{dy}{dx} = \frac{dy/d\theta}{dx/d\theta} = \frac{b\cos\theta}{-a\sin\theta} = -\frac{b\cos\theta}{a\sin\theta}$$

So the required equation is

$$\frac{y - b\sin\theta}{x - a\cos\theta} = -\frac{b\cos\theta}{a\sin\theta}$$

$$\Rightarrow \quad bx\cos\theta + ay\sin\theta = ab(\cos^2\theta + \sin^2\theta) = ab$$

$$\Rightarrow \quad \boxed{\frac{x\cos\theta}{a} + \frac{y\sin\theta}{b} = 1}$$

Exercise A.2

1 Find the lengths of the axes, the eccentricity, the foci and the directrices of:

(*i*) $\dfrac{x^2}{9} + \dfrac{y^2}{4} = 1$;

(*ii*) $4x^2 + 25y^2 = 100$

2 Show that the ellipse

$$\frac{x^2}{4 + \lambda} + \frac{y^2}{1 + \lambda} = 1$$

always has the same foci, whatever the value of λ (provided $\lambda > -1$). Sketch this family of confocal ellipses.

3 $P_1(x_1, y_1)$ is a point on the ellipse

$$\frac{x^2}{a^2} + \frac{y^2}{b^2} = 1$$

Find the gradient of the tangent to the ellipse in terms of x_1, y_1, a and b; and hence find the equations of the tangent and normal at P.

4 Find the equation of the normal to the standard ellipse at the point $(a\cos\theta, b\sin\theta)$.

5 The tangent and normal at a point on the standard ellipse meet the major axis at T, N. Prove that $OT \times ON = a^2 - b^2$.

6 Find a linear transformation which, when applied to the points of the circle $x^2 + y^2 = a^2$, transforms them into the points of the standard ellipse. Illustrate the relationship between the ellipse and the circle (known as its *auxiliary* circle).

7 A plane inclined to the horizontal at an angle α contains a hole of radius a, and a vertical beam of light shines through this hole. Show that the patch of light cast on a horizontal plane is in the shape of an ellipse whose eccentricity is $\sin\alpha$.

8 (*i*) A man is instructed to go from A to B, first picking up a stone from the straight track l. Show that for his path to have minimum length, its two parts must be equally inclined to l.

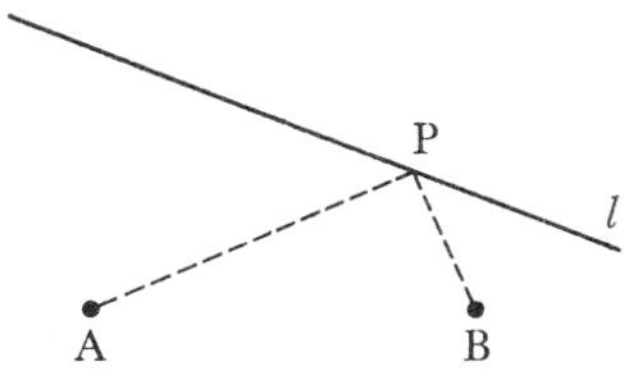

(*ii*) At which point on a tangent to an ellipse is the sum of the focal distances a minimum? Why?

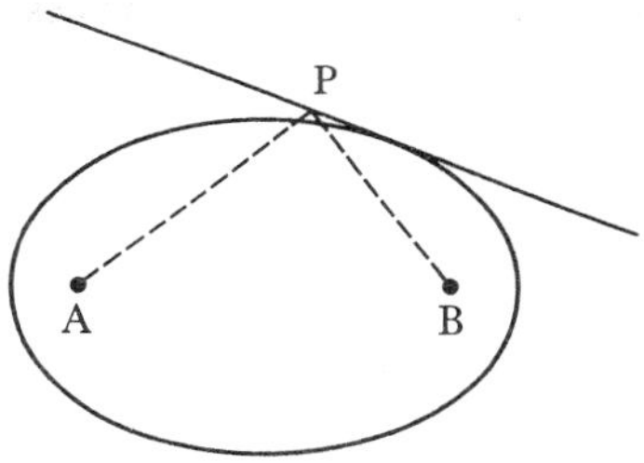

(*iii*) Hence show that the lines joining any point of an ellipse to its two foci are equally inclined to the tangent at the given point. (The *reflector property* of the ellipse.)

A.3 The hyperbola

We have seen that the hyperbola is defined as the locus of a point P such that

$$SP = ePM \quad (\text{where} \quad e > 1)$$

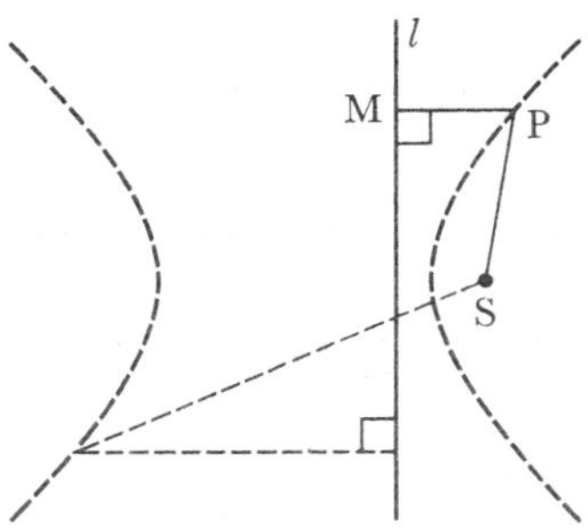

and we have already noticed that this locus has two separate branches.

As with the ellipse, we let A, A′ be the two positions of P which lie on the axis of symmetry (this time both on the same side of S), and take the origin O at their mid-point.

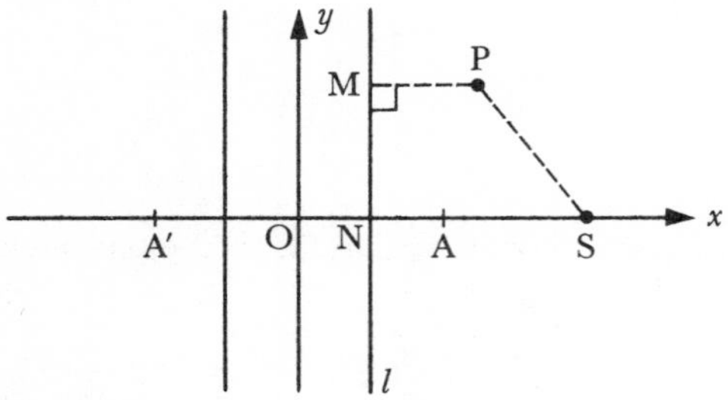

As before, we let N be the point where l meets the line of symmetry, and let

$$\text{OS} = s \qquad \text{and} \qquad \text{ON} = n$$

Then $\text{SA} = s - a \qquad \text{AN} = a - n$

$\text{SA}' = s + a \qquad \text{A}'\text{N} = a + n$

Now $\text{SA} = e\text{AN} \Rightarrow s - a = e(a - n)$

and $\text{SA}' = e\text{A}'\text{N} \Rightarrow s + a = e(a + n)$

So $s = ae$

and $a = en \Rightarrow n = \dfrac{a}{e}$

Since $e > 1$, this means that the directrix $x = a/e$ lies between O and the focus $(ae, 0)$

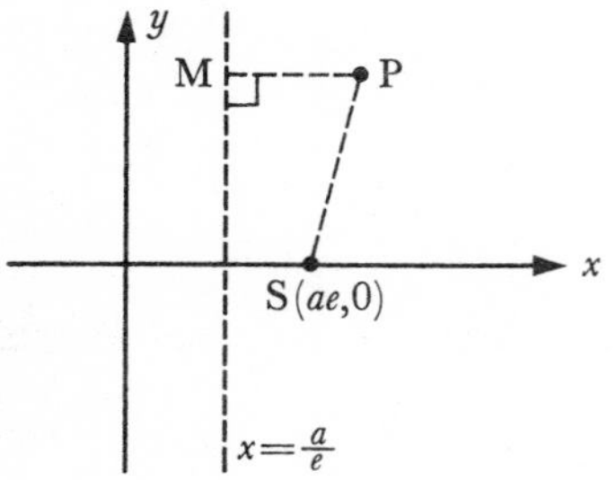

Taking P as the point (x, y)

$$\text{SP} = e\text{PM}$$

$$\Rightarrow \quad \text{SP}^2 = e^2\text{PM}^2$$

$$\Rightarrow \quad (x - ae)^2 + y^2 = e^2(x - a/e)^2$$

and this, as before, leads to

$$\frac{x^2}{a^2} + \frac{y^2}{a^2(1 - e^2)} = 1$$

But as $e > 1$ we write this as

$$\frac{x^2}{a^2} - \frac{y^2}{a^2(e^2 - 1)} = 1$$

and let $b^2 = a^2(e^2 - 1)$, so that the standard equation for the hyperbola is

$$\boxed{\frac{x^2}{a^2} - \frac{y^2}{b^2} = 1}$$

As before, we see (by symmetry) that there must be another focus at $(-ae, 0)$, and another directrix $x = -a/e$.
Furthermore, if $x = a \sec \theta$ and $y = b \tan \theta$, then

$$\frac{x^2}{a^2} - \frac{y^2}{b^2} = \sec^2 \theta - \tan^2 \theta = 1$$

So $(a \sec \theta, b \tan \theta)$ is a parametric representation of a point on the hyperbola.

A characteristic feature of the hyperbola, not shared with either the parabola or ellipse, is its pair of asymptotes.

For $$\frac{x^2}{a^2} - \frac{y^2}{b^2} = 1$$

$$\Rightarrow \quad \left(\frac{x}{a} + \frac{y}{b}\right)\left(\frac{x}{a} - \frac{y}{b}\right) = 1$$

So $$\frac{x}{a} - \frac{y}{b} = \frac{1}{x/a + y/b}$$

Now as $x \to +\infty$ and $y \to +\infty$ (or $x \to -\infty$ and $y \to -\infty$) we see that

$$\frac{x}{a} - \frac{y}{b} = \frac{1}{x/a + y/b} \to 0$$

and $\frac{x}{a} - \frac{y}{b} = 0$ is an asymptote.

Similarly $$\frac{x}{a} + \frac{y}{b} = \frac{1}{x/a - y/b}$$

So as $x \to +\infty$ and $y \to -\infty$ (or $x \to -\infty$ and $y \to +\infty$), we see that

$$\frac{x}{a} + \frac{y}{b} = \frac{1}{x/a - y/b} \to 0$$

$$\Rightarrow \quad \frac{x}{a} + \frac{y}{b} = 0 \quad \text{is an asymptote.}$$

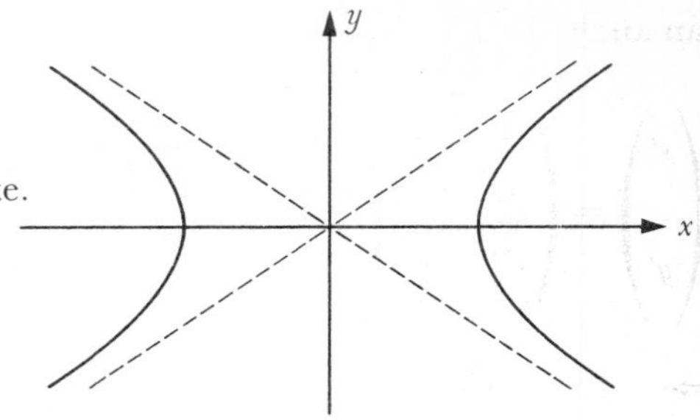

Hence the two lines

$$\frac{x}{a} \pm \frac{y}{b} = 0$$

are asymptotes, to which points on the hyperbola approach more and more closely as they get further and further from its centre.

The rectangular hyperbola

The hyperbola $\dfrac{x^2}{a^2} - \dfrac{y^2}{b^2} = 1$

has a pair of asymptotes $\dfrac{x}{a} = \pm\dfrac{y}{b}$.

If $a = b$, these become the pair of lines $x = \pm y$. Because these are perpendicular, we call the corresponding curve a *rectangular hyperbola*.

The equation of such a hyperbola is clearly

$$\frac{x^2}{a^2} - \frac{y^2}{a^2} = 1$$

$$\Rightarrow \quad x^2 - y^2 = a^2$$

and since $\quad b^2 = a^2(e^2 - 1)$

we see that $\quad a^2 = a^2(e^2 - 1)$

$$\Rightarrow \quad 1 = e^2 - 1$$

$$\Rightarrow \quad e = \sqrt{2}$$

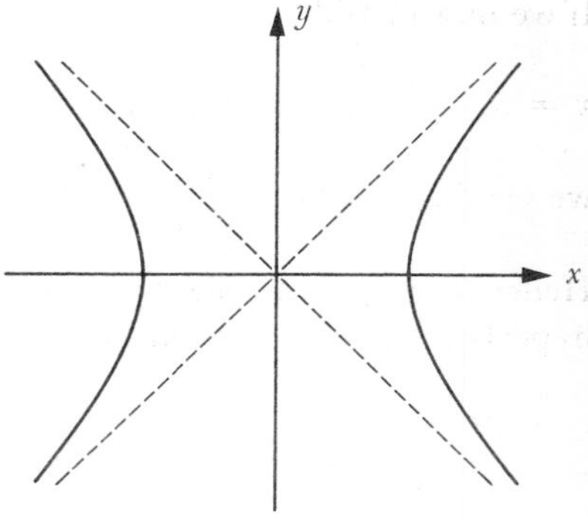

Hence the rectangular hyperbola has eccentricity $\sqrt{2}$ and when referred to its axes of symmetry has the standard equation

$$x^2 - y^2 = a^2$$

It is, however, frequently convenient to consider the equation of the rectangular hyperbola referred to its asymptotes as axes, and this can be done by means of the linear transformation which rotates the hyperbola through an angle $+\frac{\pi}{4}$:

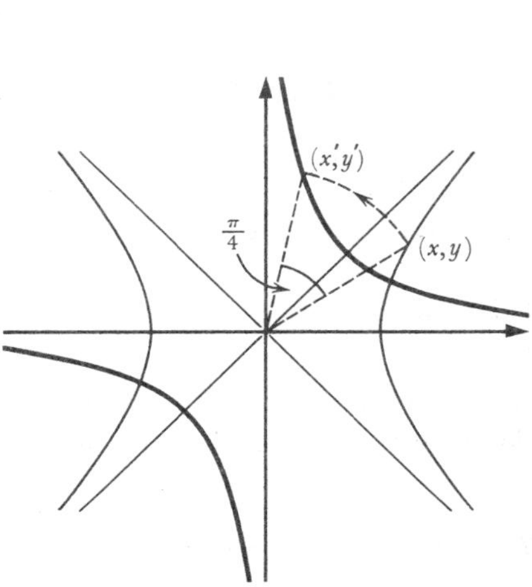

$$\begin{pmatrix} x' \\ y' \end{pmatrix} = \begin{pmatrix} \frac{1}{\sqrt{2}} & \frac{-1}{\sqrt{2}} \\ \frac{1}{\sqrt{2}} & \frac{1}{\sqrt{2}} \end{pmatrix} \begin{pmatrix} x \\ y \end{pmatrix}$$

$$\Rightarrow \quad x' = \frac{1}{\sqrt{2}}(x - y)$$

$$y' = \frac{1}{\sqrt{2}}(x + y)$$

Hence the equation $x^2 - y^2 = a^2$

is transformed into $2x'y' = a^2$

$$\Rightarrow \quad x'y' = \frac{1}{2}a^2$$

We now revert to the use of (x, y) rather than (x', y') and let $c^2 = \frac{1}{2}a^2$, obtaining

$$\boxed{xy = c^2}$$

as the standard form of the rectangular hyperbola when referred to its asymptotes.

If we now consider

$x = ct, \quad y = \frac{c}{t}$ (where t is a parameter),

we see that $xy = ct \times \frac{c}{t} = c^2$.

Hence a very convenient parametric representation of the rectangular hyperbola $xy = c^2$ is given by

$$\boxed{x = ct, \quad y = \frac{c}{t}}$$

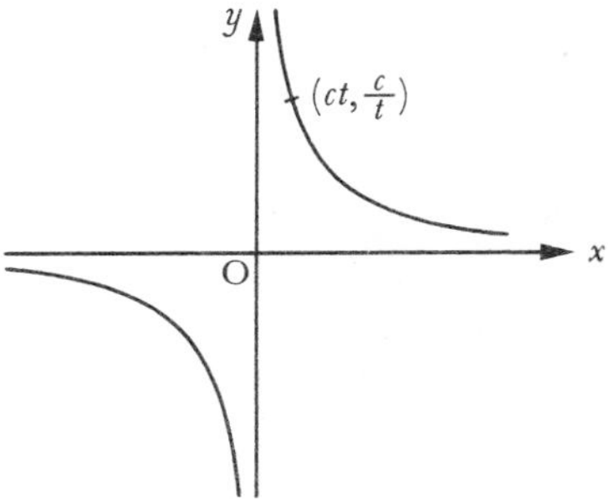

Exercise A.3

1 Find the asymptotes, eccentricity, foci, and directrices of

(*i*) $\frac{x^2}{4} - y^2 = 1$, (*ii*) $4x^2 - 9y^2 = 36$.

2 Show that if $-4 < \lambda < -1$, the curve

$$\frac{x^2}{4+\lambda} + \frac{y^2}{1+\lambda} = 1$$

represents a hyperbola, and that the position of its foci is independent of the value of λ. Compare with exercise A.2, no. 2, and sketch these two families of curves in one diagram.

3 Find the equation of the tangent to the hyperbola

$$\frac{x^2}{a^2} - \frac{y^2}{b^2} = 1$$

at the point P $(a \sec \theta, b \tan \theta)$.

If this tangent meets the two asymptotes at Q, R, show that P is the mid-point of QR.

4 (*i*) If P is a point on a hyperbola with foci S, S′ and major axis of length $2a$, use the focus-directrix property to prove that

$$|SP - S'P| = 2a$$

(*ii*) An explosion takes place at an unknown point and three observers at different points with carefully synchronised watches note the exact time at which its sound reaches them. Use (*i*) to show how this information can be used to locate the position of P. [A method used by artillery in 'sound-ranging', the same principle being the basis of many sea and air navigational systems.]

5 A rectangular hyperbola is represented by

$$x = ct, \quad y = \frac{c}{t}$$

Find the equation of
(*i*) the tangent at the point with parameter t;
(*ii*) the normal at the same point;
(*iii*) the chord joining points t_1 and t_2.

6 The tangent at a variable point P of the rectangular hyperbola $xy = c^2$ meets the two asymptotes at points Q, R. Prove that ΔOQR has constant area $2c^2$ and that the locus of its centroid is the hyperbola $xy = \frac{4}{9}c^2$.

7 P_1, P_2, P_3, P_4 are points on the standard rectangular hyperbola which have parameters t_1, t_2, t_3, t_4. If P_1P_2 is perpendicular to P_3P_4, show that $t_1t_2t_3t_4 = -1$. Hence show that if P_1P_2 is perpendicular to P_3P_4, then P_1P_3 is perpendicular to P_2P_4, and P_1P_4 is perpendicular to P_2P_3. Finally show that if three points lie on a rectangular hyperbola, so does their orthocentre.

8 The foot of the perpendicular from O to the tangent at a variable point P on $xy = c^2$ is Q.
(*i*) Prove that OP×OQ is constant.
(*ii*) Find the equation of the locus of Q and sketch it.

A.4 The sections of a cone

We have seen that the eccentricity e of an ellipse lies between 0 and 1. Moreover, if we take the particular ellipse for which $b = a$, which is clearly a circle, we see that

$$b^2 = a^2(1 - e^2)$$

$$\Rightarrow \quad a^2 = a^2(1 - e^2) \quad \Rightarrow \quad e = 0$$

Hence we can regard a circle as an ellipse whose eccentricity is zero.
Further, we saw that
for the parabola, $e = 1$;
and for the hyperbola, $e > 1$.

But we began this study by looking at the ways in which a plane meets a circular cone and noticed that the curve of intersection changes with the inclination of the plane. It now appears that this is related to the above changes in eccentricity.

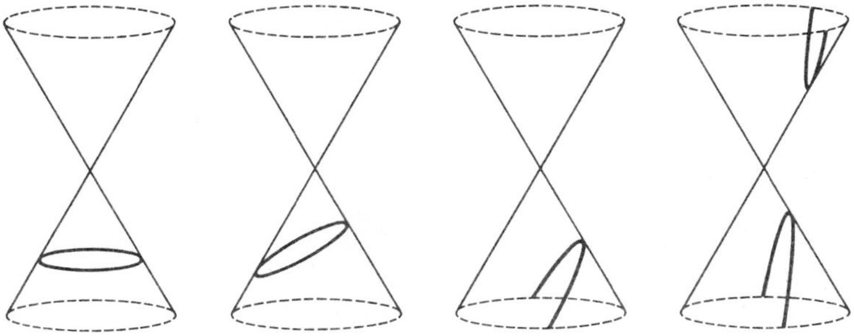

$e = 0$ Circle | $0 < e < 1$ Ellipse | $e = 1$ Parabola | $e > 1$ Hyperbola

We have, however, not yet *proved* that any of these oblique sections of a circular cone is a parabola, an ellipse or a hyperbola in the sense in which we have defined them by means of the focus-directrix properties. Our final objective is to remedy this in the case of the ellipse, the proofs for the parabola and hyperbola being broadly similar.

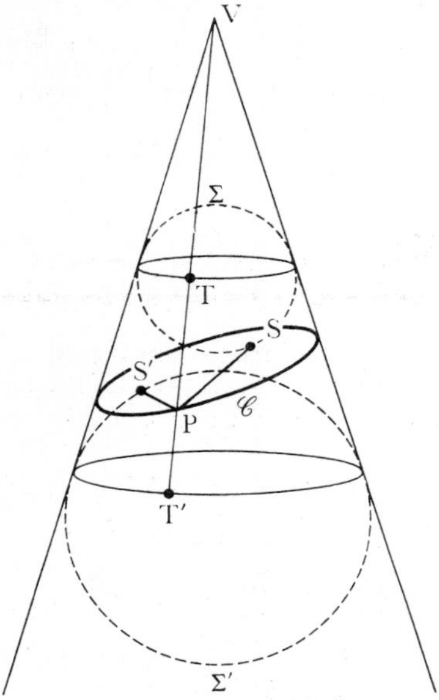

In the above figure, a circular cone with vertex V is cut by an oblique plane. 𝒞 is the curve of intersection, and P is any point of 𝒞.

We first draw two spheres Σ and Σ′, each of which touches *both* the inside of the cone (with which it has a circle of contact) *and* the given plane (at S, S′ respectively). Furthermore, the line VP will meet these two circles of contact in two points T and T′.

If we now concentrate upon the point P and the sphere Σ, we see that PS and PT are both tangents to this sphere, S and T being their points of contact. So $PS = PT$

Similarly (using sphere Σ′), $PS' = PT'$

So $PS + PS' = PT + PT' = TT'$

But TT′ is a length which is independent of the position of P.

Hence $PS + PS' =$ constant, and so 𝒞, which is the locus of P, must be an ellipse with foci S and S′.

Answers to exercises

Exercise 0.1

1 (*i*) $p \Leftarrow q$ (*ii*) $p \Rightarrow q$ (*iii*) $p \Leftarrow q$ (*iv*) $p \Leftrightarrow q$
2 (*i*) $p \not\Rightarrow q$ (*ii*) $p \not\Leftarrow q$ (*iii*) $p \not\Rightarrow q$ (*iv*) $p \not\Rightarrow q, p \not\Leftarrow q$
3 (*i*) $p' \Rightarrow q'$ (*ii*) $p' \Leftarrow q'$ (*iii*) $p' \Rightarrow q'$ (*iv*) $p' \Leftrightarrow q'$
4 (*i*) $p' \Leftarrow q'$ (*ii*) $p' \Rightarrow q'$ (*iii*) $p' \Leftrightarrow q'$

Exercise 0.2a

1 (*i*) $\{-1, 0, 1, 2, 3\}$ (*ii*) $\{1, 2, 3\}$ (*iii*) $\{5, 6, 7, \ldots\}$
(*iv*) $\{1, 2\}$ (*v*) $\varnothing$
2 (*i*) $\mathscr{E}$ (*ii*) $\varnothing$ (*iii*) A

Exercise 0.2b

1 (*i*) A (*ii*) $\varnothing$ (*iii*) $\mathscr{E}$ (*iv*) A (*v*) A (*vi*) A (*vii*) A (*viii*) A (*ix*) $\mathscr{E}$
(*x*) $\varnothing$
5 (*i*) AB (*ii*) $\varnothing$ (*iii*) B' (*iv*) A
6 (*i*) $BC + CA + AB$ (*ii*) $ABC + A'B'C'$ (*iii*) $A + BC$
(*iv*) ABC'

Exercise 0.3a

2 OA: $(2, 2\tfrac{1}{2})$, $\sqrt{41}$, $\tfrac{5}{4}$
OB: $(1\tfrac{1}{2}, 0)$, 3, 0
OC: $(0, -1)$, 2, Indeterminate
OD: $(-1, \tfrac{1}{2})$, $\sqrt{5}$, $-\tfrac{1}{2}$
OB, OC

3 AB: $(3\frac{1}{2}, 2\frac{1}{2})$, $\sqrt{26}$, 5
AC: $(2, 1\frac{1}{2})$, $\sqrt{65}$, $\frac{7}{4}$
AD: $(1, 3)$, $\sqrt{52}$, $\frac{2}{3}$
BC: $(1\frac{1}{2}, -1)$, $\sqrt{13}$, $\frac{2}{3}$
BD: $(\frac{1}{2}, \frac{1}{2})$, $\sqrt{26}$, $-\frac{1}{5}$
CD: $(-1, -\frac{1}{2})$, $\sqrt{13}$, $-\frac{3}{2}$
AD, BC are parallel
AB, BD are perpendicular
CD perpendicular to AD, BC

Exercise 0.3b

11 (*i*) $(1, 3)$ (*ii*) $(-2, 6)$ (*iii*) $(-1, -1)$ (*iv*) $(-1, 1), (2, 4)$
(*v*) $(0, 0), (2, 8), (-2, -8)$

Exercise 0.4a

1 (*i*) $x - 3y = 0$ (*ii*) $x - y = 0$ (*iii*) $x - 9y = 0$
2 (*i*) $x + y - 10 = 0$ (*ii*) $y = 1$ (*iii*) $2x - y - 5 = 0$
3 $x - 2y - 1 = 0$, $4x + y - 25 = 0$, $2x + 5y - 23 = 0$,
G: $(5\frac{2}{3}, 2\frac{1}{3})$
4 $x - y - 2 = 0$, $x = 5$, $x + 2y - 11 = 0$,
H: $(5, 3)$
5 $x - y - 4 = 0$, $x = 6$, $x + 2y - 10 = 0$,
X: $(6, 2)$
6 $x + y = 8$

Exercise 0.4b

3 $(3, 1), (4, 2), (4, 0)$
4 $(2, -2), (-7, -2), (-1, 1)$
5 $(\frac{3}{2}, -\frac{3}{2}), (4, -4), (2, 0)$

Exercise 0.5

1 (*i*) $x^2 + y^2 = 16$ (*ii*) $x^2 + y^2 - 2x - 4y - 4 = 0$
(*iii*) $x^2 + y^2 - 6x + 8y = 0$ (*iv*) $(x - a)^2 + (y - b)^2 = r^2$
2 (*i*) circle; centre $(0, 0)$, radius $\sqrt{10}$ (*ii*) circle; centre $(2, 1)$, radius 1
(*iii*) circle; centre $(-3, 4)$, radius 5
(*iv*) circle; centre $(-1, -3)$, radius 4 (*v*) parabola
(*vi*) rectangular hyperbola
3 Centre $(-g, -f)$, radius $\sqrt{(g^2 + f^2 - c)}$
4 (*i*) $x^2 + y^2 + \frac{10}{3}x + 1 = 0$
circle; centre $(-\frac{5}{3}, 0)$, radius $\frac{4}{3}$
(*ii*) $x = 0$; perpendicular bisector of AB
(*iii*) $x^2 + y^2 = 1$; circle on AB as diameter
(*iv*) $x = -1$; straight line perpendicular to AB
(*v*) $y^2 = 2x - 1$; parabola

Exercise 0.6

1 Identity
2 Reflection in Oy
3 Reflection in O
4 Reflection in $y = x$
5 Magnification $\times 2$
6 Singular
7 Anticlockwise rotation through 90°
8 Clockwise rotation through 90° and magnification $\times 2$
9 Singular
10 Shear parallel Oy

Exercise 0.7

1 (*i*) C (*ii*) B (*iii*) A

2 (*i*) I (*ii*) $\begin{pmatrix} -1 & 0 \\ 0 & -1 \end{pmatrix}$

3 (*i*) $\begin{pmatrix} 0 & 1 \\ -1 & 0 \end{pmatrix}$ $\begin{pmatrix} 0 & -1 \\ 1 & 0 \end{pmatrix}$ (*ii*) I

5 No

Exercise 0.8

1 (*i*) $\begin{pmatrix} \frac{1}{2} & 0 \\ 0 & 1 \end{pmatrix}$ (*ii*) $\begin{pmatrix} \frac{1}{3} & 0 \\ 0 & \frac{1}{3} \end{pmatrix}$ (*iii*) $\begin{pmatrix} -1 & 0 \\ 0 & 1 \end{pmatrix}$

(*iv*) $\begin{pmatrix} 0 & -1 \\ 1 & 0 \end{pmatrix}$ (*v*) None (*vi*) $\begin{pmatrix} 0 & 1 \\ 1 & 0 \end{pmatrix}$

2 (*i*) I (*ii*) A

3 (*i*) 5 (*ii*) 2 (*iii*) 0

4 $\begin{pmatrix} \frac{3}{2} & -\frac{5}{2} \\ -1 & 2 \end{pmatrix}$ (*i*) $-7, 6$ (*ii*) $2, -1$ (*iii*) $-6\frac{1}{2}, 5$ (*iv*) $1\frac{1}{2}, -2$

5 (*i*) $(-34, 15)$ (*ii*) $\left(-\frac{2}{7}, \frac{15}{7}\right)$ (*iii*) None (*iv*) $\left(\frac{7}{4}, -\frac{19}{4}\right)$

Exercise 1.1a

1 (*i*) 12.56, 0.785, 314, 31 400 (*ii*) 0 (*iii*) $f(3) = 28.26$
(*iv*) $f(\frac{1}{3}) = 0.349$

2 (*i*) 5 m, 20 m, 45 m (*ii*) 50 m s^{-1}

3 (*i*) 2 s, 4s (*ii*) $\frac{1}{4}$ m, 2.25 m

4 (*i*) 3.6 km, 7.2 km, 36 km (*ii*) 25 m, 100 m

Exercise 1.1b

1 (*i*) $fg: x \mapsto x^3 + 1$ (*ii*) $gf: x \mapsto (x+1)^3$
(*iii*) $f^{-1}: x \mapsto x - 1$ (*iv*) $g^{-1}: x \mapsto \sqrt[3]{x}$
(*v*) $(fg)^{-1}: x \mapsto \sqrt[3]{(x-1)}$ (*vi*) $g^{-1}f^{-1}: x \mapsto \sqrt[3]{(x-1)}$

2 (*i*) $fg: x \mapsto 3x^2$ (*ii*) $gf: x \mapsto (3x)^2$ (*iii*) $f^{-1}: x \mapsto \frac{1}{3}x$
(*iv*) $g^{-1}: x \mapsto \sqrt{x}$ (*v*) $(fg)^{-1}: x \mapsto \sqrt{\frac{x}{3}}$ (*vi*) $g^{-1}f^{-1}: x \mapsto \sqrt{\frac{x}{3}}$
(if domain of g is $x \geqslant 0$)

3 $(fg)^{-1} \equiv g^{-1}f^{-1}$, provided both exist

Exercise 1.2

1 (*i*) -2 (*ii*) 7 (*iii*) 5.5 (*iv*) -5
2 (*i*) 7 (*ii*) 7 (*iii*) 7 (*iv*) 7, even
3 (*i*) – (*ii*) $\frac{1}{9}$ (*iii*) 0.16 (*iv*) 1, even
4 (*i*) 0 (*ii*) 27 (*iii*) 15.625 (*iv*) -1, odd
5 (*i*) 0 (*ii*) 1.732 (*iii*) 1.58 (*iv*) –
6 (*i*) 0 (*ii*) $\frac{1}{3}$ (*iii*) 0.4 (*iv*) -1, odd
7 (*i*) 0 (*ii*) 0 (*iii*) 0.5 (*iv*) 0
8 (*i*) 0 (*ii*) $\frac{1}{9}$ (*iii*) 0.16 (*iv*) 1, even
9 (*i*) 1 (*ii*) 1 (*iii*) 1 (*iv*) 1, even
10 (*i*) 0 (*ii*) 3 (*iii*) 2.5 (*iv*) -1, odd
11 (*i*) – (*ii*) $-\frac{1}{3}$ (*iii*) – (*iv*) 1, odd
12 (*i*) – (*ii*) 31 (*iii*) – (*iv*) –
13 (*i*) – (*ii*) 1 (*iii*) – (*iv*) –
14 (*i*) No (*ii*) 0 (if defined)
15 (*i*) Even (*ii*) Odd
16 (*i*) $a = 0$ (*ii*) $b = 0$
17 (*i*) $x^3 + 2$ (*ii*) $x^3 - 3$ (*iii*) $(x - 1)^3$ (*iv*) $(x + 4)^3$ (*v*) $2x^3$
(*vi*) $-3x^3$ (*vii*) $8x^3$ (*viii*) $-x^3$

Exercise 1.3

$f(x)$ is continuous

1 Everywhere
2 Everywhere
3 $x \neq 0$
4 Everywhere
5 $x \geqslant 0$
6 $x \neq 0$
7 $x \notin Z$
8 Everywhere
9 Nowhere
10 $x = 0$

Exercise 1.4a

1 (*i*) 0 (*ii*) 0 (*iii*) 0, continuous
2 (*i*) 0 (*ii*) 0 (*iii*) discontinuous
3 (*i*) 1 (*ii*) 0 (*iii*) does not exist, discontinuous
4 (*i*) 0 (*ii*) 1 (*iii*) does not exist, discontinuous
5 (*i*) 1 (*ii*) 1 (*iii*) 1, continuous

Exercise 1.4b

1 $-\infty$, $-\infty$
2 $+\infty$, does not exist
3 $+\infty$, 0
4 does not exist ($-\infty$ and $+\infty$), $+1$
5 neither exists ($-\infty$ and $+\infty$)
6 $+\infty$, 0
7 $+\infty$, does not exist

Exercise 1.5b

1 (*i*) $\pm 2i$ (*ii*) $\pm\sqrt{3}i$
2 (*i*) i (*ii*) -1

Exercise 1.6

1 (*i*) ± 2 (*ii*) 4, 0 (*iii*) $+1, +3$ (*iv*) 2 (*v*) 3.41, 0.59
2 (*i*) $\frac{1}{2}, -1$ (*ii*) 0.781, -1.281 (*iii*) $\frac{5}{3}, -1$ (*iv*) $\frac{1}{2}\{p \pm \sqrt{(p^2 + 4q)}\}$
3 (*i*) $f(x) \geqslant -9$: least when $x = 3$
(*ii*) $f(x) \leqslant \dfrac{25}{4}$: greatest when $x = -\frac{3}{2}$
(*iii*) $f(x) \geqslant -1$: least when $x = 3$
4 (*i*) $x < 0$ and $x > 1$ (*ii*) Never (*iii*) $-1.618 \leqslant x \leqslant 0.618$
(*iv*) Always
5 (*i*) $(\frac{1}{2}, \frac{5}{2}), (\frac{5}{2}, \frac{1}{2})$ (*ii*) (1.28, 3.28), $(-0.78, 1.22)$
6 (*i*) $m < -7.46, m > -0.54$ (*ii*) $m = -7.46, -0.54$
(*iii*) $-7.46 < m < -0.54$
8 2.83 s 9 80 m, 20 m 10 26
11 (*i*) $n > 150$ (*ii*) 180 (*iii*) 70, 80 (*iv*) 75

Exercise 1.7a

1 (*i*) $2x^3 + 5x^2 - 7x + 2$ (*ii*) $x^3 + 1$ (*iii*) $x^4 - 1$
(*iv*) $4x^4 - 4x^3 - 13x^2 + 23x - 12$
2 (*i*) $2x^2 - x - 2; +9$ (*ii*) $2x^2 - 2x - 1; +4$
(*iii*) $x^4 - x^3 + x^2 - x + 1; -1$ (*iv*) $x^3 + x^2 + x + 1; 0$
(*v*) $x^2 + x + 1; 2x + 4$

Exercise 1.7b

1 (*i*) 17 (*ii*) 7 (*iii*) 7 (*iv*) -9 (*v*) 0 (*vi*) 5
2 (*i*) $(x - 1)(x - 2)(x - 3)$; 1, 2, 3 (*ii*) $x^2(x - 1)$; 0, 1
(*iii*) $(x - 1)(x + 1)^2$; ± 1 (*iv*) $(x - 2)(x + 2)(2x - 1)$; $\pm 2, \frac{1}{2}$
(*v*) $-x^2(x - 2)(x + 2)$; 0, ± 2
(*vi*) $(x + 1)(x - 1)(x - 3)(x - 5)$; -1, 1, 3, 5
3 (*ii*) 2, -10; $2x + 5$ 4 $-5, -3$; $(x + 1)^2(x - 3)$
5 (*i*) $(x - a)(x^2 + ax + a^2)$ (*ii*) $(x + a)(x^2 - ax + a^2)$

Exercise 1.7c

1 (*i*) 6, 11, 6 (*ii*) 0, -1, 0 (*iii*) $-1, -1, 1$ (*iv*) $\frac{1}{2}, -4, -2$

Exercise 1.8

1 $y \neq 0; x = 1, y = 0$
2 $y \neq 0; x = -1, y = 0$
3 $y \geqslant 2; \leqslant -2; x = 0, y = x$
4 All y; $x = 0, y = x$
5 $y \leqslant -4, > 0; x = 1, 2; y = 0$
6 $0 \leqslant y < 1; y = 1$
7 $y \leqslant 0, > 1; x = \pm 2$
8 $y > 5.83, < 0.17; x = -1, -2; y = 0$

Exercise 1.9a

1 (*i*) 16π (*ii*) 32π (*iii*) 20 (*iv*) 16
2 (*i*) $f(10, 5) = 500\pi = 1\,571$ m^3 (*ii*) $\phi(1, 1, 3) = 13$
3 (*i*) $2\pi r^2 + 2\pi rh$ (*ii*) 24π (*iii*) $\frac{3}{2}\pi c^2$
4 (*i*) lbh (*ii*) $4(l + b + h)$

Miscellaneous problems 1

1 $\frac{3}{4}$ m **4** $-x + 2$
5 $\{P(a) - P(b)\}x + \{aP(b) - bP(a)\}$
6 2.98 **7** $\frac{1}{2}(\sqrt{5} - 1) = 0.618$

Exercise 2.1a

1 (*i*) $\frac{1}{3}$ (*ii*) $\frac{1}{16}$ (*iii*) $\frac{8}{27}$ (*iv*) $-\frac{1}{5}$ (*v*) 64 (*vi*) $\frac{1}{64}$ (*vii*) 9 (*viii*) 1 (*ix*) 1 (*x*) 0
2 (*i*) 729 (*ii*) $\frac{1}{4}$ (*iii*) 64 (*iv*) $\frac{1}{100\,000}$ (*v*) $\frac{1}{125}$
3 (*i*) 729 (*ii*) $\frac{1}{64}$ (*iii*) 1 000 000 (*iv*) 16 (*v*) 1

Exercise 2.1b

1 (*i*) 5 (*ii*) 2 (*iii*) 3 (*iv*) 4 (*v*) 27 (*vi*) 32 (*vii*) 1 (*viii*) $\frac{1}{2}$ (*ix*) $\frac{1}{2}$ (*x*) $\frac{1}{27}$ (*xi*) $\frac{1}{32}$ (*xii*) $\frac{1}{8}$
2 (*i*) $\frac{1}{5}$ (*ii*) $\sqrt{2}$

Exercise 2.2

1 (*i*) 6.3 (*ii*) 20.0 (*iii*) 79.4
2 (*i*) 0.70 (*ii*) 1.43 (*iii*) 2.08
4 (*i*) 1.58 (*ii*) 1.58 (*iii*) 1.58

Exercise 2.3

1 (*i*) 10.25% (*ii*) 15.76% (*iii*) 21.55%
2 (*i*) 121 (*ii*) 133 (*iii*) 146 (*iv*) 19
3 23.7° C **4** (*i*) 7.84% (*ii*) 11.53%

Exercise 2.4

1 (*i*) 10^5; 5 (*ii*) $10^{3/2}$; 1.5 (*iii*) 10^0; 0 (*iv*) 10^{-2}; -2 (*v*) $10^{1/3}$; $\frac{1}{3}$ (*vi*) $10^{-1/2}$; -0.5
2 (*i*) 1.301 03 (*ii*) 2.477 12 (*iii*) $-1.522\,88$ ($\bar{2}.477\,12$) (*iv*) $-2.698\,97$ ($\bar{3}.301\,03$) (*v*) 0.778 15 (*vi*) 0.602 06 (*vii*) $-0.301\,03$ ($\bar{1}.698\,97$) (*viii*) 0.176 09
3 (*i*) 6.89 (*ii*) 2.96 (*iii*) 921 (*iv*) 118 (*v*) 6.42 (*vi*) 5.19(5) (*vii*) 0.012 5 (*viii*) 0.069 7 (*ix*) 37.3 (*x*) 0.244

Exercise 2.5

1 (*i*) 0.602 060 (*ii*) 0.698 970 (*iii*) 0.903 090 (*iv*) 0.954 242 (*v*) 1 (*vi*) 1.079 181 (*vii*) 1.176 091 (*viii*) 1.477 121 (*ix*) 2.255 272 (*x*) $-0.602\,060$ ($\bar{1}.397\,940$) (*xi*) 0.176 091 (*xii*) $-0.176\,091$ ($\bar{1}.823\,909$) (*xiii*) 0.690 105 (5) (*xiv*) 0.401 373 (*xv*) $-0.238\,560$ (5) $\{\bar{1}.761\,439\ (5)\}$
2 (*i*) 4.19 (*ii*) 19.9 (*iii*) 8.96 (*iv*) 0.477

3 (*i*) 1.404 (*ii*) 1.431 (*iii*) 5.321 (*iv*) 0.631
4 (*i*) 0.149 (*ii*) 6.21
5 5.11 cm
6 (*i*) £163 (*ii*) £259
7 (*i*) 14.2 years (*ii*) 7.26 years
8 20.5
9 (*i*) £377 (*ii*) 4.26 years
10 2 880 years
11 18.3%
12 13%

Exercise 2.7

1 (*ii*) 2.72, 1.04 (*iii*) 4.3% (*iv*) 14.5×10^6 (*v*) 16.5
2 (*ii*) 2, 0.5 (*iii*) 4.47 s (*iv*) 0.25 m
3 $P = 2\,000(\frac{4}{5})^n$
4 $y = 2.4x^{1.6}$

Miscellaneous problems 2

1 (*i*) 1, 1, 1 (*ii*) 4, 16, 16 (*iii*) 27, 19 683, 7 625 597 484 987
2 (*i*) 19% (*ii*) 27.1% (*iii*) 34.4% 6.57 min
3 1.06; 322, 384
5 (*i*) 2.44 (*ii*) 2.59 (*iii*) 2.70 (*iv*) 2.72
Approximately 2.72 (or, more precisely 2.718 282 = e)

Exercise 3.1a

3 80 m, 4 s
4 8 s
5 30, 15, −15, 0 m s^{-1}

Exercise 3.1b

1 (*i*) 19.5, 19.95, 20 m s^{-1} (*ii*) 10 − 5h, 10 m s^{-1}
(*iii*) −30 m s^{-1} (*iv*) $40 - 10t$ m s^{-1}
2 (*i*) 20.5, 20.05, 20 m s^{-1}
3 (*i*) −0.09 m (*ii*) 29.91, −0.09 m s^{-1} (*iii*) $10t - \frac{1}{100}t^2$ m s^{-1}

Exercise 3.2

1 (*i*) 6 (*ii*) 0 (*iii*) −7 (*iv*) 24 (*v*) 0 (*vi*) $-\frac{3}{4}$ (*vii*) −2
2 (*i*) $2x$; $x > 0$, $x < 0$, $x = 0$
(*ii*) $6x^2$; $|x| > 0$, never, $x = 0$
(*iii*) $8x + 1$; $x > -\frac{1}{8}$, $x < -\frac{1}{8}$, $x = -\frac{1}{8}$
(*iv*) $12x^3 + 12$; $x > -1$, $x < -1$, $x = -1$
(*v*) 0; never, never, always
(*vi*) $-\dfrac{3}{x^2}$ $(x \neq 0)$; never, always, never
(*vii*) $-\dfrac{2}{x^3}$ $(x \neq 0)$; $x < 0$, $x > 0$, never

Exercise 3.3a

1 $5x^4$
2 $8x^7$
3 $\frac{4}{3}x^{1/3}$
4 $\frac{1}{3}x^{-2/3}$
5 $\frac{3}{2}\sqrt{x}$
6 $-\dfrac{3}{x^4}$

7 $-\dfrac{4}{x^5}$ **8** $-\frac{1}{2}x^{-3/2}$ **9** $-\frac{3}{2}x^{-5/2}$

10 $-\frac{1}{3}x^{-4/3}$

Exercise 3.3b

1 $4x^3 + 3x^2$ **2** $5x^4 - 2x$ **3** $12x^3 + 2$

4 $20x^4 - 15x^2$ **5** $x^{-1/2}$ **6** $1 - \dfrac{1}{x^2}$

7 $-x^{-4}$ **8** $4x^3 + 4x$ **9** $2x + 2x^{-3}$

10 $x^{-2/3}$

Exercise 3.4a

1 (*i*) $15x^2, 30x$ (*ii*) $4x^3 + 1, 12x^2$ (*iii*) $4x^3 - 2x, 12x^2 - 2$
(*iv*) $5x^9, 45x^8$

2 (*i*) $-6x^{-7}, 42x^{-8}$ (*ii*) $-42x^{-4}, 168x^{-5}$ (*iii*) $-2x^{-3}, 6x^{-4}$
(*iv*) $-15x^{-4}, 60x^{-5}$

3 (*i*) $\frac{1}{3}x^{-2/3}, -\frac{2}{9}x^{-5/3}$ (*ii*) $\frac{1}{2}x^{-1/2}, -\frac{1}{4}x^{-3/2}$
(*iii*) $3x^{1/2}, \frac{3}{2}x^{-1/2}$ (*iv*) $-\frac{1}{2}x^{-3/2}, \frac{3}{4}x^{-5/2}$

4 (*i*) $6x^5 + 6x^2, 30x^4 + 12x$ (*ii*) $2 - \dfrac{3}{x^2}, \dfrac{6}{x^3}$
(*iii*) $\frac{1}{4}x^{-1/2} - \frac{1}{4}x^{-3/2}, -\frac{1}{8}x^{-3/2} + \frac{3}{8}x^{-5/2}$
(*iv*) $1 + x^{-1/2}, -\frac{1}{2}x^{-3/2}$

Exercise 3.4b

1 (*i*) $2x - 2, 2$ (*ii*) $x > 1; x < 1; x = 1$
(*iii*) Always; never; never

2 (*i*) $3x^2 - 3, 6x$ (*ii*) $|x| > 1; |x| < 1; x = \pm 1$
(*iii*) $x > 0; x < 0; x = 0$

3 (*i*) $6x - 6x^2, 6 - 12x$ (*ii*) $0 < x < 1; x < 0, x > 1; x = 0, 1$
(*iii*) $x < \frac{1}{2}; x > \frac{1}{2}; x = \frac{1}{2}$

4 (*i*) $4x^3 - 16x, 12x^2 - 16$
(*ii*) $-2 < x < 0, x > 2; x < -2, 0 < x < 2; x = 0, \pm 2$
(*iii*) $|x| > \dfrac{2}{\sqrt{3}}; |x| < \dfrac{2}{\sqrt{3}}; x = \pm\dfrac{2}{\sqrt{3}}$

5 (*i*) $3x^2 - 6x - 9, 6x - 6$
(*ii*) $x < -1, x > 3; -1 < x < 3; x = -1, 3$
(*iii*) $x > 1; x < 1; x = 1$

Exercise 3.5

1 (*i*) $6x$ (*ii*) $-\dfrac{2}{x^2}$

2 (*i*) $3x^2 + 6x, 6x + 6$ (*ii*) $4, 0$ (*iii*) $-\dfrac{2}{x^3}, \dfrac{6}{x^4}$
(*iv*) $-\frac{1}{2}x^{-3/2}, \frac{3}{4}x^{-5/2}$

3 (*i*) $4x^3$, $12x^2$ (*ii*) $6x$, 6 (*iii*) $x^{-1/2}$, $-\frac{1}{2}x^{-3/2}$ (*iv*) $-\frac{1}{x^2}, \frac{2}{x^3}$

4 (*i*) 4; 75° 58′; $y = 4x - 4$; $x + 4y = 18$
(*ii*) 1; 45°; $y = x - 2$; $y = -x$
(*iii*) -1; 135°; $x + y = 1$; $y = x - 1$
(*iv*) $-\frac{1}{2}$; 153° 26′; $x + 2y + 4 = 0$; $y = 2x + 3$

5 $8x + y + 2 = 0$, $x - y = \pm 16$

6 $9x - y = 16$, $(-4, -52)$

7 250 m, 75° 58′

8 (*i*) 9 000 m (*ii*) 18° 26′, 0°, −18° 26′

9 (*i*) (2, 4) (*ii*) 40° 36′

Exercise 3.6

1 35 m s^{-1}, 20 m s^{-2} 2 5 m s^{-1}, −1 m s^{-2}; 32 m

3 (*i*) 30 m s^{-1}, 42 m s^{-2} (*ii*) 6.75 m

4 4 s, 11 m s^{-2} 5 ±12 m s^{-2}

7 (*i*) 180 m (*ii*) 15 s (*iii*) $24 - \frac{8}{5}t$, $-\frac{8}{5}$

Exercise 3.7

1 (*i*) $\frac{1}{2}x^4 + c$ (*ii*) $x - \frac{1}{2}x^2 + c$ (*iii*) $x^3 + 2x + 4$ (*iv*) $-\frac{1}{x} + \frac{3}{2}$

2 (*i*) $y = 2x^2 - 5x + c$ (*ii*) $y = 2x^2 - 5x + 4$ (*iii*) $y = 3x - 1$
(*iv*) $y = -\frac{1}{2}x^2 + \frac{3}{2}$

3 (*i*) $5t^2 + c$ (*ii*) $2t^4 - 2t^3 + c$ (*iii*) $t^3 + 2t^2 + 2$ (*iv*) $t - t^2 + 3$

4 (*i*) $3t^2 - 8t + 6$, $t^3 - 4t^2 + 6t + 4$
(*ii*) $-10t + 12$, $-5t^2 + 12t - 2$

5 6.75 m, 4 m s^{-1} 6 3 s, 9 m s^{-2}; 162 m

7 (*i*) 1.6 h (*ii*) 102.4 km (*iii*) 96 km h^{-1}

8 (*i*) $1 - \frac{1}{9}t$, $\frac{1}{2}t^2 - \frac{1}{54}t^3 + 6$ (*ii*) 18 s, 27 s

9 (*i*) $12t - 48$ (*ii*) $2t^3 - 24t^2 + 42t$
1 s, 7 s; 2.1 s

10 100 m s^{-1}, 1 333 m 11 4.5 m, 5 s

12 (*i*) 5 m s^{-2} (*ii*) 15 s (*iii*) 56.25 m s^{-1}

Exercise 3.8

1 (*i*) $f(1) = -9$, min. (*ii*) $f(1) = 4$, max.
(*iii*) $f(-1) = 2$, max; $f(1) = -2$, min. (*iv*) None

2 (*i*) (2, 4), min; (−2, −4), max.
Nonc
(*ii*) (1, 3), min. p.i. at $(-2^{1/3}, 0)$
(*iii*) (0, 0), min; (±1, 2), max.
p.i. at $\left(\pm\sqrt{\frac{3}{5}}, \frac{162}{125}\right)$

3 (*i*) (1, −1), min; (−2, 26), max.
(*ii*) (2, 1), min; (−1, 28), max. (*iii*) +1, −3

4 (*i*) $y = x^3 - x^2 - x + 1$
(*ii*) $(-\frac{1}{3}, \frac{32}{27})$, max; $(1, 0)$, min.

5 $800\ \text{m}^2$

6 $128\ \text{cm}^3$

7 $x^2 + \dfrac{16}{x}$, 2m

8 $4 \times 12 \times 6$ (cm)

9 $6\frac{2}{3} \times 10 \times 4$ (cm)

10 $\frac{2}{27}\ \text{m}^3$

11 £2 400; $20 \times 20 \times 5$ (m)

12 84 cm, 66 cm

13 $2a^2, \sqrt{3}a^2$

14 £$5\frac{13}{16}$ min^{-1} (*ii*) £5

15 $\dfrac{4a}{25}, \dfrac{3a}{25}$

16 $\dfrac{1}{\sqrt{3}} = 0.577$

Exercise 3.9a

1 (*i*) $24x(3x^2 + 1)^3$ (*ii*) $180x(9x^2 + 4)^9$ (*iii*) $-12x(1 - x^2)^5$
(*iv*) $5(2x - 1)(x^2 - x)^4$ (*v*) $-6x(3x^2 - 1)^{-2}$ (*vi*) $-\dfrac{2x}{(x^2 - 1)^2}$
(*vii*) $-\dfrac{6x}{(x^2 + 2)^4}$ (*viii*) $\dfrac{x}{\sqrt{(1 + x^2)}}$ (*ix*) $\dfrac{2x}{3}(x^2 - 1)^{-2/3}$
(*x*) $\frac{1}{2}(1 - x)^{-3/2}$ (*xi*) $\dfrac{5}{\sqrt{x}}(\sqrt{x} + 1)^9$ (*xii*) $2anx(ax^2 + b)^{n-1}$

2 (*i*) $12t^2(t^3 + 1)^3$ (*ii*) $-30t(1 - 3t^2)^4$ (*iii*) $-\dfrac{1}{2\sqrt{(1 - t)}}$
(*iv*) $-\dfrac{t}{(t^2 + 1)^{3/2}}$

Exercise 3.9b

1 (*i*) $(1 + 6x)(1 + x)^5$ (*ii*) $2x(9x + 1)(3x + 1)^3$
(*iii*) $(5x + 7)(x + 1)(x + 2)^2$ (*iv*) $(32x^3 + 1)(2x^3 + 1)^4$
(*v*) $\dfrac{2}{(x + 1)^2}$ (*vi*) $\dfrac{1 - 2x^2}{(2x^2 + 1)^2}$ (*vii*) $-\dfrac{4x}{(1 + x^2)^2}$
(*viii*) $\dfrac{1 - 3x}{(3x + 1)^3}$ (*ix*) $\dfrac{1}{\sqrt{x}(1 - \sqrt{x})^2}$ (*x*) $-\dfrac{1}{(1 + x)\sqrt{(1 - x^2)}}$

2 (*i*) Max. at $(0, 0)$; Min. at $(2, 4)$
(*ii*) Max. at $(\sqrt{3}, -\frac{3}{2}\sqrt{3})$; Min. at $(-\sqrt{3}, \frac{3}{2}\sqrt{3})$

Exercise 3.10a

1 (*i*) $3.77\ \text{m}^2\ \text{s}^{-1}$ (*ii*) $0.53\ \text{m s}^{-1}$

2 $0.113\ \text{m s}^{-1}$

3 $\dfrac{2x}{3}$, $2\ \text{m s}^{-1}$

4 $\frac{4}{3}\ \text{m s}^{-1}$

5 $0.025\,5\ \text{cm s}^{-1}$

6 $0.5\ \text{atm s}^{-1}$

Exercise 3.10b

1 (*i*) $\dfrac{x}{y}$, 3 (*ii*) $-\dfrac{x + 2y}{2x + y}$, -1 (*iii*) $-\dfrac{x^2}{y^2}$, $-\dfrac{1}{4}$

2 (*i*) $\frac{1}{3t}$ (*ii*) $-\frac{1}{4t^2}$ (*iii*) $\frac{2t}{3t^2-1}$

3 $t = 4$; $x = 40$; 76°

4 (*i*) $y^2 = x^3 + 3x^2$

(*iv*) $\frac{3(t^2-1)}{2t}$; $(-2, \pm 2)$

Exercise 3.11a

1 (*i*) 8 120 (*ii*) 3.007 (*iii*) 3.987

2 $2\pi = 6.28$ m

3 $\delta V = \pi x(2r - x)\,\delta x$; 1.06 cm

4 4%, 6% **5** 2p, 3p **6** $\frac{1}{2}$

Exercise 3.11b

2 1.879 **3** 2.055 **4** 0.66

Exercise 3.12

1 80 m, 50 m s^{-1} **2** 0.2 m s^{-2}, 2.88 km

3 4 m s^{-2}, 5 s **4** 1 m s^{-2}, 22 m s^{-1}

5 $4\frac{3}{4}$ min **6** 1.45 m s^{-2}, 417 m

7 $\frac{5}{12}$ m s^{-2}, $40\frac{5}{6}$ m **8** 7.1 s, 5.2 s

Miscellaneous problems 3

2 $(0, 0)$, $(\pm\sqrt{3}a, 9a^4)$

3 $\frac{h(a-x)}{a}$ **4** $h = r, 2r, 2r + a$

5 1.259 9 **6** 1.47 **7** $2\times3^{-3/4}$ **9** 1

Exercise 4.1

1 35 **2** 250 **3** 22

4 3.75 **5** 10.5 **6** 4

Exercise 4.2

1 $\frac{1}{3}$

Exercise 4.4

1 (*i*) $\frac{1}{3}x^3 + c$ (*ii*) $x^3 - 2x^2 + 2x + c$ (*iii*) $\frac{1}{3}x^3 - \frac{1}{2}x^4 + c$
(*iv*) $\frac{1}{4}x^4 - \frac{2}{3}x^3 + c$

2 (*i*) $\frac{1}{3}t^3 - t^2 + 3t + c$ (*ii*) $\frac{1}{3}t^3 + \frac{5}{2}t^2 + c$
(*iii*) $\frac{1}{3}t^3 + \frac{3}{2}t^2 + 2t + c$ (*iv*) $\frac{4}{3}t^3 + 2t^2 + t + c$

3 (*i*) $\frac{1}{6}u^6 + \frac{1}{2}u^2 + c$ (*ii*) $\frac{1}{5}u^5 + \frac{2}{3}u^3 + u + c$
(*iii*) $\frac{1}{3}u^6 - \frac{2}{3}u^3 + c$ (*iv*) $\frac{1}{9}u^9 + \frac{4}{5}u^5 + 4u + c$

4 (*i*) $\frac{2}{7}x^{7/2} + c$ (*ii*) $\frac{3}{4}x^{4/3} + c$ (*iii*) $\frac{2}{5}x^{5/2} + \frac{2}{3}x^{3/2} + c$
(*iv*) $\frac{1}{2}x^2 + 4x^{3/2} + 9x + c$

5 (*i*) $-\frac{3}{x} + c$ (*ii*) $-\frac{1}{2x^2} - \frac{1}{x^3} + c$

(*iii*) $\frac{2}{3}x^{3/2} + 2x^{1/2} + c$ (*iv*) $-\frac{2}{\sqrt{x}} - \frac{5}{x} + c$

6 (*i*) $\frac{1}{2}ax^2 + bx + c$ (*ii*) $\frac{1}{3}ax^3 + \frac{1}{2}bx^2 + cx + d$
(*iii*) $\frac{2}{3}ax^{3/2} + bx + c$ (*iv*) $\frac{2}{3}ax^{3/2} + 2bx^{1/2} + c$

7 $\frac{x^{n+1}}{n+1} + c \ (n \neq -1)$

8 $\frac{n}{n+1}u^{1/n+1} + c \ (n \neq -1)$

Exercise 4·5

1 (*i*) $8\frac{2}{3}$ (*ii*) 44 (*iii*) $6\frac{1}{5}$ (*iv*) $2\frac{1}{4}$
2 (*i*) $4\frac{2}{3}$ (*ii*) $11\frac{1}{4}$ (*iii*) $\frac{1}{2}$ (*iv*) $\frac{1}{3}$
3 (*i*) $3\frac{3}{4}$ (*ii*) $\frac{1}{6}$ (*iii*) $3\frac{5}{6}$ (*iv*) $1\frac{1}{3}$
4 (*i*) $10\frac{2}{3}$ (*ii*) $1\frac{1}{3}$ (*iii*) $2\frac{2}{3}$ (*iv*) $\frac{4}{3}$

Exercise 4.6

1 (*i*) $-10\frac{2}{3}$ (*ii*) $5\frac{1}{3}$ 2 (*i*) 0 (*ii*) 8
3 (*i*) 45 (*ii*) -45 (*iii*) 0 (*iv*) -80
4 (*i*) $2\frac{14}{15}$ (*ii*) $3\frac{1}{6}$

Exercise 4.7

1 (*i*) 259 m^2 (*ii*) 256 m^2
2 (*i*) 116 m (*ii*) 117 m
3 (*i*) 2.96 m s^{-1} (*ii*) 2.85 m s^{-1}
4 (*i*) 0.708 3 (2.2%) (*ii*) 0.694 5 (0.2%)
5 (*i*) 0.693 8 (0.1%) (*ii*) 0.693 1 (0.0%)
6 (*i*) 3.104 (1.2%) (*ii*) 3.127 (0.5%)
7 3.143 8 (*i*) 0.845 (*ii*) 0.812

Exercise 4.8

1 (*i*) $\frac{7\pi}{3}$ (*ii*) $\frac{28\pi}{15}$ (*iii*) $\frac{\pi}{6}$ (*iv*) 8π (*v*) $\frac{2\pi}{15}$ (*vi*) $\frac{\pi}{30}$
2 (*i*) 3 (*ii*) $\frac{14}{3}$ (*iii*) 12 (*iv*) $\frac{28}{3}$ (*v*) $\frac{8}{3}$
3 (*i*) $\frac{28\pi}{3}$ (*ii*) $\frac{15\pi}{2}$ (*iii*) 32π (*iv*) 6π (*v*) $\frac{32\pi}{5}$
4 $\frac{1}{3}\pi r^2 h$ 5 $\frac{4}{3}\pi r^3$ 6 $\frac{128}{3}$, 128π
7 (0, 0), (2, 4); $\frac{32\pi}{5}$ 8 $\frac{5}{3}, \frac{20\pi}{7}$
9 $\frac{2}{3}\pi a^3$ 10 $\frac{16}{3}, \frac{96\pi}{5}$
11 162 m^3 12 $\frac{256\pi}{3} = 268$ cm^3

Exercise 4.9

1 (*i*) $\frac{3}{2}$ (*ii*) 44 (*iii*) 1.22

2 (*i*) 3 142 kg (*ii*) 750 kg m^{-3} **3** 7.5 cm

Exercise 4.10

1 (*i*) $\frac{1}{6}(x + 1)^6 + c$ (*ii*) $\dfrac{1}{2(1 - 2x)} + c$

(*iii*) $-\frac{2}{3}(2 - x)^{3/2} + c$ (*iv*) $\frac{1}{5}(2x + 1)^{5/2} + c$

(*v*) $\frac{3}{10}(2x - 3)(1 + x)^{2/3} + c$

2 (*i*) $\frac{1}{10}(x^2 + 1)^5 + c$ (*ii*) $\frac{1}{30}(2x^3 + 3)^5 + c$

(*iii*) $-\frac{1}{3}(1 - x^2)^{3/2} + c$ (*iv*) $2\sqrt{(x + x^2)} + c$

(*v*) $\frac{3}{4}(x^2 - 1)^{2/3} + c$

3 (*i*) $\frac{1}{5}$ (*ii*) $\frac{1}{35}$ (*iii*) $\frac{1}{3}(10\sqrt{10} - 1) = 10.2$

(*iv*) $\frac{2}{3}\sqrt{7} = 1.76$ (*v*) $\frac{4}{3}(3\sqrt{3} - 2\sqrt{2}) = 3.157$

Miscellaneous problems 4

1 Between 13.8 and 14.8

2 (*i*) 0 (*ii*) 4 **3** $\dfrac{8}{15}, \dfrac{\pi}{12}$

5 $(4, \pm 4), \dfrac{32\pi}{5}$ **6** $\frac{1}{6}\pi h^3$

7 $\dfrac{4\pi ab^2}{3}$ **8** Approx. 400 000 tonne

Exercise 5.1a

1 (*i*) 0.57, 0.82 (*ii*) 0.22(5), −0.97 (*iii*) −0.56, −0.83

(*iv*) −0.83, 0.56 (*v*) 0.64, 0.77 (*vi*) −0.53, 0.85

2 (*i*) 0.70 (*ii*) −0.23 (*iii*) 0.67 (*iv*) −1.48 (*v*) 0.84 (*vi*) −0.62

3 (*i*) $\sin\theta, -\cos\theta, -\tan\theta$ (*ii*) $-\sin\theta, -\cos\theta, +\tan\theta$

(*iii*) $-\cos\theta, \sin\theta, -\dfrac{1}{\tan\theta}$

Exercise 5.1b

1 (*i*) $\dfrac{\sqrt{3}}{2}, -\dfrac{1}{2}, -\sqrt{3}$ (*ii*) $-\dfrac{1}{\sqrt{2}}, -\dfrac{1}{\sqrt{2}}, +1$

(*iii*) $-\dfrac{1}{2}, \dfrac{\sqrt{3}}{2}, -\dfrac{1}{\sqrt{3}}$

Exercise 5.2

1 (*i*) 21° 43′, 158° 17′ (*ii*) 90° (*iii*) 227° 44′, 312° 16′

2 (*i*) 63° 15′, 296° 45′ (*ii*) 127° 36′, 232° 24′ (*iii*) 180°

3 (*i*) 135°, 315° (*ii*) 0°, 180°, 360° (*iii*) 69° 49′, 249° 49′

4 (*i*) 0°, 180°, 270°, 360° (*ii*) 0°, 60°, 300°, 360°
(*iii*) 30°, 150°, 210°, 330°

5 (*i*) 136° 50′, 223° 10′ (*ii*) 243° 4′
(*iii*) 12° 14′, 102° 14′, 192° 14′, 282° 14′

Exercise 5·3

2 (*i*) $\sin\theta$ (*ii*) $\tan\theta$ (*iii*) $\sec\theta$ (*iv*) $\operatorname{cosec}\theta$

3 (*i*) $\cot^2\theta$ (*ii*) 2 (*iii*) $\sec\theta - 1$ (*iv*) $\sin\theta\cos\theta$ (*v*) $\cot^2\theta$
(*vi*) $\sec^2\theta$ (*vii*) $\sec^2\theta$ (*viii*) $\sin\theta + \cos\theta$

4 (*i*) 60°, 300° (*ii*) 135°, 315° (*iii*) 0°, 120°, 240°, 360°
(*iv*) 0°, 45°, 180°, 225°, 360° (*v*) 41° 49′, 138° 11′
(*vi*) 60°, 300°

5 (*i*) $\frac{5}{3}, \frac{4}{3}$

Exercise 5·4

1 (*i*) $b = 2.54$ $c = 4.77$ $C = 70°$
(*ii*) $a = 6.71$ $c = 4.22$ $C = 37° 14'$

2 (*i*) $A = 41° 24'$ $B = 55° 46'$ $C = 82° 49'$
(*ii*) $A = 26° 44'$ $B = 36° 32'$ $C = 116° 44'$

3 (*i*) $a = 3.09\,(5)$ $B = 48°$ $C = 82°$
(*ii*) $c = 6.58$ $A = 44° 2'$ $B = 32° 18'$

4 $A = 56° 26'$, $C = 93° 34'$, $c = 5.99$ or
$A = 123° 34'$, $C = 26° 26'$, $c = 2.67$

5 8° 36′, 154 n.m.

6 (*i*) 60° (*ii*) 13 cm

8 (*i*) 33° 33′ (*ii*) 4.92 cm

Exercise 5·5

1 (*i*) $\dfrac{\sqrt{3}+1}{2\sqrt{2}} = \dfrac{1}{4}(\sqrt{6}+\sqrt{2})$; $\dfrac{1-\sqrt{3}}{2\sqrt{2}} = -\dfrac{1}{4}(\sqrt{6}-\sqrt{2})$;
$\dfrac{\sqrt{3}+1}{-1-\sqrt{3}} = -2-\sqrt{3}$

(*ii*) $\dfrac{\sqrt{3}-1}{2\sqrt{2}} = \dfrac{1}{4}(\sqrt{6}-\sqrt{2})$; $\dfrac{\sqrt{3}+1}{2\sqrt{2}} = \dfrac{1}{4}(\sqrt{6}+\sqrt{2})$;
$\dfrac{\sqrt{3}-1}{\sqrt{3}+1} = 2-\sqrt{3}$

2 (*i*) $\cos x$, $-\sin x$, $-\cot x$
(*ii*) $\sin x$, $-\cos x$, $-\tan x$

3 (*i*) $\dfrac{1}{\sqrt{2}}(\cos x + \sin x)$, $\dfrac{1}{\sqrt{2}}(\cos x - \sin x)$, $\dfrac{1+\tan x}{1-\tan x}$

(*ii*) $\frac{1}{2}(\sqrt{3}\cos x - \sin x)$, $\frac{1}{2}(\cos x + \sqrt{3}\sin x)$, $\dfrac{\sqrt{3}-\tan x}{1+\sqrt{3}\tan x}$

4 (*i*) $\sin 60°$ (*ii*) $\cos 50°$ (*iii*) $\sin(2\alpha - \beta)$
(*iv*) $\cos(2\theta - 3\phi)$ (*v*) $\tan(60° + \theta)$ (*vi*) $\tan(\theta - 45°)$

5 (*i*) $2\sin A\cos B$ (*ii*) $-2\sin A\sin B$ (*iii*) $\cot B$
(*iv*) $\tan A - \tan B$ (*v*) $\cot B - \tan A$ (*vi*) -1

6 $\dfrac{\tan A + \tan B + \tan C - \tan A\tan B\tan C}{1 - \tan B\tan C - \tan C\tan A - \tan A\tan B}$
(*i*) 45°

Exercise 5.6

1 (*i*) $\sin 30° = \frac{1}{2}$ (*ii*) $\cos 30° = \dfrac{\sqrt{3}}{2}$ (*iii*) $\cos 30° = \dfrac{\sqrt{3}}{2}$
(*iv*) $\cos 30° = \dfrac{\sqrt{3}}{2}$ (*v*) 1 (*vi*) $\tan 30° = \dfrac{1}{\sqrt{3}}$

2 (*i*) $\frac{24}{25}$ (*ii*) $\frac{7}{25}$ (*iii*) $\frac{24}{7}$

3 (*i*) $\sqrt{\left(\dfrac{\sqrt{2}-1}{2\sqrt{2}}\right)}$ (*ii*) $\sqrt{\left(\dfrac{\sqrt{2}+1}{2\sqrt{2}}\right)}$ (*iii*) $\sqrt{2} - 1$

4 (*i*) $3\sin A - 4\sin^3 A$ (*ii*) $4\cos^3 A - 3\cos A$

5 (*i*) $\tan\dfrac{A}{2}$ (*ii*) $\tan x$ (*iii*) $\cos 2\theta$ (*iv*) $\tan 2A$ (*v*) $\tan\theta$ (*vi*) 2

6 (*i*) $8\cos^4 A - 8\cos^2 A + 1$ (*ii*) $\dfrac{4\tan A - 4\tan^3 A}{1 - 6\tan^2 A + \tan^4 A}$

7 $\dfrac{1 + \sqrt{3} - \sqrt{6}}{\sqrt{3} + \sqrt{2} - 1}$

8 (*i*) 30°, 90°, 150°, 270° (*ii*) 90°, 180°, 270°
(*iii*) 105°, 165°, 285°, 345°
(*iv*) 0°, 48° 35′, 131° 25′, 180°, 270°, 360°

Exercise 5.7

1 (*i*) $\sqrt{2}\sin(x + 45°)$; $\pm\sqrt{2}$
(*ii*) $\sqrt{13}\sin(x + 56°\ 19')$; $\pm\sqrt{13}$
(*iii*) $5\sin(x - 36°\ 52')$; ± 5
(*iv*) $-\sqrt{10}\sin(x - 18°\ 26')$; $\pm\sqrt{10}$

2 (*i*) 0°, 90°, 360° (*ii*) 139° 47′, 286° 35′
(*iii*) 60° 27′, 193° 17′ (*iv*) 57° 40′, 159° 12′

4

0°	10°	20°	30°	40°	50°	60°	70°	80°	
0.000	0.505	0.910	1.150	1.221	1.173	1.082	1.015	0.995	
90°	100°	110°	120°	130°	140°	150°	160°	170°	
1.000	0.975	0.864	0.650	0.359	0.065	−0.149	−0.226	−0.158	
180°	190°	200°	210°	220°	230°	240°	250°	260°	
0.000	0.158	0.226	0.149	−0.650	−0.359	−0.650	−0.864	−0.975	
270°	280°	290°	300°	310°	320°	330°	340°	350°	360°
−1.000	−0.995	−1.015	−1.092	−1.173	−1.221	−1.15	−0.910	−0.505	0.000

Exercise 5.8

1 (*i*) $2 \sin 4\theta \cos \theta$ (*ii*) $2 \sin A \cos 3A$
(*iii*) $2 \cos x \cos 60° = \cos x$ (*iv*) $-\sqrt{2} \sin x$

2 (*i*) $\cot \dfrac{A-B}{2}$ (*ii*) $\tan \dfrac{A+B}{2}$ (*iii*) $-\cot 2A$ (*iv*) $\cot \dfrac{A}{2}$

5 (*i*) 0°, 30°, 60°, 120°, 150°, 180° (*ii*) 45°, 120°, 135°
(*iii*) 0°, 90°, 180° (*iv*) 0°, 10°, 50°, 90°, 130°, 170°, 180°

Exercise 5.9a

1 (*i*) $\dfrac{\pi}{18}$ (*ii*) $\dfrac{\pi}{5}$ (*iii*) $\dfrac{2\pi}{3}$ (*iv*) π (*v*) $\dfrac{3\pi}{2}$ (*vi*) $\dfrac{3\pi}{4}$
(*vii*) $\dfrac{\pi}{1\,080}$ (*viii*) $\dfrac{\pi}{64\,800}$

2 (*i*) 90° (*ii*) 270° (*iii*) 150° (*iv*) 120° (*v*) 135° (*vi*) 15°
(*vii*) 12° (*viii*) 18°

3 (*i*) $+1$ (*ii*) $\dfrac{1}{\sqrt{2}}$ (*iii*) $\sqrt{3}$ (*iv*) $\frac{1}{2}$ (*v*) -1 (*vi*) $\dfrac{1}{\sqrt{2}}$
(*vii*) $-\frac{1}{2}$ (*viii*) 0 (*ix*) 0 (*x*) $-\frac{1}{2}$

4 (*i*) $\cos x$ (*ii*) $-\cos x$ (*iii*) $-\cos x$ (*iv*) $-\sin x$
(*v*) $\cot x$ (*vi*) $\tan x$

5 (*i*) $\dfrac{\pi}{6}, \dfrac{5\pi}{6}$ (*ii*) $\dfrac{\pi}{4}, \dfrac{3\pi}{4}$ (*iii*) π (*iv*) $\dfrac{\pi}{3}, \dfrac{5\pi}{3}$ (*v*) $\dfrac{\pi}{4}, \dfrac{5\pi}{4}$ (*vi*) $\dfrac{2\pi}{3}, \dfrac{5\pi}{3}$

6 (*i*) 0.698 (*ii*) 0.807 (*iii*) 3.213 (*iv*) 6.886

7 (*i*) 29° 48′ (*ii*) 99° 7′ (*iii*) 294° 16′ (*iv*) 38′

8 (*i*) 6 m, 9 m^2 (*ii*) 1.571 m, 1.571 m^2 (*iii*) 1.745 m, 3.490 m^2

9 (*i*) 20.9 cm (*ii*) 43.3 cm (*iii*) 253 cm^2

10 $3\frac{1}{3}$ m, 33 m^2

11 (*i*) $\dfrac{2\pi r}{l}$ (*ii*) πrl 12 1.86

Exercise 5.9b

3 480 km, 1 390 000 km

Exercise 5.10

1 (*i*) $4;\ 2\pi = 6.28$ s; $\dfrac{1}{2\pi} = 0.159\ s^{-1}$; -1.66

(*ii*) $2;\ \dfrac{\pi}{5} = 0.628$ s; $\dfrac{5}{\pi} = 1.59\ s^{-1}$; 1.82

(*iii*) $5;\ \dfrac{\pi}{50} = 0.062\,8$ s; $\dfrac{50}{\pi} = 15.9\ s^{-1}$; -4.4

(*iv*) $a;\ \dfrac{1}{n};\ n;\ a \sin 4n\pi\ (= 0, \text{ if } n \in Z)$

2 (*i*) $10 \sin t$ (*ii*) $3 \sin \pi t$ (*iii*) $\frac{1}{2} \sin 200\pi t$

Exercise 5.11a

2 (*i*) $2\cos 2x$ (*ii*) $-3\sin 3x$ (*iii*) $2\pi\cos 2\pi x$ (*iv*) $-6\sin\dfrac{3x}{2}$

(*v*) $2\sin x\cos x$ (*vi*) $-3\cos^2 x\sin x$ (*vii*) $2x\cos 2x + \sin 2x$

(*viii*) $-\dfrac{1}{x^2}\cos x - \dfrac{1}{x}\sin x$ (*ix*) $2\cos 2x$ (*x*) $12\sin^2 4x\cos 4x$

(*xi*) $\dfrac{\cos x}{2\sqrt{\sin x}}$ (*xii*) $\dfrac{2x\sin x - x^2\cos x}{\sin^2 x}$

3 (*ii*) $\dfrac{x^2}{9} + \dfrac{y^2}{4} = 1$ (*iii*) $-\dfrac{2}{3}\cot\theta$

4 (*i*) $\dfrac{x^2}{a^2} + \dfrac{y^2}{b^2} = 1$ (*ii*) $-\dfrac{b}{a}\cot\theta$

5 (*i*) $3\cos t - \sin t$ m s^{-1}, $-3\sin t - \cos t$ m s^{-2}
(*ii*) 1.25 s (*iii*) -3.16 m s^{-2}, 5.16 m

6 (*i*) a, $2a$ (*ii*) $\dfrac{3a}{2}$

7 1.445, 3.142; 10 cm

8 (*ii*) $\sqrt{5} = 2.236$ cm (*iii*) -2.19 m s^{-1}

9 (*i*) $\dfrac{\pi}{3}$, max; $-\dfrac{\pi}{3}$, min. (*ii*) $\dfrac{\pi}{2}, \dfrac{3\pi}{2}$, max; $\dfrac{7\pi}{6}, \dfrac{11\pi}{6}$, min.

10 (*i*) 0.861 6 (*ii*) 0.711 2

11 0.739 (42° 21′)

12 223 s

13 $\frac{1}{2}(\sqrt{5} - 1) = 0.618$

Exercise 5.11b

1 (*i*) $3\sec^2 3x$ (*ii*) $4\sec 4x\tan 4x$ (*iii*) $-\frac{1}{2}\operatorname{cosec}^2\dfrac{x}{2}$

(*iv*) $-2\operatorname{cosec} 2x\cot 2x$ (*v*) $x\sec^2 x + \tan x$ (*vi*) $2\tan x\sec^2 x$

(*vii*) $3\sec^3 x\tan x$ (*viii*) $\dfrac{x\sec x\tan x - \sec x}{x^2}$

(*ix*) $-6\cot 3x\operatorname{cosec}^2 3x$ (*x*) $\dfrac{\sec^2 x}{2\sqrt{\tan x}}$

3 (*i*) $\operatorname{cosec}\theta$; $x^2 - y^2 = 1$; $\dfrac{x}{y}$

(*ii*) $\dfrac{b}{a}\sec\theta$; $\dfrac{x^2}{a^2} - \dfrac{y^2}{b^2} = 1$; $\dfrac{b^2x}{a^2y}$

4 $20\sec\theta - 5\tan\theta + 25$; 14° 29′, 44.4 s

Exercise 5.12

1 (*i*) $-\frac{1}{2}\cos 2x + c$ (*ii*) $2\sin\dfrac{x}{2} + c$ (*iii*) $-\frac{1}{3}$ (*iv*) $\frac{3}{2}$

2 (*i*) $\frac{1}{2}(x - \sin x\cos x) + c$ (*ii*) $\frac{1}{2}(x + \sin x) + c$ (*iii*) $\dfrac{\pi}{2}$ (*iv*) $\frac{1}{2}$

3 (*i*) $-\frac{1}{6}\cos 3x - \frac{1}{2}\cos x + c$ (*ii*) $\frac{1}{6}\sin 3x + \frac{1}{2}\sin x + c$
(*iii*) $\frac{1}{2}\sin x - \frac{1}{6}\sin 3x + c$

4 (*i*) $\frac{1}{2}\sin x^2 + c$ (*ii*) $-\frac{1}{3}\cos x^3 + c$ (*iii*) $\frac{1}{3}\sin^3 x + c$
(*iv*) $-\frac{1}{4}\cos^4 x + c$

5 (*i*) $\tan x + c$ (*ii*) $\sec x. + c$ (*iii*) $\tan x - x + c$
(*iv*) $-\cot x - x + c$

6 (*i*) 4π (*ii*) $9\pi^2$

7 $4 + 6\pi$

8 $\frac{1}{2}$

9 $\frac{1}{4} + \frac{\pi}{8}$

10 (*i*) π (*ii*) $\frac{1}{2}\pi - 1$

Exercise 5.13a

1 (*i*) $\frac{\pi}{6}, \frac{\pi}{3}, 0.464$ (*ii*) $-\frac{\pi}{2}, \pi, \frac{3\pi}{4}$ (*iii*) 0.775, 0.796, 0.611

2 $\sin^{-1} x + \cos^{-1} x = \frac{\pi}{2}$

3 (*i*) $\tan^{-1} 1 = \frac{\pi}{4}$ (*ii*) $\tan^{-1} \frac{x + y}{1 - xy}$

Exercise 5.13b

1 (*i*) $\frac{\pi}{6}$ (*ii*) $\frac{\pi}{12}$ (*iii*) $\frac{\pi}{4}$

2 (*i*) $\tan^{-1}(x + 1) + c$ (*ii*) $\sin^{-1}(x - 1) + c$
(*iii*) $\sin^{-1}\sqrt{x} - \sqrt{(x - x^2)} + c$

3 (*i*) $\frac{1}{2}\sin^{-1}\frac{2x}{3} + c$ (*ii*) $\frac{1}{2}\tan^{-1} 2x + c$

(*iii*) $\frac{1}{2}\left\{a^2 \sin^{-1}\frac{x}{a} + x\sqrt{(a^2 - x^2)}\right\} + c$

4 (*i*) $\sec^{-1} x + c$ (*ii*) $\frac{1}{2}\left(\tan^{-1} x + \frac{x}{1 + x^2}\right) + c$

Exercise 5.14

5 $x = r\cos\theta, y = r\sin\theta$

6 (*i*) $r = 2\sin\theta$ (*ii*) $r = \sec\theta\tan\theta$ (*iii*) $r^2 = 2\operatorname{cosec} 2\theta$

(*iv*) $r = \frac{1}{\cos\theta + \sin\theta}$

7 (*i*) $x^2 + y^2 = ax$ (*ii*) $y = a$ (*iii*) $(x^2 + y^2)^2 = 2a^2xy$
(*iv*) $x^2 - y^2 = a^2$

Miscellaneous problems 5

1 1 m^2

2 (*ii*) 18°, 90°, 162°, 234°, 306°

6 273 km

8 Max. at $\frac{\pi}{4}, \pi, \frac{7\pi}{4}$

Min. at $0, \frac{3\pi}{4}, \frac{5\pi}{4}, 2\pi$

Six inflections

9 (*i*) $a(\theta - \sin\theta), a(1 - \cos\theta)$

(*ii*) $a(1 - \cos\theta), a\sin\theta, \cot\frac{\theta}{2}$

10 (*i*) $-\tan t$ (*ii*) $x\sin t + y\cos t = \sin t\cos t$

(*iii*) $(\cos t, 0), (0, \sin t)$ (*iv*) 1

Exercise 6.1a

1 (*i*) 8, 9, 10 (*ii*) 29, 37, 46 (*iii*) 34, 55, 89

(*iv*) 128, 256, 512 (*v*) 5, −3, 6 (*vi*) 40 320, 362 880, 3 628 800

2 (*i*) 1, 8, 27, 64 (*ii*) $\frac{1}{2}, \frac{2}{3}, \frac{3}{4}, \frac{4}{5}$ (*iii*) 0, 2, 0, 2 (*iv*) −1, 2, −3, 4

3 (*i*) 7 (*ii*) 132 (*iii*) $n + 1$ (*iv*) $n(n - 1)$

Exercise 6.1b

1 (*i*) 30 (*ii*) $1\frac{17}{60}$ (*iii*) 255 (*iv*) 0 (*v*) $1\frac{63}{64}$ (*vi*) 20

(*vii*) 98 (*viii*) −10

2 (*i*) $\sum_{r=1}^{25} r^2$ (*ii*) $\sum_{r=2}^{100} \frac{1}{r}$ (*iii*) $\sum_{r=1}^{10} r^3$ (*iv*) $\sum_{r=0}^{50} 3^r$

(*v*) $\sum_{r=1}^{100} (-1)^{r-1} \times \frac{1}{r}$ (*vi*) $\sum_{r=1}^{n} r(r + 1)$

Exercise 6.2

1 (*i*) $4r - 2, 2n^2$ (*ii*) $\frac{1}{2}(3r - 1), \frac{1}{4}n(3n + 1)$

(*iii*) $5 - 2r, 4n - n^2$

2 (*i*) 50, 2 500 (*ii*) 41, 2 091 (*iii*) 31, −155

3 (*i*) 1 275 (*ii*) 1 717

4 220 m

5 1 683, 3 367

6 34

7 $\frac{3}{4}$

8 8, 4, 540

9 (*i*) $4n + 8$ (*ii*) 4

Exercise 6.3

1 (*i*) 1 458, 2 186 (*ii*) $\frac{1}{81}, 4\frac{40}{81}$ (*iii*) −256, −170

(*iv*) $x^9, \frac{x^{10} - 1}{x - 1}$ (*v*) $(-1)^{n-1}x^n, x\frac{1 - (-x)^n}{1 + x}$

2 (*i*) 3 (*ii*) $5\frac{2}{5}$ (*iii*) 10 (*iv*) $\frac{1}{1 + x}$

3 (*i*) $\frac{4}{33}$ (*ii*) $\frac{41}{333}$

4 £183 000 000 000 000 000 (approximately!)

5 £1 260

6 £96.1(5), £444

7 7s; $17\frac{6}{7}$ m

8 $\frac{4}{5}$; $r = -\frac{1}{2}$; No

Exercise 6.4a

1 $\frac{1}{4}n(n+1)(n+2)(n+3)$; 210

2 $\dfrac{n}{n+1}$; $\dfrac{4}{5}$; 1

Exercise 6.4b

1 $\frac{1}{3}n(4n^2-1)$; 166 650

2 (*i*) $r(r+2)$; $\frac{1}{6}n(n+1)(2n+7)$
(*ii*) $2r(r-1)$; $\frac{1}{3}n(n+1)(4n-1)$
(*iii*) $3r(3r-2)$; $\frac{3}{2}n(n+1)(2n-1)$

Exercise 6.6

1 (*i*) 1, 3, 5, 7, 9, 11 (*ii*) 2, 3, 5, 8, 12, 17
(*iii*) 4, 13, 40, 121, 364, 1 093 (*iv*) -1, 2, -4, 8, -16, 32

2 1, 2, 6, 24, 120, 720; $n!$

3 5, 13, 35, 97, 275

4 3.162 281

5 $\sqrt[3]{10}$; 2.15

Exercise 6.7

1 21, 34, 55, 89

2 $\frac{1}{2}(1+\sqrt{5})$

Exercise 6.8

1 39 916 800

2 40 320

3 720

4 (*i*) 720 (*ii*) 1 000

5 (*i*) 60 (*ii*) 20 (*iii*) 10

6 (*i*) 1 024 (*ii*) 243 (*iii*) 32

7 4 368

8 (*i*) 220 (*ii*) 243

9 (*i*) $\frac{1}{2}n(n-1)$ (*ii*) $\frac{1}{6}n(n-1)(n-2)$
(*iii*) $\frac{1}{24}n(n-1)(n-2)(n-3)$

10 5 775

11 (*i*) ${}^nC_r$ (*ii*) 2^n

12 (*i*) ${}^nC_r$ (*ii*) ${}^{n+1}C_r$

13 90, 6

14 (*i*) 15 120 (*ii*) 59 049

15 (*i*) 120 (*ii*) 60

Exercise 6.9

1 (*i*) $a^4+4a^3b+6a^2b^2+4ab^3+b^4$
(*ii*) $8x^3+12x^2y+6xy^2+y^3$
(*iii*) $p^5-10p^4q+40p^3q^2-80p^2q^3+80pq^4-32q^5$
(*iv*) $x^4-4x^2+6-\dfrac{4}{x^2}+\dfrac{1}{x^4}$
(*v*) $64x^6-192x^5y+240x^4y^2-160x^3y^3+60x^2y^4-12xy^5+y^6$
(*vi*) $81x^8-216x^6y^2+216x^4y^4-96x^2y^6+16y^8$

2 (*i*) $1+8x+28x^2$ (*ii*) $1+12x+60x^2$
(*iii*) $1-\frac{7}{2}x+\frac{21}{4}x^2$ (*iv*) $32-240x+720x^2$

(*i*) $28x^2$ (*ii*) $-96x^3$ (*iii*) $40x^4$ (*iv*) 12 (*v*) 15

4 (*i*) 1.105 (*ii*) 0.941 48 (*iii*) 257.026 (*iv*) 0.990 04

5 (*i*) $1 - 5x + 5x^2$ (*ii*) $1 + 12x + 78x^2$ (*iii*) $16 - 32x - 8x^2$

Miscellaneous problems 6

2 $u_5 = 16$; but $u_n \neq 2^{n-1}$ $(u_5 = 31)$

3 $\frac{1}{2}n(n^2 + 1)$ **4** $20n^2$ **5** £20 000 (approx.)

6 (*i*) 1.6 cm²; infinite (*ii*) 0.4 cm²; infinite

7 $$\frac{nx^{n+2} - (n+1)x^{n+1} + x}{(x-1)^2}$$

$$\frac{n^2x^{n+3} - (2n^2 + 2n - 1)x^{n+2} - (n+1)^2x^{n+1} - x^2 - x}{(x-1)^3}$$

8 $$\frac{\sin\frac{n+1}{2}x \sin\frac{nx}{2}}{\sin\frac{x}{2}}$$

9 (*i*) 2^n (*ii*) 0 (*iii*) $n\,2^{n-1}$ (*iv*) ${}^{2n}C_n = \dfrac{(2n)!}{(n!)^2}$

10 (*i*) $u_1 = 0, u_2 = 1, u_3 = 2, u_4 = 9, u_5 = 44$
(*ii*) $u_{n+1} + 1 = n(u_n + u_{n-1})$
(*iii*) $u_n = nu_{n-1} + (-1)^n$

Exercise 7.1

1 $\frac{1}{13}, \frac{1}{2}, \frac{4}{13}$ **2** $\frac{1}{6}, \frac{1}{2}, \frac{2}{3}$
3 $\frac{1}{4}, \frac{1}{4}, \frac{1}{2}$ **4** $\frac{1}{8}, \frac{1}{2}, 0$
5 $\frac{1}{36}, \frac{5}{18}, \frac{25}{36}$ **6** $\frac{1}{9}, \frac{8}{9}, \frac{2}{27}$
7 $\frac{3}{8}, \frac{5}{16}, \frac{1}{4}$ **8** $\frac{1}{10}, \frac{2}{5}$ **9** $\frac{1}{49}, \frac{1}{7}, \frac{12}{49}$

Exercise 7.2a

1 (*i*) $\frac{1}{6}, \frac{1}{2}, \frac{1}{3}$ (*ii*) $\frac{1}{6}, 0, \frac{1}{6}$ (*iii*) $\frac{2}{3}, \frac{1}{2}, \frac{2}{3}$

3 (*i*) ***C***, ***A***; ***A***, ***B*** (*ii*) None

4 (*i*) $\frac{5}{6}, \frac{1}{2}, \frac{2}{3}$ (*ii*) $\frac{1}{3}, \frac{1}{2}, \frac{1}{3}$ (*iii*) $\frac{5}{6}, 1, 1$

6 (*i*) None (*ii*) ***C′***, ***A′***; ***A′***, ***B′***

7 (*i*) $\frac{1}{8}$ (*ii*) $\frac{1}{10}$ **8** $\frac{4}{7}$

Exercise 7.2b

1 $\frac{1}{6}, \frac{1}{6}; \frac{1}{36}$. Independent

2 $\frac{1}{4}, \frac{1}{4}; \frac{1}{16}$. Independent

3 $\frac{1}{4}, \frac{1}{4}; \frac{1}{17}$. Not independent

4 $\frac{1}{4}, \frac{1}{5}; \frac{1}{20}$. Independent

5 $\frac{1}{4}, \frac{1}{10}; \frac{1}{20}$. Not independent

6 $\frac{15}{52}$ **7** 13 **8** $6, \frac{4}{27}$

Exercise 7.3a

1 (*i*) $\frac{1}{13}, \frac{12}{13}, \frac{1}{2}, \frac{1}{2}$ (*ii*) $\frac{1}{26}, \frac{1}{26}, \frac{6}{13}, \frac{6}{13}$ (*iii*) $\frac{1}{13}, \frac{1}{13}, \frac{12}{13}, \frac{12}{13}$
(*iv*) $\frac{1}{2}, \frac{1}{2}, \frac{1}{2}, \frac{1}{2}$
All independent

2 (*i*) $\frac{1}{6}, \frac{5}{6}, \frac{1}{2}, \frac{1}{2}$ (*ii*) $\frac{1}{6}, 0, \frac{1}{3}, \frac{1}{2}$ (*iii*) $\frac{1}{3}, 0, \frac{2}{3}, 1$ (*iv*) $1, \frac{2}{5}, 0, \frac{3}{5}$
None independent

Exercise 7.3b

2 $\frac{1}{8}, \frac{3}{8}, \frac{3}{8}, \frac{1}{8}$
3 $\frac{29}{40}$
4 $\frac{3}{10}, \frac{3}{10}, \frac{1}{15}, \frac{7}{15}, \frac{8}{15}$
5 $\frac{117}{125}$
6 $\frac{1}{15}, \frac{31}{90}, \frac{3}{10}$
7 $\frac{1}{64}, \frac{9}{64}, \frac{11}{32}$
8 (*ii*) $\frac{1}{3}, \frac{1}{3}, \frac{1}{12}, \frac{1}{4}$ (*iii*) $\frac{5}{12}, \frac{7}{12}$ (*iv*) $\frac{4}{5}, \frac{1}{5}, \frac{4}{7}, \frac{3}{7}$
9 $\frac{2}{3}$
10 (*i*) 62% (*ii*) $\frac{1}{31}$ (*iii*) $\frac{9}{19}$
11 (*i*) $\frac{3}{40}$ (*ii*) $\frac{1}{3}$ (*iii*) $\frac{6}{37}$
12 $\frac{1}{3}, \frac{1}{4}, \frac{5}{12}; \frac{1}{9}$

Exercise 7.4a

1 $\frac{1}{36}, \frac{2}{36}, \frac{3}{36}, \frac{4}{36}, \frac{5}{36}, \frac{6}{36}, \frac{5}{36}, \frac{4}{36}, \frac{3}{36}, \frac{2}{36}, \frac{1}{36}$
2 $\frac{3}{18}, \frac{5}{18}, \frac{4}{18}, \frac{3}{18}, \frac{2}{18}, \frac{1}{18}$

Exercise 7.4b

1 (*i*) 0.5 0.5
(*ii*) 0.25 0.50 0.25
(*iii*) 0.125 0.375 0.375 0.125
(*iv*) 0.062 0.250 0.375 0.250 0.062
(*v*) 0.031 0.156 0.312 0.312 0.156 0.031
(*vi*) 0.016 0.094 0.234 0.312 0.234 0.094 0.016
2 0.000 3, 0.006 4, 0.051 2, 0.205 0, 0.409 6, 0.327 7
3 0.26(5) **4** 0.34 **5** (*i*) 0.26 (*ii*) 0.66
6 (*i*) 0.107 (*ii*) 0.376 (*iii*) 0.302 (*iv*) 0.323
7 (*i*) 0.38 (*ii*) 0.14. Less.
8 (*i*) 0.367 (*ii*) 0.389
9 (*i*) 0.236 (*ii*) 0.576 (*iii*) 0.026
10 (*i*) 1.07% (*ii*) 2.14%

Exercise 7.4c

1 $\frac{1}{9}$ **2** 0.82, 0.15 **3** 0.77, 0.125 **4** 7

Exercise 7.5

1 $3\frac{1}{2}$, 7 **2** $15\frac{1}{6}$p **3** 1.5
4 1.5, 1.875, 1.875 **5** $4\frac{17}{36}$
6 $3\frac{7}{11}$ **7** $2\frac{19}{36}$ **8** Jones, by $6\frac{1}{4}$p
9 £$\frac{3}{128} \approx 2\frac{1}{2}$p

Exercise 7.6

1 (*ii*) 6; 6; 6 (*iii*) 5, 7; 1 (*iv*) 1.2; 1.10
2 (*ii*) 7; 7; 7 (*iii*) 6, 8; 1 (*iv*) 1.2; 1.10
3 (*ii*) 6, 7, 8; 7, 7 (*iii*) 6, 8; 1 (*iv*) 3; 1.73
4 (*ii*) 9; 8; 7.5 (*iii*) 6, 9; 1.5 (*iv*) 3.45; 1.86
5 (*ii*) 5; 6.5; 6.8 (*iii*) 5, 8; 1.5 (*iv*) 4.36; 2.09
6 (*ii*) 6, 9; 8, 7.6 (*iii*) 6, 9; 1.5 (*iv*) 2.64; 1.62
7 67, 18.8
8 (*ii*) 2; 4.0 (*iii*) 9.812 m s^{-2}; 0.004 m s^{-2}
9 (*i*) 695 480 km (*ii*) 280 km
10 $y = 0.80\,x - 3.48$

Exercise 7.7

1 (*i*) 55; 26 (*ii*) 42 (*iii*) 47 **2** 39, 69, 56
3 101, 10 **4** 7.1, 2.44 **5** 44, 1.46 **6** 37.05, 2.77
7 48.5, 16.7 **8** 44.6, 18.6 **9** 100.5, 15.1
10 (*ii*) 27.6; 10.3, 47.4 (*iii*) 35.0; 22.9
11 (*ii*) 26.4; 9.7, 45.6 (*iii*) 33.6; 21.9
12 (*ii*) 28.8; 11.0, 49.3 (*iii*) 36.3; 23.6

Miscellaneous problems 7

1 0.014 5; 0.000 25 **2** 0.067 **3** 0.047 **4** $\frac{29}{84}$
5 0.84 **6** 0.41, 0.57, 23 **7** 5 **8** $\frac{496}{729}, \frac{73}{729}, \frac{160}{729}$
9 0.167, 0.333; 0.042, 0.375; 0.007, 0.367
10 (*i*) $\frac{1}{2}, \frac{1}{2}, \frac{1}{2}$ (*ii*) $\frac{1}{4}, \frac{1}{4}, \frac{1}{4}$ (*iii*) $\frac{1}{4}$
11 5.98 roubles
12 (*i*) $\frac{1}{10}, \frac{9}{10}, \left(\frac{9}{10}\right)^{10} = 0.348$
(*ii*) $\frac{1}{100}, \frac{99}{100}, \left(\frac{99}{100}\right)^{100} = 0.363$
(*iii*) Approximately 0.368 $\left(\text{more precisely, } 0.367\,9\ldots = \frac{1}{e}\right)$

Exercise 8.1a

1 (*i*) ***AE*** (*ii*) ***AE*** (*iii*) ***AE*** (*iv*) ***AE***
2 (*i*) ***BA*** + ***AC***, ***BD*** + ***DC***, ***BE*** + ***EC***
(*ii*) ***BA*** + ***AD*** + ***DC***, ***BD*** + ***DA*** + ***AC***,
BA + ***AE*** + ***EC***, ***BE*** + ***EA*** + ***AC***,
BD + ***DE*** + ***EC***, ***BE*** + ***ED*** + ***DC***

Exercise 8.1b

1 (*i*) 13 cm, 026° (*ii*) 5.9 cm, 16°
(*iii*) 7.2 cm, 326° (*iv, v, vi*) 11.8 cm, 008°

2 (*i*) 6 cm, 180° (*ii*) 8 cm, 225° (*iii*) 4 cm, 090°
(*iv*) 5.7 cm, 273° (*v*) 7.2 cm, 033° (*vi*) 11.2 cm, 060°
(*vii*) 5.7 cm, 093° (*viii*) 7.2 cm, 214° (*ix*) 11.2 cm, 240°
(*x*) 1.69 cm, 102°

Exercise 8.1c

1 (*i*) 8 cm (*ii*) 24 cm (*iii*) 24 cm, all due N
2 (*i*) 9 cm (*ii*) 12 cm (*iii*) 21 cm, all due E
3 (*i*) 2 cm due N (*ii*) 1.5 cm due E (*iii*) 2.5 cm, 037°
(*iv*) 5 cm, 037° (*v*) 2.5 cm, 037°

Exercise 8.2a

2 (*i*) $3\boldsymbol{i} + 2\boldsymbol{j}$, $\boldsymbol{i} + 3\boldsymbol{j}$, $-2\boldsymbol{j}$, $\frac{1}{2}\boldsymbol{i} + \frac{1}{2}\boldsymbol{j}$
(*ii*) $(2, 0)$, $(\frac{1}{2}, 0)$, $(-3, 0)$, $(-1, 1)$
3 (*i*) $\boldsymbol{b} - \boldsymbol{a}$, $\boldsymbol{a} - \boldsymbol{b}$ (*ii*) $\boldsymbol{a} + \boldsymbol{b}$ (*iii*) $\frac{1}{2}(\boldsymbol{a} + \boldsymbol{b})$
4 (*i*) $\frac{1}{2}(\boldsymbol{a} + \boldsymbol{b})$, $\frac{1}{2}(\boldsymbol{b} + \boldsymbol{c})$, $\frac{1}{2}(\boldsymbol{c} + \boldsymbol{d})$, $\frac{1}{2}(\boldsymbol{d} + \boldsymbol{a})$ (*ii*) $\frac{1}{2}(\boldsymbol{c} - \boldsymbol{a})$
(*iii*) $\frac{1}{4}(\boldsymbol{a} + \boldsymbol{b} + \boldsymbol{c} + \boldsymbol{d})$
5 (*i*) $\boldsymbol{a} + \boldsymbol{c}$, $\boldsymbol{b} + \boldsymbol{c}$, $\boldsymbol{a} + \boldsymbol{b}$, $\boldsymbol{a} + \boldsymbol{b} + \boldsymbol{c}$ (*ii*) $\boldsymbol{c} - \boldsymbol{a}$, $\boldsymbol{c} - \boldsymbol{a}$
(*iii*) $\frac{1}{2}(\boldsymbol{a} + \boldsymbol{b} + \boldsymbol{c})$
6 (*i*) $\frac{1}{2}\boldsymbol{a}$, $\frac{1}{2}\boldsymbol{b}$, $\frac{1}{2}\boldsymbol{c}$ (*ii*) $\frac{1}{2}(\boldsymbol{b} + \boldsymbol{c})$, $\frac{1}{2}(\boldsymbol{c} + \boldsymbol{a})$, $\frac{1}{2}(\boldsymbol{a} + \boldsymbol{b})$
(*iii*) $\frac{1}{4}(\boldsymbol{a} + \boldsymbol{b} + \boldsymbol{c})$

Exercise 8.2b

1 (*i*) $\sqrt{5}$, 027°; $\sqrt{10}$, 342°; $\sqrt{29}$, 158°
(*ii*) $2.12\boldsymbol{i} + 2.12\boldsymbol{j}$; $3.54\boldsymbol{i} - 3.54\boldsymbol{j}$; $-9.56\boldsymbol{i} + 2.92\boldsymbol{j}$
2 (*i*) $\sqrt{14}$ (*ii*) $\sqrt{29}$ (*iii*) $\sqrt{83}$ (*iv*) $\sqrt{3}$
3 $-3\boldsymbol{i} - \boldsymbol{j}$, $-2\boldsymbol{i} + 3\boldsymbol{j}$
5 $(1 + \frac{2}{3}t)\boldsymbol{i} + \frac{2}{3}t\boldsymbol{j} + (2 + \frac{1}{3}t)\boldsymbol{k}$
$\frac{2}{3}t\boldsymbol{i} + (\frac{2}{3}t - 1)\boldsymbol{j} + (\frac{1}{3}t + 1)\boldsymbol{k}$
6 $\boldsymbol{r} = (-\frac{6}{5} + \frac{1}{5}t)\boldsymbol{i} + (\frac{16}{5} - \frac{1}{5}t)\boldsymbol{j} + \boldsymbol{k}$ $t = 16$

Exercise 8.3

1 $2t\boldsymbol{i} - 5\boldsymbol{j} + 2t\boldsymbol{k}$, $2\boldsymbol{i} + 2\boldsymbol{k}$; $\sqrt{57}$, $2\sqrt{2}$
2 n
3 $2n\pi + \frac{1}{2}\pi$
4 $3\pi \approx 6.28$ s;

$$-4\boldsymbol{i}\cos\frac{2t}{3} - 4\boldsymbol{j}\sin\frac{2t}{3} + 8\boldsymbol{k}\cos\frac{4t}{3}$$

$4\sqrt{5} \approx 8.94$ m s^{-2}
6 7 m s^{-1}, N 38° 13′ E; 10 m s^{-1}
7 (*i*) $2t\boldsymbol{i} + 6t\boldsymbol{j}$, $(4t + \frac{1}{2}t^2)\boldsymbol{i} + (-8t + \frac{1}{2}t^2)\boldsymbol{j}$
(*ii*) 5 s (*iii*) 6 s
8 $4x^2 - y^2 = 4a^2$ $(a\sqrt{2}, 2a)$

Exercise 8.4

1 (*i*) $25t\boldsymbol{i} + (15t - 5t^2)\boldsymbol{j}$ (*ii*) $25\boldsymbol{i} + (15 - 10t)\boldsymbol{j}$ (*iii*) 3 s
(*iv*) 75 m (*v*) 11.25 m
2 (*i*) $10t\boldsymbol{i} + (10\sqrt{3}t - 5t^2)\boldsymbol{j}$; $10\boldsymbol{i} + (10\sqrt{3} - 10t)\boldsymbol{j}$
(*ii*) 10.35 m s^{-1}, 15° downwards;
(*iii*) After 1.73 s
3 (*i*) 45 m (*ii*) 90 m
4 40 m s^{-1}
6 19.6 m
7 $\frac{1}{2}$, 3; 392, 294 m

Exercise 8.5

1 (*i*) 4.5 m s^{-2} (*ii*) 0.4 m s^{-2} (*iii*) 16 m s^{-2}
(*iv*) 4 m s^{-2} (*v*) $\frac{64\pi^2}{9} = 70.2$ m s^{-2} (*vi*) $\frac{2\pi^2}{9} = 2.19$ m s^{-2}
2 5 m s^{-2}; 500 m
3 50 m s^{-1}

Exercise 8.6

1 337° 23′, 30 min
2 48° 36′, 45 s
3 160° 4 $99\frac{1}{2}$° 5 NW
6 36° 52′, 54 min
7 v km h^{-1}, NE, $112\frac{1}{2}$°
8 27 m s^{-1}, 266°
9 $\boldsymbol{r} = 2t\boldsymbol{i} + (1 - t)\boldsymbol{j} + t^2\boldsymbol{k}$, $t = \frac{1}{2}$
10 (*i*) $-60(6\boldsymbol{i} + 5\boldsymbol{j} + \boldsymbol{k})$
(*ii*) $(30 - 6t)\boldsymbol{i} + (30 - 5t)\boldsymbol{j} + (2 - t)\boldsymbol{k}$
(*iii*) 320 s
11 (*i*) 3.92 m s^{-1}, 233° (*ii*) 5.09 m s^{-1}, 199°

Exercise 8.8a

1 (*i*) 6 N (*ii*) 120 N (*iii*) 0.15 N
2 (*i*) 3 m s^{-2} (*ii*) 0.6 N (*iii*) 0.002 5 N
3 (*i*) 20 kg (*ii*) 2.5×10^3 kg (*iii*) 0.015 kg
4 (*i*) 8 m s^{-2} (*ii*) 24 N
5 (*i*) 8 m s^{-2}, 48 N (*ii*) 32 m s^{-2}, 96 N (*iii*) 4 m s^{-2}, 12 N
6 8.44×10^6 m s^{-1}
7 (*i*) $\boldsymbol{0}$, $10\boldsymbol{i}$, $-10\boldsymbol{j}$, $-40\boldsymbol{j}$
(*ii*) $20\boldsymbol{i} - 20\boldsymbol{j}$, $10\boldsymbol{i} - 20\boldsymbol{j}$, $-10\boldsymbol{j}$, $-40\boldsymbol{j}$
(*iii*) $100\boldsymbol{i} - 500\boldsymbol{j}$, $10\boldsymbol{i} - 100\boldsymbol{j}$, $-10\boldsymbol{j}$, $-40\boldsymbol{j}$
8 1.10, 1.20, 1.35, 1.55, 1.85, 2.30, 3.10 m s^{-2}; 85 km h^{-1}

Exercise 8.8b

1 $10\boldsymbol{i} + 5\boldsymbol{j}$, $2\boldsymbol{i} + \boldsymbol{j}$, $-10\boldsymbol{i} - 5\boldsymbol{j}$
2 $6\boldsymbol{i} + 6\boldsymbol{j} + 3\boldsymbol{k}$, $3\boldsymbol{i} + 3\boldsymbol{j} + 1.5\boldsymbol{k}$, $-6\boldsymbol{i} - 6\boldsymbol{j} - 3\boldsymbol{k}$
3 36.1 N, 146° 18′; 3.61 m s^{-2}, 146° 18′; 36.1 N, 326° 18′
4 14.0 N, 255° 22′; 3.5 m s^{-2}, 255° 22′; 14.0, 75° 22′
5 7.37 N, 298° 40′; 0.37 m s^{-2}, 298° 40′; 7.37 N, 118° 40′
6 0.54 N, 249° 12′; 2.7 m s^{-2}, 249° 12′; 0.54 N, 69° 12′
7 $2\boldsymbol{i} + \boldsymbol{j}$, $28\boldsymbol{i} + 16\boldsymbol{j}$, $11\boldsymbol{i} + 6\boldsymbol{j}$
8 10^7 m s^{-1} at 37° to its original direction

Exercise 8.8c

1 (*i*) 11 800 N (*ii*) 7 800 N (*iii*) 9 800 N (*iv*) 9 800 N
2 (*i*) 2.2 m s^{-2} (*ii*) 0.2 m s^{-2}
3 3.6 N, 4.8 m s^{-2}
4 75.1 N, 63.0 N
5 157 N, 118 N
6 3.74 N ($\sqrt{14}$), 1.41 N ($\sqrt{2}$), 2.83 N ($\sqrt{8}$)
7 (*i*) 1.73 m s^{-2} (*ii*) 5° 49′
8 1 055 N, 1 225 N, 660 N
9 (*i*) 1.84×10^5 N (*ii*) 9.71×10^4 N
10 (*i*) 4.18 N (*ii*) 0.82 N
11 $3\frac{1}{2}$ h, 30°

Exercise 8.9a

1 (*i*) 0.5 m s^{-2} (*ii*) 0.3 m s^{-2}
2 (*i*) 0.45 m s^{-2} (*ii*) 0.25 m s^{-2}
3 (*i*) 0.55 m s^{-2} (*ii*) 0.35 m s^{-2}
4 (*i*) 0.4 m s^{-2}, 490 N (*ii*) 0.38 m s^{-2}, 497 N or 483 N
(*iii*) 2.98 m s^{-2}, 467 N, or 3.72 m s^{-2}, 454 N
(*iv*) 2.98 m s^{-2}, or 3.73 m s^{-2}, 460 N
5 86.4 N, 0.83 m s^{-2}
6 13.1 m s^{-1}
8 52° 9 92.6 km
11 (*i*) 3.35 m s^{-2} at 26° 34′ with Ox, no
(*ii*) 2.68 m s^{-2}, $3/\sqrt{5}$ N

Exercise 8.9b

1 (*i*) 1.6 m s^{-2}, 400 N (*ii*) 1.25 m s^{-2}, 312 N (*iii*) 1.04 m s^{-2}, 460 N
2 6 m s^{-2}, 12 N
3 2 m s^{-2}, 24 N 4 $\frac{25}{7}$ m s^{-2} ≈ 3.6, $\frac{180}{7} \approx 26$ N
5 $\frac{1}{7}(20\sqrt{3} - 15) = 2.8$ m s^{-2}, 23 N
6 $\frac{20}{9} = 2.2$ m s^{-2}, $24\frac{4}{9} \approx 24$ N, $31\frac{1}{9} \approx 31$ N
7 2.4 m s^{-2}; 14.1, 9.4 N
8 (*i*) 1.5 m s^{-2}, 400 N (*ii*) 1.1 m s^{-2}, 300 N (*iii*) 0.94 m s^{-2}, 460 N

Exercise 8.10a

1 $1\ \mathrm{m\,s^{-2}}$, $2\ \mathrm{m}$
2 (*i*) $10\ \mathrm{m\,s^{-2}}$ (*ii*) 1 (*iii*) 20 m
3 60 N $\mu = 0.4$ 4 0.1
5 $18.2\ \mathrm{m\,s^{-2}}$, $15.4\ \mathrm{m\,s^{-2}}$
6 $5 - \sqrt{3} \approx 3.3\ \mathrm{m\,s^{-2}}$, $8.1\ \mathrm{m\,s^{-1}}$
7 $g(\sin\alpha - \mu\cos\alpha)$
8 $\dfrac{(M - \mu m)g}{M + m}$, $\dfrac{(\mu + 1)Mmg}{M + m}$
9 1.5 m, $2.45\ \mathrm{m\,s^{-1}}$

Exercise 8.10b

1 (*i*) 78.4 N (*ii*) 73.4 N (*iii*) 118 N (*iv*) 72.8 N
2 (*i*) 0.2 (*ii*) 96 N
3 (*i*) 46.0 N (*ii*) 11.9 N
4 (*i*) 44.0 N (*ii*) 11.4 N
5 (*i*) $W\tan\lambda$ (*ii*) $\dfrac{W\sin\lambda}{\cos(\lambda - \theta)}$ (*iii*) $W\sin\lambda$
6 (*i*) $\dfrac{W\sin(\lambda + \alpha)}{\cos\lambda}$ (*ii*) $\dfrac{W\sin(\lambda + \alpha)}{\cos(\lambda - \theta)}$ (*iii*) $W\sin(\lambda + \alpha)$
7 (*i*) $\dfrac{W\sin(\lambda - \alpha)}{\cos\lambda}$ (*ii*) $\dfrac{W\sin(\lambda - \alpha)}{\cos(\lambda - \theta)}$ (*iii*) $W\sin(\lambda - \alpha)$
8 $2.21\ \mathrm{s^{-1}}$; 1.06 s

Miscellaneous problems 8

3 $\boldsymbol{F} = m\dfrac{\mathrm{d}^2\boldsymbol{r}}{\mathrm{d}t^2}$
4 (*i*) 954 m (*ii*) 0.053 *W* (*iii*) 1.24 km
5 $r = \dfrac{1}{2 + \cos\theta}$
8 $\boldsymbol{s} = \boldsymbol{a} + t\boldsymbol{u} + \frac{1}{2}t^2\boldsymbol{g}$
$v = \sqrt{\left(\dfrac{b^2g^2}{u^2} + \dfrac{u^2}{4}\right)}$; $u = \sqrt{2bg}$
9 $\dfrac{aU}{V}$, $\dfrac{2a}{U}\sin^{-1}\dfrac{U}{V}$
10 $\dfrac{2a}{V^2 - v^2}\{V + \sqrt{(V^2 - v^2)}\}$ 11 $\frac{1}{4}$

Exercise A.1

1 $y = \frac{1}{2}x^2 + 1$

Exercise A.2

1 (*i*) $6, 4; \dfrac{\sqrt{5}}{3}; (\pm\sqrt{5}, 0); x = \pm\dfrac{9}{\sqrt{5}}$

(*ii*) $10, 4; \dfrac{\sqrt{21}}{5}; (\pm\sqrt{21}, 0); x = \pm\dfrac{25}{\sqrt{21}}$

2 $(\pm\sqrt{3}, 0)$

3 $-\dfrac{b^2x_1}{a^2y_1}; \dfrac{xx_1}{a^2} + \dfrac{yy_1}{b^2} = 1; \dfrac{a^2x}{x_1} - \dfrac{b^2y}{y_1} = a^2 - b^2$

4 $\dfrac{ax}{\cos\theta} - \dfrac{by}{\sin\theta} = a^2 - b^2$

6 $x' = x, y' = \dfrac{by}{a}$

Exercise A.3

1 (*i*) $x = \pm 2y; \dfrac{\sqrt{5}}{2}; (\pm\sqrt{5}, 0); x = \pm\dfrac{4}{\sqrt{5}}$

(*ii*) $2x = \pm 3y; \dfrac{\sqrt{13}}{3}, (\pm\sqrt{13}, 0); x = \pm\dfrac{9}{\sqrt{13}}$

2 $(\pm\sqrt{3}, 0)$ 3 $\dfrac{x\sec\theta}{a} - \dfrac{y\tan\theta}{b} = 1$

5 (*i*) $x + t^2y = 2ct$

(*ii*) $t^2x - y = c\left(t^3 - \dfrac{1}{t}\right)$

(*iii*) $x + t_1t_2y = c(t_1 + t_2)$

6 $(x^2 + y^2)^2 = 4c^2xy$ or $r^2 = 2c^2\sin 2\theta$ (a *lemniscate*)